Applications of GIS and Remote Sensing in Soil Environment Monitoring

Applications of GIS and Remote Sensing in Soil Environment Monitoring

Editors

Antonio Ganga
Blaž Repe
Mario Elia

Basel • Beijing • Wuhan • Barcelona • Belgrade • Novi Sad • Cluj • Manchester

Editors

Antonio Ganga
Department of Architecture,
Design and Urban Planning
University of Sassari
Sassari
Italy

Blaž Repe
Geography Department
University of Ljubljana
Ljubljana
Slovenia

Mario Elia
Department of Soil, Plant and
Food Sciences (DISSPA)
University of Bari
Bari
Italy

Editorial Office
MDPI
St. Alban-Anlage 66
4052 Basel, Switzerland

This is a reprint of articles from the Special Issue published online in the open access journal *Sustainability* (ISSN 2071-1050) (available at: www.mdpi.com/journal/sustainability/special_issues/ AGRSISEM).

For citation purposes, cite each article independently as indicated on the article page online and as indicated below:

Lastname, A.A.; Lastname, B.B. Article Title. *Journal Name* **Year**, *Volume Number*, Page Range.

ISBN 978-3-0365-9479-8 (Hbk)
ISBN 978-3-0365-9478-1 (PDF)
doi.org/10.3390/books978-3-0365-9478-1

Cover image courtesy of Antonio Ganga

Contents

About the Editors

Antonio Ganga

Antonio Ganga works as a researcher at the Department of Architecture, Design and Urban Planning at the University of Sassari (ITA). He possesses a Master's degree in Landscape Architecture and a PhD in Pedology, both from the University of Sassari. He deals with spatial analysis in environmental data management. Additional research areas include investigating intervention techniques aimed at improving degraded soils, exploring the potential for urban wastewater and industrial by-products as soil amendments, and conducting integrated ethnopedological surveys to better understand local knowledge of soils. The author previously served as a Visiting Professor at the Faculdade de Ciência Agronômicas, Universidade Estadual Paulista (UNESP) Câmpus de Botucatu in Brazil, and currently maintains collaborative ties with the institution. He served as a Guest Speaker at the Faculty of Agronomy, Horticulture, and Bioengineering within the Department of Soil Science and Microbiology at the University of Life Sciences in Poznan, Poland. Moreover, His is a Topical Advisory Panel Member and Guest Editor for the journal *Sustainability*, specifically for the Special Issue entitled "Applications of GIS and Remote Sensing in Soil Environment Monitoring". He is a member of the Directive board of AMFM GIS Italia, (Automated Mapping/Facilities Management/Geographic Information Systems). He is also a member of the Scientific Committee of Asita, which is the Italian Federation of Scientific Associations for Territorial and Environmental Information.

Blaž Repe

Blaž Repe is an associate professor at the Geography Department, Faculty of Arts, University of Ljubljana, and, since 2007, he has been teaching GIS, digital cartography, Soil geography, Biogeography, Geographical decision support, Physical geography and Applicative physical geography. In 2007, he defended his PhD thesis in the field of soil geography, GIS and digital soil mapping (Geographical soil map and its use in geography). He is the author of many papers, several textbooks for universities and high schools as well as various teaching materials for all levels of education. In 2017, with his fellow soil scientists, he published a Springer monographic publication, Soils of Slovenia, and, in 2018, he translated the 2015 version of WRB soil classification into the Slovenian language. He was a guest lecturer in Slovakia, Croatia, Bosnia and Herzegovina as well as Poland. His knowledge was broadened while visiting the universities in Bratislava (Slovakia) and Freiburg (Germany). Since 2009, he has been a member of Steering Committee of Slovene Soil Science Society. At the moment, and among other things, he is dealing with soil classification, digital soil mapping and invasive alien plant species.

Mario Elia

Dr. Mario Elia, graduated in Forestry science and is a fire scientist of the Department of Soil, Plant and Food Sciences (DISSPA) at the University of Bari, Italy. His research interests lie in the field of Fire Ecology and Forest Fire Management, focusing mainly in the areas of fire ecological modeling, GIS, remote sensing and ecosystem conservation at multiple scales. He has collected experience in modelling data through research projects and scientific collaborations worldwide. Dr. Elia experienced managing projects, leading research studies, writing projects and papers, developing board studies and collaborating to national and international congresses and seminars. He has worked as visiting researcher at the Toledo University (Oh, USA), Oregon State University (Or, USA) and University of Lisbon (POR).

Preface

Effective land resource management requires the monitoring of environmental characteristics, for which soil data, both temporal and spatial, are of immense importance. Remote sensing and GIS applications facilitate the efficient management of these data, thus enabling the development of predictive and valid models to control soil erosion and land degradation. Further research is needed to improve the study of soil dynamics in various environmental biosystems, providing a greater understanding of erosion processes and the effectiveness of conservation techniques, in line with contemporary and sustainable practices. Furthermore, recent research on soil dynamics offers a promising opportunity to improve our understanding of the link between soil erosion and natural disasters such as landslides, floods, slope instability, biodiversity loss and climate change.

This publication brings together recent studies that contribute to improving techniques and knowledge in these contexts, especially with regard to soil management and monitoring through the use of original techniques, which are essential for the future efficient management of this limited resource.

Antonio Ganga, Blaž Repe, and Mario Elia
Editors

Editorial

Applications of GIS and Remote Sensing in Soil Environment Monitoring

Antonio Ganga [1],*, Mario Elia [2] and Blaž Repe [3]

1 Department of Architecture Design and Planning, University of Sassari, 07100 Sassari, Italy
2 Department of Agricultural and Environmental Science, University of Bari, 70121 Bari, Italy; mario.elia@uniba.it
3 Department of Geography, Faculty of Arts, University of Ljubljana, Aškerčeva 2, 1000 Ljubljana, Slovenia; blaz.repe@ff.uni-lj.si
* Correspondence: aganga@uniss.it

Citation: Ganga, A.; Elia, M.; Repe, B. Applications of GIS and Remote Sensing in Soil Environment Monitoring. *Sustainability* **2023**, *15*, 13705. https://doi.org/10.3390/su151813705

Received: 7 September 2023
Accepted: 12 September 2023
Published: 14 September 2023

Monitoring plays an essential role in the efficient and sustainable management of the environment. Accurate and rapid procedures and data enable the activation and implementation of public policies and initiatives with which to address emergencies and medium-term environmental depletion processes. Among these, soil consumption is one of the most important issues. Soil sealing threatens the protection of the environment and the security of food production [1,2] in a world where only 10–12% of natural soils are still available for agriculture [3]. Another important threat to the soil as a resource is the increase in the erosion process. Several studies expect soil erosion to increase in the 21st century due to global climate change and land use [4,5]. At this point, measures to mitigate the effects of soil erosion are currently on the agenda of international institutions such as the Food and Agriculture Organization [6] and the European Union [7].

On the other hand, the availability of remote sensing data with greater temporal and spatial resolution has increased recently [8]. This wealth of data, integrated with field observations, enables increasingly efficient monitoring processes. Therefore, it is essential to implement and perfect more accurate and efficient methods and models.

With this in mind, this Special Issue aims to collect various contributions dealing with ways to improve models for managing data in the environmental monitoring process, focusing on soil issues. In this volume, various aspects of environmental monitoring have been addressed in a logical framework. In some cases, the focus was on hazard detection and mapping [9–13]; in others, the focus was on the detection and description of soil properties and their contribution to land use determination [14–16]. The papers addressed the issues at different scales: regional scale [10,13,14,17]; watershed scale [9,12]; and field scale [11,17]. The GIS approach is indeed useful for implementing an analytical model at multiple scales, even for estimating soil erosion [18]. Finally, the published review [19] addressed the monitoring of salinization, another problem affecting more than 100 nations [20,21], and gave a general overview of the problem in China, one of the largest food producers with a critical problem in terms of self-sufficiency [22].

Author Contributions: Conceptualization, A.G., B.R. and M.E.; writing—original draft preparation, A.G., B.R. and M.E.; writing—review and editing, A.G., B.R. and M.E. All authors have read and agreed to the published version of the manuscript.

Conflicts of Interest: The authors declare no conflict of interest.

References

1. Rojas, R.V.; Achouri, M.; Maroulis, J.; Caon, L. Healthy Soils: A Prerequisite for Sustainable Food Security. *Environ. Earth Sci.* **2016**, *75*, 180. [CrossRef]
2. Gardi, C.; Panagos, P.; Van Liedekerke, M.; Bosco, C.; De Brogniez, D. Land Take and Food Security: Assessment of Land Take on the Agricultural Production in Europe. *J. Environ. Plan. Manag.* **2015**, *58*, 898–912. [CrossRef]
3. Gebrehiwot, K. Chapter 3—Soil Management for Food Security. In *Natural Resources Conservation and Advances for Sustainability*; Jhariya, M.K., Meena, R.S., Banerjee, A., Meena, S.N., Eds.; Elsevier: Amsterdam, The Netherlands, 2022; pp. 61–71; ISBN 978-0-12-822976-7.
4. Borrelli, P.; Robinson, D.A.; Panagos, P.; Lugato, E.; Yang, J.E.; Alewell, C.; Wuepper, D.; Montanarella, L.; Ballabio, C. Land Use and Climate Change Impacts on Global Soil Erosion by Water (2015–2070). *Proc. Natl. Acad. Sci. USA* **2020**, *117*, 21994–22001. [CrossRef] [PubMed]
5. Eekhout, J.P.C.; de Vente, J. Global Impact of Climate Change on Soil Erosion and Potential for Adaptation through Soil Conservation. *Earth-Sci. Rev.* **2022**, *226*, 103921. [CrossRef]
6. Voluntary Guidelines for Sustainable Soil Management | Policy Support and Governance | Food and Agriculture Organization of the United Nations. Available online: https://www.fao.org/policy-support/tools-and-publications/resources-details/en/c/1027927/ (accessed on 29 August 2023).
7. COMMUNICATION FROM THE COMMISSION TO THE EUROPEAN PARLIAMENT, THE COUNCIL, THE EUROPEAN ECONOMIC AND SOCIAL COMMITTEE AND THE COMMITTEE OF THE REGIONS EU Soil Strategy for 2030 Reaping the Benefits of Healthy Soils for People, Food, Nature and Climate. 2021. Available online: https://eur-lex.europa.eu/legal-content/EN/TXT/?uri=CELEX%3A52021DC0699 (accessed on 29 August 2023).
8. Ali, I.; Greifeneder, F.; Stamenkovic, J.; Neumann, M.; Notarnicola, C. Review of Machine Learning Approaches for Biomass and Soil Moisture Retrievals from Remote Sensing Data. *Remote Sens.* **2015**, *7*, 16398–16421. [CrossRef]
9. Dapin, I.G.; Ella, V.B. GIS-Based Soil Erosion Risk Assessment in the Watersheds of Bukidnon, Philippines Using the RUSLE Model. *Sustainability* **2023**, *15*, 3325. [CrossRef]
10. Ganga, A.; Elia, M.; D'Ambrosio, E.; Tripaldi, S.; Capra, G.F.; Gentile, F.; Sanesi, G. Assessing Landslide Susceptibility by Coupling Spatial Data Analysis and Logistic Model. *Sustainability* **2022**, *14*, 8426. [CrossRef]
11. Giambastiani, Y.; Giusti, R.; Gardin, L.; Cecchi, S.; Iannuccilli, M.; Romanelli, S.; Bottai, L.; Ortolani, A.; Gozzini, B. Assessing Soil Erosion by Monitoring Hilly Lakes Silting. *Sustainability* **2022**, *14*, 5649. [CrossRef]
12. Malav, L.C.; Yadav, B.; Tailor, B.L.; Pattanayak, S.; Singh, S.V.; Kumar, N.; Reddy, G.P.O.; Mina, B.L.; Dwivedi, B.S.; Jha, P.K. Mapping of Land Degradation Vulnerability in the Semi-Arid Watershed of Rajasthan, India. *Sustainability* **2022**, *14*, 10198. [CrossRef]
13. Barrena-González, J.; Lavado Contador, J.F.; Pulido Fernández, M. Mapping Soil Properties at a Regional Scale: Assessing Deterministic vs. Geostatistical Interpolation Methods at Different Soil Depths. *Sustainability* **2022**, *14*, 10049. [CrossRef]
14. Wang, Q.; Xie, J.; Yang, J.; Liu, P.; Chang, D.; Xu, W. A Model between Cohesion and Its Inter-Controlled Factors of Fine-Grained Sediments in Beichuan Debris Flow, Sichuan Province, China. *Sustainability* **2022**, *14*, 12832. [CrossRef]
15. Arciniegas-Ortega, S.; Molina, I.; Garcia-Aranda, C. Soil Order-Land Use Index Using Field-Satellite Spectroradiometry in the Ecuadorian Andean Territory for Modeling Soil Quality. *Sustainability* **2022**, *14*, 7426. [CrossRef]
16. Czigány, S.; Sarkadi, N.; Lóczy, D.; Cséplő, A.; Balogh, R.; Fábián, S.Á.; Ciglič, R.; Ferk, M.; Pirisi, G.; Imre, M.; et al. Impact of Agricultural Land Use Types on Soil Moisture Retention of Loamy Soils. *Sustainability* **2023**, *15*, 4925. [CrossRef]
17. ul Haq, Y.; Shahbaz, M.; Asif, H.M.S.; Al-Laith, A.; Alsabban, W.H. Spatial Mapping of Soil Salinity Using Machine Learning and Remote Sensing in Kot Addu, Pakistan. *Sustainability* **2023**, *15*, 12943. [CrossRef]
18. Wickenkamp, V.; Duttmann, R.; Mosimann, T. A Multiscale Approach to Predicting Soil Erosion on Cropland Using Empirical and Physically Based Soil Erosion Models in a Geographic Information System. In *Soil Erosion: Application of Physically Based Models*; Schmidt, J., Ed.; Environmental Science; Springer: Berlin/Heidelberg, Germany, 2000; pp. 109–134; ISBN 978-3-662-04295-3.
19. Ma, Y.; Tashpolat, N. Current Status and Development Trend of Soil Salinity Monitoring Research in China. *Sustainability* **2023**, *15*, 5874. [CrossRef]
20. Pereira, P.; Barceló, D.; Panagos, P. Soil and Water Threats in a Changing Environment. *Environ. Res.* **2020**, *186*, 109501. [CrossRef] [PubMed]
21. Ondrasek, G.; Rengel, Z. Environmental Salinization Processes: Detection, Implications & Solutions. *Sci. Total Environ.* **2021**, *754*, 142432. [CrossRef] [PubMed]
22. Huang, J.; Wei, W.; Cui, Q.; Xie, W. The Prospects for China's Food Security and Imports: Will China Starve the World via Imports? *J. Integr. Agric.* **2017**, *16*, 2933–2944. [CrossRef]

 *sustainability*

Article

A Model between Cohesion and Its Inter-Controlled Factors of Fine-Grained Sediments in Beichuan Debris Flow, Sichuan Province, China

Qinjun Wang [1,2,3,4,*], Jingjing Xie [2,3], Jingyi Yang [2,3], Peng Liu [2,3], Dingkun Chang [2,3] and Wentao Xu [2,3]

[1] International Research Center of Big Data for Sustainable Development Goals, Beijing 100094, China
[2] Key Laboratory of Digital Earth Science, Aerospace Information Research Institute, Chinese Academy of Sciences, Beijing 100094, China
[3] Yanqihu Campus, University of Chinese Academy of Sciences, Beijing 101408, China
[4] Key Laboratory of the Earth Observation of Hainan Province, Hainan Aerospace Information Research Institute, Sanya 572029, China
* Correspondence: wangqj@radi.ac.cn

Abstract: Cohesion is the attraction between adjacent particles within the same material, which is the main inter-controlled factor of fine-grained sediment stability, and thus plays an important role in debris flow hazard early warning. However, there is no quantitative model of cohesion and its inter-controlled factors, including effective internal friction angle, permeability coefficient and density. Therefore, establishing a quantitative model of cohesion and its inter-controlled factors is of considerable significance in debris flow hazard early warning. Taking Beichuan county in southwestern China as the study area, we carried out a series of experiments on cohesion and its inter-controlled factors. Using the value of cohesion as the dependent variable and values of normalized density, normalized logarithm of permeability coefficient and normalized effective internal friction angle as the independent variables, we established a quantitative model of cohesion and its inter-controlled factors by the least-squares multivariate statistical method. Fitting of the model showed that its determination coefficient (R^2) was 0.61, indicating that the corresponding correlation coefficient (R) was 0.78. Furthermore, t-tests of the model showed that except for the *p* value of density, which was 0.05, those of other factors were less than 0.01, indicating that cohesion was significantly correlated to its inter-controlled factors, providing a scientific basis for debris flow hazard early warning.

Keywords: debris flow; fine sediments; cohesion; Beichuan

Citation: Wang, Q.; Xie, J.; Yang, J.; Liu, P.; Chang, D.; Xu, W. A Model between Cohesion and Its Inter-Controlled Factors of Fine-Grained Sediments in Beichuan Debris Flow, Sichuan Province, China. *Sustainability* **2022**, *14*, 12832. https://doi.org/10.3390/su141912832

Academic Editors: Antonio Ganga, Blaž Repe and Mario Elia

Received: 15 August 2022
Accepted: 30 September 2022
Published: 8 October 2022

Publisher's Note: MDPI stays neutral with regard to jurisdictional claims in published maps and institutional affiliations.

1. Introduction

Debris flow is gravity sediment flow with a large amount of soils and stones caused by rainstorms or snow/ice melting. It often causes houses to collapse and results in damaged roads, electricity lines and other facilities, thus posing a serious threat to the safety of local people's lives and property [1]. For example, on August 20, 2019, catastrophic debris flow in Sichuan affected 446,000 people, leading to a direct economic loss of 15.89 billion yuan [2]. Therefore, rapid debris flow early warning plays an important role in ensuring the safety of mountainous people.

There are numerical debris flow hazard simulation methods, which are essential for the development of a hazard early warning system [3–6], such as the full three-dimensional (3D) smoothed particle hydrodynamics (SPH) method, the modified MPS method, the coupled moving particle simulation–finite element method and the liquid–gas-like phase transition model in sand flow under microgravity. In such models, quantitative debris flow parameters are important to the debris flow hazard simulations.

With particle size (represented by diameter) less than 2 mm, fine-grained sediments are Quaternary sediments and the main materials that flow in water during debris flow.

Their stability is closely related to the debris flow initial water volume [7–15]. Therefore, debris flow early warning needs to quickly detect their stability. It is mainly controlled by external and internal factors. External factors include water sources, such as rainfall, rainfall intensity, runoff and topographic conditions, e.g., slope, surface coverage and structure. Internal factors include cohesion, permeability coefficient and effective internal friction angle [16–18].

Cohesion is the attraction within a material, such as electrostatic attraction, van der Waals force, cementation and valence bonds. In the case of effective stress, cohesion is obtained by reducing the friction from the total shear strength, which is the inter-controlled factor of the debris flow stability. Its value mainly reflects the strain capacity of soil to resist external stress, and is related to the effective internal friction angle (the internal friction between soil particles, mainly including the surface friction of soil particles and the bonding force between them), the permeability coefficient (the unit flow under the unit hydraulic gradient, indicating the difficulty of fluid passing through the pore skeleton), density (mass per unit volume) and moisture (the ratio of the weight of water contained in the soil to the weight of dry soil) [19–27].

As there is no quantitative model of cohesion and its inter-controlled factors in debris flow, carrying out cohesion research and discovering its inter-controlled factors to establish a quantitative model is of great significance in debris flow hazard early warning.

2. Materials and Methods

2.1. Study Area

The study area is mainly located in Beichuan county, with geographical coordinates of 104°23′–104°31.7′ E, 31°48.5′–31°53.5′ N, covering an area of about 140 km² (Figure 1). Since the Wenchuan earthquake in 2008, six heavy debris flow disasters have occurred, resulting in nearly 10,000 people left homeless, more than half of the buildings buried and a large area of farmland flooded [28].

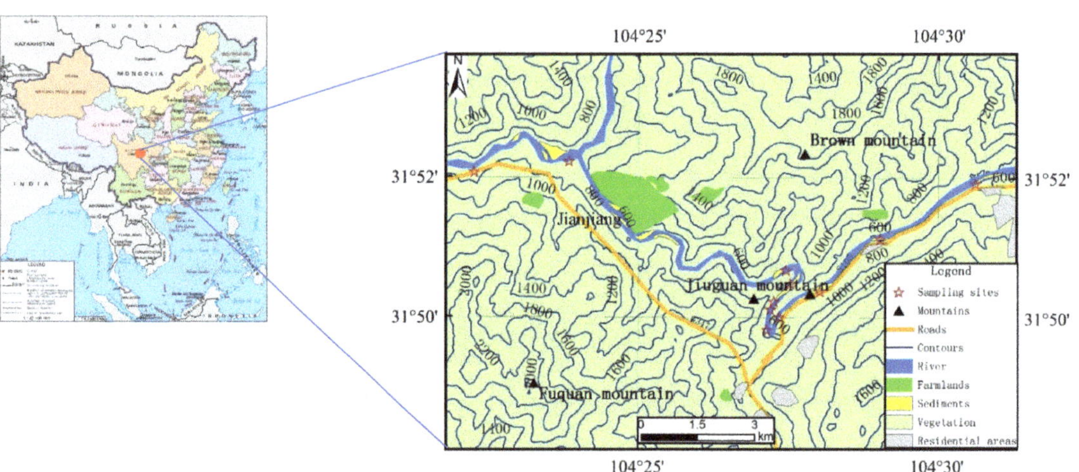

Figure 1. Location map of the study area.

2.2. Materials and Equipment

To establish a model of cohesion and its inter-controlled factors of fine-grained sediments in the study area, we collected multi-resource materials: remote sensing images, digital elevation model (DEM), soil and its parameters such as cohesion, permeability coefficient, density and particle size using the corresponding equipment listed in Table 1.

Table 1. Materials and equipment.

Materials	Equipment	Manufacturer/Provider
Remote sensing images	Gaofen (GF)	Land satellite remote sensing application center, China
Digital Elevation Model (DEM)	Advanced Spaceborne Thermal Emission and Reflection Radiometer (ASTER)	Ministry of International Trade and Industry, Japan
Soil	Ring knife (200 mL)	Longnian Hardware Tools Store, China
Cohesion	ZJ strain-controlled direct shear instrument	Nanjing soil instrument factory Company Limited (Co., Ltd.), China
Permeability coefficient	TST-55 permeameter	Zhejiang Dadi Instrument Co., Ltd., China
Density	MDJ-300A solid densitometer	Shanghai Lichen Instrument Technology Co., Ltd., China
Moisture	Electric heating constant temperature drying oven	Shanghai-southern Electric Furnace Oven Factory, China
Particle size	Microtrac S3500	American Microtrac Incorporated (Inc.)

2.3. Methods

A technical flowchart of this research is shown in Figure 2, which mainly includes the steps of data acquisition, parameter measurement experiment and model establishment.

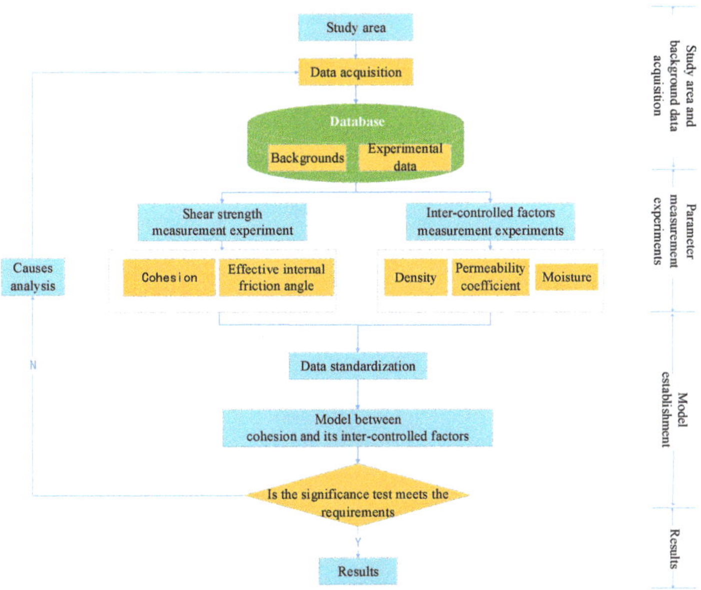

Figure 2. Technical flowchart.

2.3.1. Data Acquisition

(1) Background data

GF-2 satellite remote sensing images with a spatial resolution of 0.8 m and DEM with a spatial resolution of 30 m were acquired first. Then, a fine-grained sediment map was extracted from the GF imagery to determine the locations of sampling sites.

(2) Sample collection

From 19 to 25 March 2021, with cloudy weather and 11–15 °C temperature, 200 samples (600 mL for each sample) were collected from 11 sampling sites using ring knives according to the guide GB/T 36197-2018 [29], whose locations are shown in Figure 1.

(3) Database establishment

A database composed of remote sensing images, DEM, fine-grained sediments map, soil locations, pictures and descriptions was established according to the principles of GB/T 30319-2013 [30].

2.3.2. Parameter Measurement Experiment

Experiments were carried out according to the specification SL237-1999 [31]. Measured parameters include particle size, cohesion and its inter-controlled factors, such as effective internal friction angle, permeability coefficient, density and moisture. Detailed information about the experiments can be found in Reference [28].

(1) Particle size measurement experiment

Microtrac S3500 in Section 2.2 was used to measure particle size; the main steps include setting the sample number and parameters on the instrument, particle size automatic measurement, saving data and cleaning the pipeline.

A histogram of fine-grained sediments' particle size is shown in Figure 3. We can see that the minimum soil particle size in the study area is 0.45 um and the maximum is about 90 um, most of which is distributed in the range of 10–20 um. According to the standard for the engineering classification of soil (GB/T 50145-2007) [32], they are classified as silt loams.

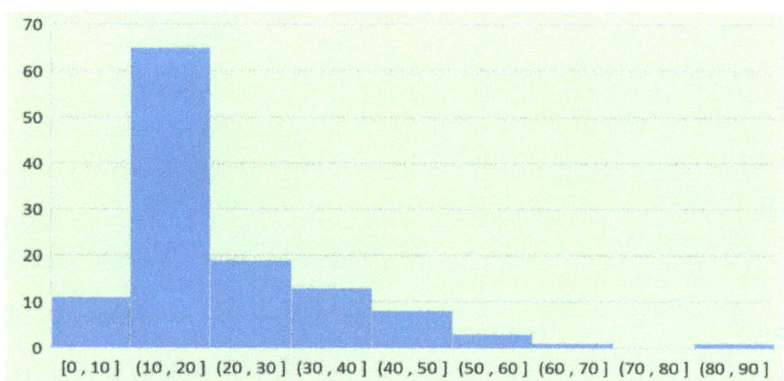

Figure 3. Histogram of fine-grained sediments' particle size.

(2) Cohesion measurement experiment

A ZJ strain-controlled direct shear instrument in Section 2.2 was used to measure the soil cohesion and effective internal friction angle. The main steps of the experiment include sample preparation, adding shear normal stress σ of 50, 100, 200 and 300 kpa to obtain shear strength τ, then calculating cohesion and effective internal friction angle by showing the 4 pairs of data in the coordinate system, in which σ is on the horizontal axis and τ is on the vertical axis.

Histograms of fine-grained sediments' cohesion and effective internal friction angle are shown in Figure 4. We can see that cohesion is distributed from 13.95 to 39.55 kPa with the main range of 17.15–26.75 kPa, and the effective internal friction angle is distributed from 16.16° to 23.68°, with the main range of 18.98–21.80°.

(3) Cohesion inter-controlled factor measurement experiments

A series of measurement experiments were carried out to acquire cohesion inter-controlled factors, such as permeability coefficient, density and moisture.

Firstly, the TST-55 permeameter in Section 2.2 was used to carry out the permeability coefficient experiment, whose main steps include sample preparation, flowing water through the sample and recording related parameters such as initial water head, starting time and the end water head. Then, we calculated the permeability coefficient according to its formula.

Number

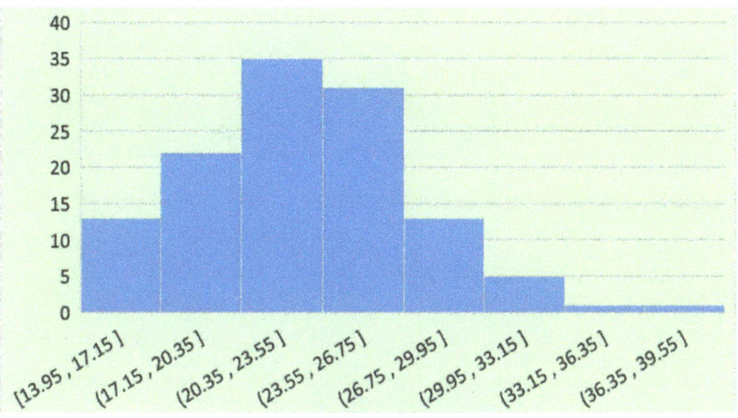

(a) Cohesion (kPa)

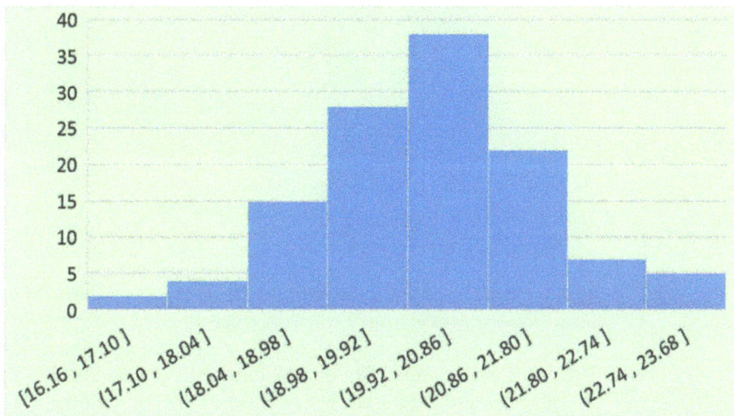

(b) Internal friction angle (°)

Figure 4. Histograms of fine-grained sediments' cohesion (**a**) and effective internal friction angle (**b**) in Beichuan.

A histogram of the fine-grained sediments' permeability coefficient is shown in Figure 5. We can see that the permeability coefficient is distributed from 0.47 to 2.85 m/d, with the main range of 1.15–2.17 m/d.

Secondly, the MDJ-300A solid densitometer in Section 2.2 is used to measure density; the main steps include sample preparation, weighing the sample and its bag, and then calculating density.

A histogram of fine-grained sediments' density is shown in Figure 6. We can see that density is distributed from 1.34 to 1.73 g/mL, with the main range of 1.34–1.54 g/mL.

Number

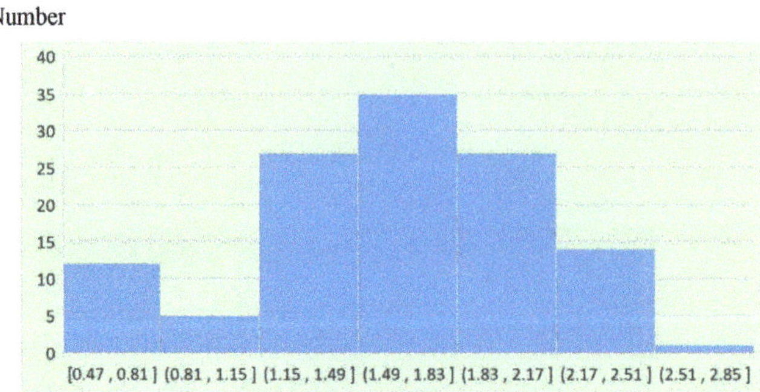

Figure 5. Histogram of fine-grained sediments' permeability coefficient in Beichuan.

Number

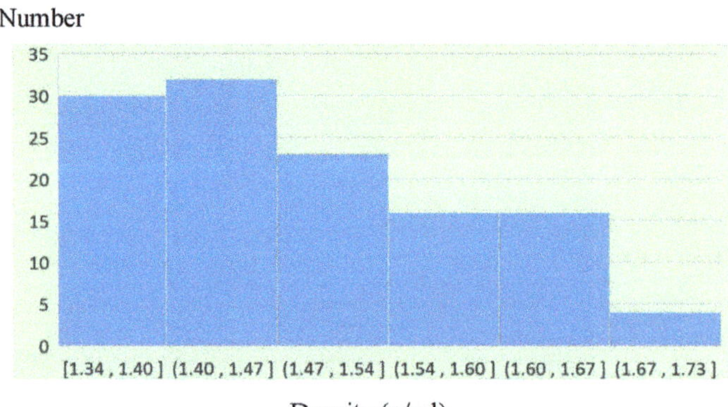

Figure 6. Histogram of fine-grained sediments' density in Beichuan.

Finally, an electric heating constant temperature drying oven in Section 2.2 was used to measure moisture. The main steps include sample preparation and weighting, drying the sample for more than 8 hours, weighting dried samples and then calculating moisture.

A histogram of fine-grained sediments' moisture is shown in Figure 7. We can see that moisture is distributed from 4.01% to 30.69%, with the main range of 4.01–12.02%.

Number

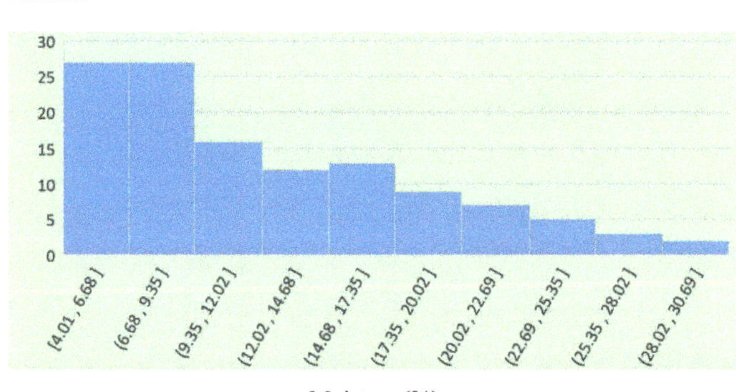

Figure 7. Histogram of fine-grained sediments' moisture in Beichuan.

2.3.3. Model

(1) Data standardization

To eliminate the impacts of magnitude dimensions and orders for different units, data standardization was carried out using Equation (1).

$$z_{ij} = \left(x_{ij} - \overline{x_i}\right)/s_i \tag{1}$$

where z_{ij} is the standardized value, x_{ij} is the measured value, $\overline{x_i}$ is the mean value and s_i is the standard deviation.

(2) Close factors to cohesion selection

As shown in Table 2, in order to determine the factors close to cohesion, the correlation coefficients between cohesion and effective internal friction angle, permeability coefficient, density and moisture were calculated.

Table 2. Coefficients between cohesion correlated to its factors.

	Effective Internal Friction Angle (°)	ln(p) (m/d)	Density (g/cm^3)	Moisture (%)
Cohesion (KPa)	−0.66	−0.58	0.36	0.32

p: permeability coefficient; ln: natural logarithm.

From Table 2, we can see that with the correlation coefficients of −0.66, −0.58, 0.36 and 0.32, the effective internal friction angle, logarithm of permeability coefficient, density and moisture, respectively, are related to cohesion. The correlation coefficients of the former two are negative, indicating that cohesion decreases with the increase in effective internal friction angle and permeability coefficient, while the others are positive, indicating that cohesion increases with the increase in density and moisture.

Figure 8 shows the fitting relationship between cohesion and its inter-controlled factors. From which, we can see that the fitting determination coefficient (R^2) between cohesion and the permeability coefficient is 0.31, which is less than 0.34 of the fitting determination coefficient between cohesion and the logarithm of permeability coefficient, logarithmic transformation on the permeability coefficient should be made before regression. We also applied logarithms on the effective internal friction angle and density, and then calculated determination coefficients between cohesion with them. The results showed that their logarithmic determination coefficients were 0.44 and 0.12, respectively, which are no more than those of 0.44 and 0.13 in their linear regression format.

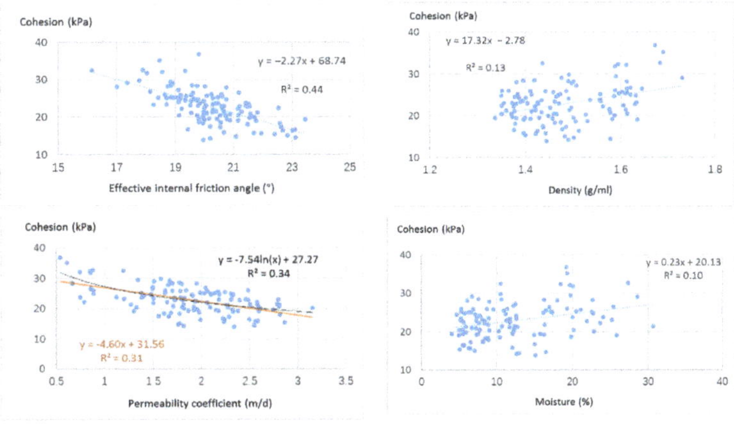

Figure 8. Fitting relationship between cohesion and its inter-controlled factors.

According to the principles of the statistical significance test, when the correlation coefficient (R) is more than 0.3 and the p value of the t-test is no more than 0.05, the factor passes the t-test. However, the t-test on the correlation between cohesion and its inter-controlled factors (Table 3) showed that the p value of moisture was 0.45, which is much more than 0.05, indicating that moisture did not pass the t-test, and it was then deleted from the cohesion inter-controlled factors.

Table 3. The t-test of the correlation between cohesion and its inter-controlled factors.

	Coefficient	p Value	Lower Limit 95.0%	Upper Limit 95.0%
Intercept	22.91	0.00	22.40	23.43
Density	0.86	0.05	−0.03	1.76
ln(p)	−1.59	0.00	−2.23	−0.96
Effective internal friction angle	−2.49	0.00	−3.04	−1.95
Moisture	−0.33	0.45	−1.20	0.54

p: permeability coefficient; ln: natural logarithm.

Therefore, close factors to cohesion are effective internal friction angle, logarithm of permeability coefficient and density.

(3) Model establishment

After selecting the close factors to cohesion, a model between them is established using the least-squares multivariate statistical method.

Firstly, a model between cohesion and each inter-controlled factor is established by the correlation fitting method, such as scatter plot analysis in Microsoft Excel. Then, a transformation on each factor is carried out according to the model between cohesion and each inter-controlled factor so that the cohesion and each inter-controlled factor are linearly correlated. Finally, a model of cohesion and its inter-controlled factors is established by the least-squares multivariate statistical method, whose principles are as follows.

By minimizing the summation of squared errors, the least-squares multivariate statistical method finds the best matching function.

$$y = f(x, w)$$

In order to determine w, the function can be solved as

$$L(y, f(x, w)) = \sum_{i=1}^{n} |y_i - f(x_i, w_i)|^2 \tag{2}$$

where w_i ($i = 1, 2, \ldots, n$) can be calculated by minimizing the function.

3. Results and Discussion

3.1. Results

Using the experimental data, the model of cohesion and standardized inter-controlled factors is established as

$$y = 22.91 + 0.62x_1 - 1.57x_2 - 2.48x_3$$

where y is cohesion, x_1 is the normalized density, x_2 is the normalized logarithm of permeability coefficient ln(p) (p is the permeability coefficient) and x_3 is the normalized effective internal friction angle.

The fitting correlation of the model is shown in Figure 9.

From Figure 9, we can see that the model's determination coefficient (R^2) is 0.61, indicating that the corresponding correlation coefficient (R) is 0.78. In the model, the absolute value of each variable coefficient indicates its closeness to cohesion. Because the absolute coefficients of normalized density, normalized logarithm of permeability coefficient and normalized effective internal friction angle are 0.62, 1.57 and 2.48, respectively, the closest parameter to cohesion is the effective internal friction angle, followed by the logarithm of permeability coefficient and density in the study area.

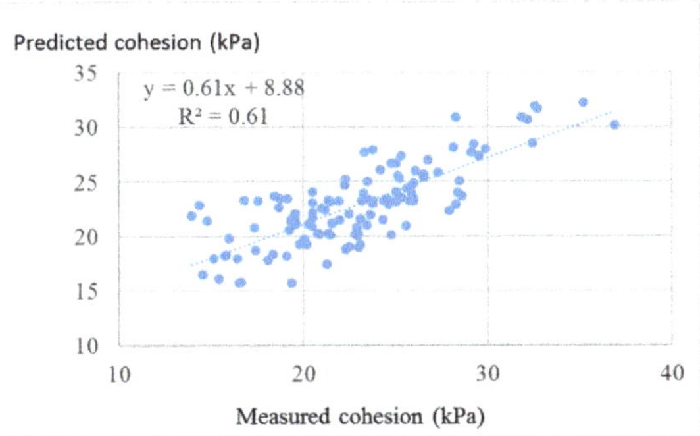

Figure 9. Fitting correlation between the measured and the predicted cohesion.

The results of the *t*-tests on each regression coefficient are shown in Table 4. We can see that except for the *p* value of density, which is 0.05, others are less than 0.01, indicating that cohesion has a significant correlation with its inter-controlled factors.

Table 4. The *t*-tests on the model.

	Coefficient	*p* Value	Lower Limit 95.0%	Upper Limit 95.0
Intercept	22.91	0.00	22.40	23.43
Density	0.62	0.05	0.01	1.22
ln(p)	−1.57	0.00	−2.20	−0.93
Effective internal friction angle	−2.48	0.00	−3.03	−1.93

p: permeability coefficient; ln: natural logarithm.

3.2. Discussion

Cohesion is negatively correlated to effective internal friction angle and logarithm of permeability coefficient and positively correlated to density; the reasons for this are analyzed as follows.

(1) The fixed shear strength of a sample makes a negative correlation between cohesion and the effective internal friction angle.

The effective internal friction angle is determined by the friction resistance and linkage between soil particles. Larger factors that lead to the increase in the soil friction angle—such as a coarser surface, more edges and corners and greater spaces between soil particles—result in the weakening of the gravity between soil particles and the reduction in cohesion [23,33].

In a sample, the correlation of cohesion and the effective internal friction angle can be expressed by

$$\tau_f = \sigma tg\varphi + c \tag{3}$$

where c is the cohesion (kPa), τ_f is the shear strength (kPa), ϕ is the effective internal friction angle (°), σ is the normal pressure (kPa) and $\sigma tg\phi$ is internal friction.

Therefore, when shear strength τ_f and normal pressure σ are fixed, the greater the effective internal friction angle is and the smaller the cohesion becomes, thus leading to a negative correlation between them.

(2) Attraction controlled by distance between soil particles makes a negative correlation between cohesion and the permeability coefficient, and a positive correlation between cohesion and density.

The value of the permeability coefficient mainly reflects the number, size and connectivity of soil pores. The greater the permeability coefficient is, the larger the distance between soil particles becomes, leading to smaller attraction and the weakening of cohesion, and thus leading to a negative correlation between cohesion and the permeability coefficient [23].

The greater the soil density is, the smaller the distance between soil particles is and the greater the mutual attraction between particles becomes, leading to greater soil cohesion. Therefore, there is a positive correlation between cohesion and density [34].

(3) The correlation between cohesion and moisture is not significant in the study area.

Many studies showed that soil moisture has some influence on cohesion, but their correlation varies with the content of water and soil components: (1) with the increase in water content, soil cohesion increases first and then decreases [35], and (2) soil cohesion differs with the soil components. For example, it is higher in red soil, mid-range in kaolin and the lowest in sandy soil [36].

Therefore, the correlation between moisture and cohesion is complex, as cohesion does not show a consistent trend with the change in moisture, thus leading to a small correlation coefficient between them. Furthermore, the soil type in the study area is silt loam, in which the correlation between cohesion and moisture is not as significant as in red soil or kaolin in other places.

4. Conclusions

We designed a series of soil experiments to establish a model of cohesion and its inter-controlled factors. The main conclusions are as follows.

(1) The cohesion inter-controlled factors of fine-grained sediments in Beichuan debris flow were discovered.

The selection of close factors to cohesion showed that with the correlation coefficients of −0.66, −0.58 and 0.36, effective internal friction angle, logarithm of permeability coefficient and density, respectively, were related to cohesion. Therefore, the cohesion inter-controlled factors of fine-grained sediments in Beichuan include effective internal friction angle, logarithm of permeability coefficient and density, as analyzed in Section 3.2.

(2) A model of cohesion and its inter-controlled factors in Beichuan debris flow was established.

A model of cohesion and its inter-controlled factors (effective internal friction angle, logarithm of permeability coefficient and density) in Beichuan debris flow was established by the least-squares multivariate statistical method. The results show that the absolute coefficients of normalized density, normalized logarithm of permeability coefficient and normalized effective internal friction angle were 0.62, 1.57 and 2.48, respectively, indicating that the closest parameter to cohesion is the effective internal friction angle, followed by the logarithm of permeability coefficient and density in the study area. The results of t-tests on each regression coefficient showed that except for the p value of density, which was 0.05, those of other factors were less than 0.01, indicating that cohesion had a significant correlation with its inter-controlled factors.

(3) The quantitative model of cohesion and its inter-controlled factors provides a scientific basis for debris flow hazard early warning.

Fine sediments with particle sizes less than 2 mm are easily transported by water, especially during high-intensity rainfall or rapid snow melt. Thus, these materials play an important role in the debris flow early warning system. Debris flow early warning needs to quickly detect the stability of these fine-grained sediments, being one of the factors controlling disaster scales.

Cohesion reflects the strain capacity of soil to resist external stress, and is closely related to soil stability. Although cohesion varies with water content, we can quickly obtain its value by the quantitative model presented in this article when the fine-grained sediments on the surface are in their natural state. Then, dangerous areas can be identified by the early warning system according to the value of soil stability estimated by cohesion.

In other words, less stability indicates higher danger, and thus provides a scientific basis for debris flow early warning.

Author Contributions: Conceptualization, Q.W. and J.X.; methodology, Q.W., J.X., J.Y. and P.L.; validation, Q.W., J.X. and P.L.; formal analysis, Q.W. and J.X.; investigation, Q.W., J.X., P.L., D.C. and W.X.; data curation, J.X., J.Y. and P.L.; writing—original draft preparation, Q.W., J.X. and J.Y.; writing—review and editing, Q.W.; visualization, D.C.; supervision, W.X.; project administration, Q.W.; funding acquisition, Q.W. All authors have read and agreed to the published version of the manuscript.

Funding: This research was funded in part by the National Natural Science Foundation of China (grant number 42071312), the Innovative Research Program of the International Research Center of Big Data for Sustainable Development Goals (grant number CBAS2022IRP03), the Hainan Hundred Special Project (grant number 31, JTT [2018]), the National Key R&D Program (grant number 2021YFB3900503), the Special Project of Strategic Leading Science and Technology of the Chinese Academy of Sciences (grant number XDA19090139), the Second Tibetan Plateau Scientific Expedition and Research (STEP) (grant number 2019QZKK0806) and the Hainan Provincial Department of Science and Technology (grant number ZDKJ2019006).

Institutional Review Board Statement: Not applicable.

Informed Consent Statement: Not applicable.

Data Availability Statement: Not applicable.

Acknowledgments: We express our acknowledgments to Chuanxin Li and Shuo Gao in China University of Geosciences (Beijing) for their hardware supporting. We also express our acknowledgments to reviewers and editors for their good comments and suggestions to make the article better. Finally, we express our acknowledgments to Yu Chen, Bihong Fu and Pilong Shi in the Aerospace Information Research Institute, Chinese Academy of Sciences for their hardware supporting.

Conflicts of Interest: The authors declare that they have no conflict of interest.

References

1. Liu, P.; Wei, Y.M.; Wang, Q.J.; Xie, J.J.; Chen, Y.; Li, Z.C.; Zhou, H.Y. A research on landslides automatic extraction model based on the improved mask R-CNN. *ISPRS Int. J. Geo-Inf.* **2021**, *10*, 168. [CrossRef]
2. Xiao, J.B. The Emergency Management Department of China Announced the National Top Ten Disasters in 2019. Available online: http://society.people.com.cn/n1/2020/0112/c1008-31544517.html,2020 (accessed on 2 January 2022).
3. Zhu, C.; Huang, Y.; Zhan, L.T. SPH-based simulation of flow process of a landslide at Hongao landfill in China. *Nat. Hazards* **2018**, *93*, 1113–1126. [CrossRef]
4. Huang, Y.; Zhu, C.Q. Simulation of flow slides in municipal solid waste dumps using a modified MPS method. *Nat. Hazards* **2014**, *74*, 491–508. [CrossRef]
5. Zhu, C.Q.; Chen, Z.Y.; Huang, Y. Coupled moving particle simulation–finite-element method analysis of fluid–structure interaction in geodisasters. *Int. J. Geomech.* **2021**, *21*, 4021081.1–4021081.11. [CrossRef]
6. Huang, Y.; Zhu, C.Q.; Xiang, X.; Mao, W.W. Liquid-gas-like phase transition in sand flow under microgravity. *Microgravity Sci. Technol.* **2015**, *27*, 155–170. [CrossRef]
7. Wu, Q.; Xu, L.R.; Zhou, K.; Liu, Z.Q. Starting analysis of loose deposits of gully debris flow. *J. Nat. Disasters* **2015**, *24*, 89–97.
8. Xu, X.C.; Chen, J.P.; Shan, B. Study of grain distribution characteristics of solid accumulation in debris flow. *Yangtze River* **2015**, *46*, 51–54.
9. Ma, M. Stability analysis of high slope of accumulation at the outlet of pressure flood discharge and sediment discharge tunnel of Jiudianxia Water Control Project. *Water Conserv. Plan. Des.* **2019**, *4*, 64–67.
10. Tang, M.G.; Xu, Q.; Li, J.Q.; Luo, J.; Kuang, Y. An experimental study of the failure mechanism of shallow landslides after earthquake triggered by rainfall. *Hydrogeol. Eng. Geol.* **2016**, *43*, 128–135.
11. Jonathan, W.F.R.; Reuben, A.H.; Brian, S.I. Particle size distribution of main-channel-bed sediments along the upper Mississippi River, USA. *Geomorphology* **2016**, *264*, 118–131.
12. Xie, J.; Wang, M.; Liu, K.; Coulthard, T.J. Modeling sediment movement and channel response to rainfall variability after a major earthquake. *Geomorphology* **2018**, *320*, 18–32. [CrossRef]
13. Marcel, H.; Velio, C.; Coraline, B.; Guo, X.; Berti, M.; Graf, C.; Hübl, J.; Miyata, S.; Smith, J.B.; Yin, H.-Y. Debris-flow monitoring and warning: Review and examples. *Earth-Sci. Rev.* **2019**, *199*, 102981. [CrossRef]
14. Mohd, A.; Aminuddin, A.; Ghania Reza, M.; Ngai, W.C. Stable channel analysis with sediment transport for rivers in Malaysia: A case study of the Muda, Kurau, and Langat rivers. *Int. J. Sediment Res.* **2020**, *35*, 455–466. [CrossRef]

15. Philip, K.L.; Lalit, K.; Richard, K. Monitoring river channel dynamics using remote sensing and GIS techniques. *Geomorphology* **2019**, *325*, 92–102.
16. Miao, Q.H.; Yang, D.W.; Yang, H.B.; Li, Z. Establishing a rainfall threshold for flash flood warnings in China's mountainous areas based on a distributed hydrological model. *J. Hydrol.* **2016**, *541*, 371–386. [CrossRef]
17. Wang, Q.J.; Wei, Y.M.; Chen, Y.; Lin, Q.Z. Hyperspectral Soil Dispersion Model for the Source Region of the Zhouqu Debris Flow, Gansu, China. *IEEE J. Sel. Top. Appl. Earth Obs. Remote Sens.* **2016**, *9*, 876–883. [CrossRef]
18. Fan, R.L.; Zhang, L.M.; Wang, H.J.; Fan, X.M. Evolution of debris flow activities in Gaojiagou Ravine during 2008–2016 after the Wenchuan earthquake. *Eng. Geol.* **2018**, *235*, 1–10. [CrossRef]
19. Hu, W.; Scaringi, G.; Xu, Q.; Pei, Z.; Van Asch, T.W.; Hicher, P.-Y. Sensitivity of the initiation and runout of flowslides in loose granular deposits to the content of small particles: An insight from flume tests. *Eng. Geol.* **2017**, *231*, 34–44. [CrossRef]
20. Liu, Z.W. Numerical stimulation analysis for the effects of soil mass strength parameters on the tunnels with side slopes. *J. Water Resour. Archit. Eng.* **2014**, *12*, 176–180.
21. Song, Y.S.; Xu, X.; Yang, S.; Gao, M. Influence of water content on shear strength of granite residual soil in Huangdao area. *J. Shandong Univ. Sci. Technol. (Nat. Sci.)* **2019**, *38*, 33–40.
22. Dong, S.; Wang, H.Y.; Li, J.; Ma, F.; Bai, X. Effects of water content and compaction degree on mechanical characteristics of compacted silty soil. *J. Guangxi Univ. (Nat. Sci. Ed.)* **2020**, *45*, 978–985.
23. Feng, H.K. Grey correlation analysis of influencing factors of soil shear strength parameters. *Subgrade Work.* **2008**, *4*, 166–167.
24. Sun, J.; Hong, B.N.; Liu, X.; Li, J.W. Sensitivity analysis on stability of embankment slope based on cohesion and friction angle. *J. Water Resour. Archit. Eng.* **2013**, *11*, 99–104.
25. Zhang, K.; Li, M.Z.; Yang, B.B. Research on effect of water content and dry density on shear strength of remolded loess. *J. Anhui Univ. Sci. Technol. (Nat. Sci.)* **2016**, *36*, 74–79.
26. Zhou, C.M.; Cheng, Y.; Wang, Y.; Wang, Q.H.; Fu, M.X. Study on influencing factors of shear strength parameters of compacted loess. *J. Disaster Prev. Mitig. Eng.* **2018**, *38*, 258–264.
27. Wang, C.Y.; Sun, Z.H.; Bian, H.B.; Lu, X.Y.; Qiu, X.M. Significance analysis of factors affecting the cohesion of silty clay. *J. Shandong Agric. Univ. (Nat. Sci. Ed.)* **2020**, *51*, 646–650.
28. Wang, Q.J.; Xie, J.J.; Yang, J.Y.; Liu, P.; Chang, D.K.; Xu, X.T. Research on permeability coefficient of fine sediments in debris-flow gullies, southwestern China. *Soil Syst.* **2022**, *6*, 29. [CrossRef]
29. Xu, M.; Yang, L.Z.; Wang, D.H.; Chen, M.J.; Chu, W.Y. *(GB/T 36197-2018)*; National Standard of the People's Republic of China: Technical Guide for Soil Sampling of Soil Quality. China National Standardization Administration Committee: Beijing, China, 2018.
30. Li, C.M.; Wang, D.; Liu, X.L.; Tan, M.J.; Liu, Y.; Jin, Z.G.; Yin, Y.; Wu, Y.J.; Qiu, R.Q.; Sun, W.; et al. *(GB/T 30319-2013)*; National Standard of the People's Republic of China: Basic Provisions for Basic Geographic Information Database. China National Standardization Administration Committee: Beijing, China, 2013.
31. Sheng, S.X.; Dou, Y.; Tao, X.Z.; Zhu, S.Z.; Xu, B.M.; Li, Q.Y.; Guo, X.L.; He, X.M. *(SL237-1999)*; National Standard of the People's Republic of China: Geotechnical Test Code. China Water Resources and Hydropower Press: Beijing, China, 1999.
32. Fan, M.Q.; Teng, Y.J.; Wu, W.; Liu, Y.H.; Luo, M.Y.; Wang, Y.; Xin, H.B.; Zhang, W.; Wen, Y.F.; Gong, B.W.; et al. *(GB/T 50145-2007)*; Engineering Classification Standard for Soil. China Planning Press: Beijing, China, 2008.
33. Wang, Y.; Akeju, O.V. Quantifying the cross-correlation between effective cohesion and friction angle of soil from limited site-specific data. *Soils Found.* **2016**, *56*, 1055–1070. [CrossRef]
34. Sun, X.D.; Wang, D. Analysis of cohesion value of soil. *Liaoning Build. Mater.* **2010**, *3*, 39–42.
35. Yan, M.K. The Effect of Water Content on Apparent Cohesion of Sand and Its Engineering Application. Master's Thesis, Chang'an University, Xi'an China, 2018.
36. Wang, H.; Cheng, H.; Huang, G.M.; Xu, T.X.; Zhao, J.M.; Jiang, H. Analysis of impact of constitutive factors and environmental factor (water) on soil cohesion. *J. Nanchang Inst. Technol.* **2019**, *38*, 60–66.

 *sustainability*

Article

Mapping Soil Properties at a Regional Scale: Assessing Deterministic vs. Geostatistical Interpolation Methods at Different Soil Depths

Jesús Barrena-González, Joaquín Francisco Lavado Contador * and Manuel Pulido Fernández

Instituto Universitario de Investigación para el Desarrollo Territorial Sostenible (INTERRA), Universidad de Extremadura, 10071 Cáceres, Spain
* Correspondence: frlavado@unex.es

Abstract: To determine which interpolation technique is the most suitable for each case study is an essential task for a correct soil mapping, particularly in studies performed at a regional scale. So, our main goal was to identify the most accurate method for mapping 12 soil variables at three different depth intervals: 0–5, 5–10 and >10 cm. For doing that, we have compared nine interpolation methods (deterministic and geostatistical), drawing soil maps of the Spanish region of Extremadura ($41,635$ km^2 in size) from more than 400 sampling sites in total (e.g., more than 500 for pH for the depth of 0–5 cm). We used the coefficient of determination (R^2), the mean error (ME) and the root mean square error ($RMSE$) as statistical parameters to assess the accuracy of each interpolation method. The results indicated that the most accurate method varied depending on the property and depth of study. In soil properties such as clay, EBK (Empirical Bayesian Kriging) was the most accurate for 0–5 cm layer ($R^2 = 0.767$ and $RMSE = 3.318$). However, for 5–10 cm in depth, it was the IDW (Inverse Distance Weighted) method with R^2 and $RMSE$ values of 0.689 and 5.131, respectively. In other properties such as pH, the CRS (Completely Regularized Spline) method was the best for 0–5 cm in depth ($R^2 = 0.834$ and $RMSE = 0.333$), while EBK was the best for predicting values below 10 cm ($R^2 = 0.825$ and $RMSE = 0.399$). According to our findings, we concluded that it is necessary to choose the most accurate interpolation method for a proper soil mapping.

Keywords: soil mapping; interpolation methods; different depths; regional scale

Citation: Barrena-González, J.; Lavado Contador, J.F.; Pulido Fernández, M. Mapping Soil Properties at a Regional Scale: Assessing Deterministic vs. Geostatistical Interpolation Methods at Different Soil Depths. *Sustainability* **2022**, *14*, 10049. https://doi.org/10.3390/su141610049

Academic Editors: Antonio Ganga, Blaž Repe and Mario Elia

Received: 19 July 2022
Accepted: 11 August 2022
Published: 13 August 2022

Publisher's Note: MDPI stays neutral with regard to jurisdictional claims in published maps and institutional affiliations.

1. Introduction

The interpolation methods have been a breakthrough in soil science, expanding the interest of study areas and saving time and money on complex field works. Many of them have been developed particularly for mapping soil properties [1–4] such as texture [2], nitrogen [5], phosphorus [6] and organic matter content [7] through different techniques [8–10]. They can be divided into two main groups: deterministic and geostatistical. Deterministic methods include Inverse Distance Weighting (IDW), Radial Basis Function (RBF), Global Polynomial Interpolation (GPI), Local Polynomial Interpolation (LPI) or Splines. On the other hand, kriging and its variants such as ordinary, simple, empirical, universal, etc., are included in the group of geostatistical techniques. In recent years, however, so-called hybrid techniques have been used, which consist of the fusion of a linear model and a non-linear interpolation model (e.g., Regression Kriging (RK), or Geographically Weighted Regression (GWR)) [11,12].

Given the variety of available interpolation methods, choosing one is a complex and critical task that entails variations in study results, as also do the nature and context of the data (e.g., experimental design, sampling density or topographic characteristics) [13]. Therefore, working on the identification of the most appropriate interpolation method in each case is essential for a proper soil mapping.

Currently, there is much controversy about which interpolation method is ideal for different properties and case studies. In this regard, the working spatial scale is one of the most important issues when it is used in one method or another. At the regional scale, many studies applied interpolation techniques and obtained different results depending on the variables interpolated [14–19]. Shen et al. [12], for example, found that ordinary kriging provided the best results for predicting total soil phosphorus content, while Chen et al. [20] identified IDW as the best method for predicting and monitoring soil moisture at regional scale. In other cases in which soil carbon [21] or soil moisture content [22] are predicted, hybrid methods such as RK and GWR were preferred.

The application of these techniques has not been used only to surface soil properties but also to map them at different depth intervals [23–25]. The most common depth interval is 0–10 cm, even when in some environments with shallow soils the adequate interval should be 0–5 cm in depth [26,27]. For those works that used different depth intervals, a large variety of methods have been used so far [28]. However, geostatistical techniques are the most widely used [2,29,30].

In addition, there is not yet a clear consensus about which technique is the most appropriate to reflect properties at different depths, frequently depending on individual cases. In some studies, for example, IDW seems to be an appropriate method for mapping variables such as P, K or SOM in the subsoil [31,32], while in others it is the kriging technique that works the best for topsoil [33]. Conversely, in some cases RBF reported better results than IDW or kriging techniques [34].

In regions such as Extremadura (SW Spain), shallow soils occupy 70% of the total area. This characteristic implies that soil mapping at different depth intervals is considered necessary. Nevertheless, so far in this region, there are no works focused on mapping soil properties at different depths. Beyond the soil mapping developed by the European Soil Data Centre (ESDAC) [35–37], there are no studies that focus on mapping soil properties. Some of these studies were focused on mapping quality indices at fixed 0–30 cm intervals, mainly using geostatistical techniques [38]. Some studies compared different interpolation techniques but focused on the quantification of soil losses or on some specific properties without considering different depths and at restricted spatial scales [39,40]. Other studies at a regional scale produced maps of soil aridity [41], bioclimatic indices [42] or sensitivity to land degradation [43].

The scarcity of investigations attending the mapping of soil properties at regional scales is evident. This highlights the need to provide accurate mapping techniques for soil properties. Therefore, the main objective of this work was to identify the most suitable interpolation method for the studied variables at different depths.

2. Materials and Methods

2.1. Study Area

The study has been carried out in the Autonomous Region of Extremadura, located in the southwest of Spain (Figure 1). Extremadura has a surface area of 41,635 km^2, which represents 8.25% of the total area of Spain. The orography of Extremadura is characterised by the presence of two large river basins (the Tagus and Guadiana rivers) and three parallel mountain ranges (Sistema Central, Montes de Toledo and Sierra Morena), which interrupt the dominance of the extensive peneplains. Precambrian and Paleozoic slates and granites are the dominant rocks on which our soils are developed. Cambisol, Leptosol and Regosol soil types occupy 70% of the total surface area, which is mainly used for livestock rearing. On the other hand, Fluvisol or Luvisol soil types predominate in the river valleys, mainly dedicated to agricultural activities [35]. The climate is Mediterranean, with mild winters and hot and dry summers. Yearly total average rainfall is below 600 mm, except in the mountain ranges where values over 1000 mm are reached. The average temperature in the region is around 16–17 °C with lower values in the vicinity of the mountain ranges.

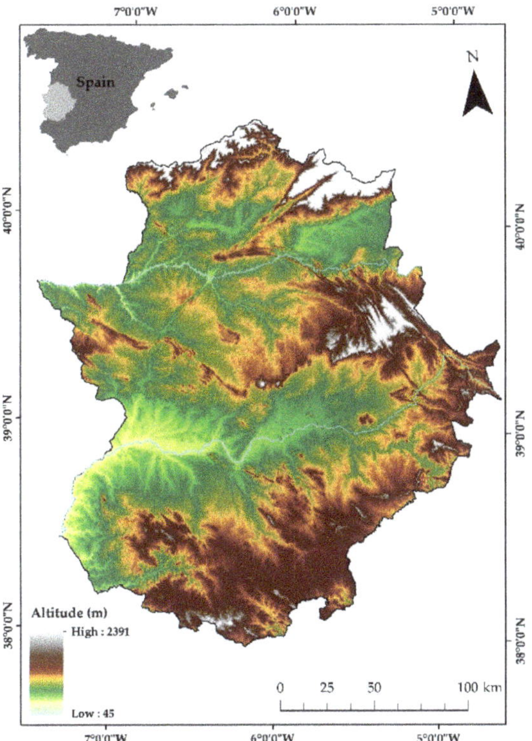

Figure 1. Location map of the study area.

2.2. Dataset Characteristics and Variables Selection

Various data sources contributed to the dataset used in this study. For all sources, data available for any of the three studied soil depths (0–5, 5–10 and >10 cm, respectively) were considered, disregarding depth intervals that do not comply with those proposed in the study. The >10 cm class refers to soils mostly not exceeding 35 or 40 cm depth. This region is dominated by Cambisol and especially Leptosol soils, most of which are less than 40 cm in depth. Data have been structured as a point soil-sample database, where most of the records come from different research projects developed by the Geo-Environmental Research Group (GIGA) of the University of Extremadura over the last 20 years, to whom authors belong. Some data from the book *"Estudio de los Suelos de la Tierra de Barros"* [36] have also been used. Additionally, other data come from the Spanish Soil Properties Database, created by the Centre for Energy, Environmental and Technological Research (CIEMAT), from the Soil Catalogue of Extremadura in soil profiles carried out at the end of the 1990s, as well from the Soil Grids platform [37]. In one of the data sources used, it was not possible to verify which analysis method was used. However, in the other sources, the method was found to be the same. The total number of sampling points varied depending on the soil property and the depth where soil samples were taken (Figure 2, Table 1). For instance, more than 500 sampling points are available for some properties, such as pH at 0–5 cm depth and 94 for >10 cm depth phosphorus.

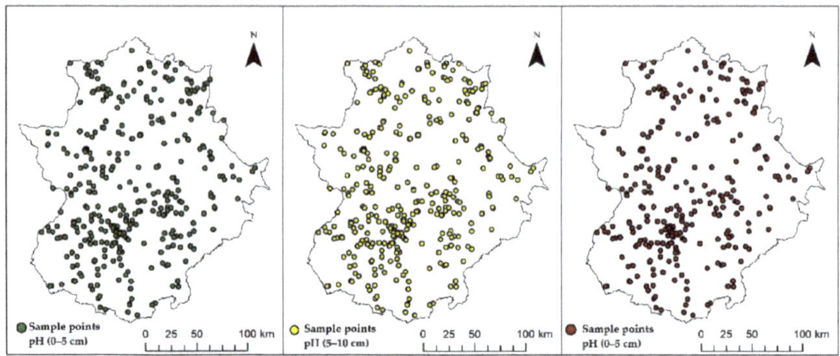

Figure 2. Sampling points available for pH at 0–5, 5–10 and >10 cm, respectively.

Table 1. Methods used for the analysis of the studied properties and distribution of the number of samples by property and soil depth interval.

Soil Property	Depth (cm)	n *	Unit	Method
Clay	0–5	495		
	5–10	393		
	>10	326		
Silt	0–5	424	%	Soil Survey Laboratory Methods Manual [38]
	5–10	392		
	>10	328		
Sand	0–5	333		
	5–10	294		
	>10	239		
pH	0–5	509		1:2.5 soil/water
	5–10	449		
	>10	349		
CEC	0–5	445		
	5–10	402		
	>10	306		
Calcium	0–5	314		
	5–10	262		
	>10	171	Cmol kg^{-1}	MAPA [39]
Magnesium	0–5	319		
	5–10	226		
	>10	178		
Sodium	0–5	284		
	5–10	262		
	>10	185		
Available N	0–5	265		
	5–10	254	%	Dumas [40]
	>10	205		
Available P	0–5	176		
	5–10	155	ppm	Olsen et al. [41]
	>10	94		
Potassium	0–5	296		Ammonium acetate at pH 7 (USDA) [42]
	5–10	243	Cmol kg^{-1}	
	>10	190		
SOM	0–5	420		Walkley and Black wet combustion [43]
	5–10	419	%	
	>10	303		

* n corresponds to the total number of samples for each property.

2.3. Data Preprocessing and Statistical Analysis

The first step of the analysis consisted of filtering the outliers from the original dataset. In order to do this the cluster and outlier analysis tool (Anselin Local Moran's I) integrated in ArcGIS 10.5 software was applied [44,45]. This analysis provides a simple and adequate identification of statistically significant outliers, with 95% confidence in relation to neighbouring data.

Once the outliers were identified and removed from the original dataset, the remaining data were randomly partitioned into training and validation datasets, amounting, respectively, to 80% and 20% of the data available for each one of the interpolation methods selected. This sequence of work was carried out for the 12 soil properties and three depths considered.

Statistical parameters such as mean, minimum, maximum and coefficient of variation were calculated to characterise the filtered dataset. In order to reflect the differences observed between some statistics among interpolation methods, bar and whisker plots were used. All the statistical analyses were carried out using Statistica v. 6.0 (StatSoft, Tulsa, OK, USA) [46] and Microsoft Office Excel software packages.

2.4. Interpolation Methods

Nine interpolation methods were considered as spatially predictive techniques for each of the variables selected in the study. Six deterministic and three geostatistical methods were chosen, all of them integrated into the Geostatistical Analyst extension of ArcGIS 10.5 (ESRI, Redlands, CA, USA) [47].

The deterministic methods were:

1. Inverse Distance Weighting (IDW), which works on the assumption that closer points are more similar to each other than those further away. This method uses the values surrounding some unmeasured point to predict, giving greater weight to the closer points [48].
2. Four Radial Basis Functions: Completely Regularized Spline (CRS), Spline With Tension (SWT), Multiquadric (M-Q) and Inverse Multiquadric (IM-Q). These are exact interpolation techniques, i.e., the surface must pass through each of the measured values. Each function will give an interpolation surface with a different shape and different results [49].
3. Local Polynomial Interpolation (LPI), which works by fitting many polynomials within the specified overlapping neighbourhoods as opposed to Global Polynomial Interpolation (GPI) [50].

Contrariwise, the geostatistical techniques assume spatial autocorrelation as part of the interpolation process. The geostatistical methods used in this study were:

1. Ordinary Kriging (OK), which works on the assumption that a constant mean is unknown throughout the process [51].
2. Simple Kriging (SK), which assumes stationarity from the beginning of a known mean. Hence, mathematically speaking, it is the simplest but least general method [52].
3. Empirical Bayesian Kriging (EBK) tends to maximum likelihood estimation. This allows us to measure the evidence and the uncertainty of the emulator [49].

It should be noted that in each of the interpolation methods used, the parameters of the model were modified in order to obtain the lowest mean square error as the output result.

2.5. Assessment of the Interpolation Methods Reliability

From the cross-validation, the mean error (ME), the coefficient of determination (R^2) and the root mean square error ($RMSE$) were used to assess the accuracy of the interpolation methods. In order to consider the most reliable method, the lowest $RMSE$ and the highest R^2 were taken into account [53]. In some cases, where these parameters were equal, the ME value closest to 0 was used to determine the most reliable method.

ME gives an absolute value that determines the degree of bias obtained in the estimates. Higher values indicate larger differences between predicted and observed values [54]. ME values are calculated as follows:

$$ME = \frac{\sum_{i=1}^{n}(\hat{Z}\,(s_i) - z\,(s_i))}{n} \tag{1}$$

where, $z\,(s_i)$ is the observed value at point s_i, $\hat{Z}\,(s_i)$ is the predicted value at point s_i and n the number of samples.

The coefficient of determination or R-squared (R^2) assesses the ability of a model to predict an outcome of a regression. Thus, the coefficient of determination indicates how well the data fit the model. Values vary between positive 1 and 0, values close to 1 indicate a better fitness of the model [55]. The equation is given as follows:

$$R^2 = \frac{[\sum_{i=1}^{n}(P_i - P_{ave})(Q_i - Q_{ave})]^2}{\sum_{i=1}^{n}(P_i - P_{ave})^2 \sum_{i=1}^{n}(Q_i - Q_{ave})^2}$$

Being P_{ave} the average of the estimated value; Q_{ave} is the average of measured value; and n is the points number used for estimation.

The root mean square error ($RMSE$) was used to quantify the accuracy of the interpolation model used. Low $RMSE$ values indicate higher reliability of the model. The $RMSE$ is calculated as follows:

$$RMSE = \sqrt{\frac{\sum_{i=1}^{n}[\hat{Z}\,(s_i) - z\,(s_i)]^2}{n}} \tag{2}$$

Being, $z\,(s_i)$ the observed value at point s_i, $\hat{Z}\,(s_i)$ the predicted value at point s_i, and n the number of samples.

3. Results

3.1. Descriptive Statistics

Table 2 shows the descriptive statistics of each of the studied variables. A trend can be observed of clay, silt, and sand content, as well as pH, cation exchange capacity, calcium, and magnesium to increase with depth, showing other variables, i.e., nitrogen, phosphorus, potassium and organic matter, an opposite behaviour. The coefficient of variation (CV) reflects the variability of soil properties was low for pH, with values not exceeding 17%, being high for clay and silt content, as well as for some chemical properties such as cation exchange capacity, phosphorus, potassium, or soil organic matter. The CV was particularly high for calcium, magnesium, sodium, and nitrogen.

3.2. Optimal Interpolation Method by Soil Property

3.2.1. Particle Size Distribution

Model statistics for the three studied soil depths, obtained with the validation dataset when modelling particle size distribution, are shown in Table 3. In Appendix A, Figure A1 shows the regional cartographic representation of the best performing methods for clay, sand, and silt content. Regarding the clay content, the more accurate methods were EBK for 0–5 cm soil depth and IDW for larger depths. $RMSE$ values were below the average reflected in Table 1, showing an increasing trend with depth (3.554%, 5.131% and 5.205%). In turn, R^2 values decreased (0.767, 0.689 and 0.648). It is worth noting that the differences in $RMSE$ between the most and least accurate methods were about 1% or higher being R^2 differences minor. In the case of silt content, IDW was the most accurate method for 0–5 and 5–10 cm depth, performing the OK method better for greater depths. As in the case of clay content, $RMSE$ values were below the average measured for the reference dataset. Model accuracy decreased when increasing soil depth (6.22%, 5.103% and 5.605%), showing an opposite behaviour to that observed for clay content, while R^2 increased (0.831, 0.897 and 0.921). With regards to the sand content, the best performing methods were CRS at 0–5 cm depth, IDW for 5–10 cm depth and SK at >10 cm depth. *RMSE* data increased in this case with increasing soil depth (4.627%, 5.217% and 6.354%), and the same is true for R^2 values (0.739, 0.816 and 0.799), contrasting with the trend observed in the clay model.

Table 2. Descriptive statistics of the studied variables.

Variable	Depth	N	Mean	Min.	Max.	CV (%)
Clay (%)	0–5 cm	495	15	1	50	55
	5–10 cm	393	19	2.	81	65
	>10 cm	326	22.811	3	77	59.
Silt (%)	0–5 cm	424	33	1	70	46
	5–10 cm	392	33	1	71	49
	>10 cm	328	31	3	74	54
Sand (%)	0–5 cm	333	47	6	91	27
	5–10 cm	294	43	4	89	32
	>10 cm	239	41	2	85	38
pH (1:2.5)	0–5 cm	509	5.93	4.00	8.20	14
	5–10 cm	449	6.01	3.70	9.1	15
	>10 cm	349	6.14	4.20	9.30	16
CEC (Cmol kg^{-1})	0–5 cm	445	15.52	0.81	80.46	67
	5–10 cm	402	15.57	2.00	69.66	76
	>10 cm	306	16.34	2.00	61.60	70
Ca (Cmol kg^{-1})	0–5 cm	314	6.78	0.41	61.87	116
	5–10 cm	262	6.90	0.17	62.00	142
	>10 cm	171	8.85	0.18	55.59	137
Mg (Cmol kg^{-1})	0–5 cm	319	1.80	0.07	12.30	87
	5–10 cm	226	1.77	0.05	25.00	93
	>10 cm	178	2.29	0.03	16.62	104
Na (Cmol kg^{-1})	0–5 cm	284	0.65	0.07	2.92	100
	5–10 cm	262	0.40	0.04	2.13	144
	>10 cm	185	0.43	0.02	2.54	129
N (%)	0–5 cm	265	0.23	0.02	0.95	87
	5–10 cm	254	0.13	0.01	0.38	85
	>10 cm	205	0.09	0.00	0.26	97
P (ppm)	0–5 cm	176	22.10	0.40	69.00	53
	5–10 cm	155	15.12	0.40	67.00	44
	>10 cm	94	11.84	0.40	48.30	51.
K (Cmol kg^{-1})	0–5 cm	296	0.57	0.03	3.40	77.
	5–10 cm	243	0.42	0.02	2.12	87
	>10 cm	190	0.42	0.02	2.91	95
SOM (%)	0–5 cm	420	3.62	0.10	20.80	76
	5–10 cm	419	1.86	0.10	14.50	98
	>10 cm	303	1.21	0.10	6.00	82

3.2.2. Selected Chemical and Sorption Complex Properties

Table 4 shows the main model statistics of the interpolation methods used for the selected chemical variables. The maps that represent the regional pattern of these variables with the more accurate interpolation methods are shown in Figure A2. For the pH, the best performing method at 0–5 cm depth was CRS (R^2 = 0.834 and *RMSE* = 0.333). The EBK algorithm provided very similar results to the former, but with slightly higher RMSE and lower R^2. At 5–10 cm depth, OK method performed better (R^2 = 0.823 and *RMSE* = 0.328), followed by SWT with slightly higher R^2 and *RMSE*. At soil depths greater than 10 cm, EBK was the most accurate method (R^2 = 0.825 and *RMSE* = 0.399). Overall, results obtained at each of the three depths were comparable, with RMSE values that varied below 0.07.

Regarding the Cation Exchange Capacity (CEC), IDW was the most accurate method at 0–5 cm depth, with *RMSE* of 3.778 and R^2 of 0.854. At 5–10 cm, OK was the method that provided the best results, with *RMSE* and R^2 accounting to 4.533 and 0.702, respectively. The same was true at soil depths greater than 10 cm, where EBK was the best performing method, with *RMSE* of 4.415 and an R^2 of 0.697. In the case of Calcium (Ca), unlike the previous variables, it was a geostatistical method, EBK, that provided the best results at 0–5 cm depth, with *RMSE* of 0.864 and R^2 of 0.977. At 5–10 cm depth, on the other hand, it was the IDW method that showed better results, with *RMSE* of 1.879, one point higher than the topmost soil layer, but still showing a good fitness with an R^2 of 0.973. The *RMSE* increased at deeper depths (>10 cm), where EBK was identified as the most accurate method but still retaining a high R^2 value of 0.964. Magnesium (Mg) models for the 0–5 cm soil depth showed the SK method as the best performing one with *RMSE* of 0.741 and an R^2 of 0.569, indicating poor model fitness. Nevertheless, with increasing depths, results improve considerably, with the OK method being better at 5–10 cm that reduced the *RMSE* to 0.534 and improved the R^2 to 0.770. The same was true at >10 cm depth, where the IDW method provided an *RMSE* that increased again to 0.781 while still maintaining a good R^2 value of 0.708. Finally, for Sodium (Na) content, it was the CRS method that produced the best results at 0–5 cm depth, with *RMSE* and R^2 that amounted to 0.167 and 0.888, respectively. For the 5–10 cm soil layer, CRS was also the method showing the best results, although both *RMSE* and R^2 decreased to 0.138 and 0.511, respectively. At >10 cm soil layer, although IDW was the most accurate method, the model validation statistics obtained were poor (*RMSE* = 0.196 and R^2 = 0.044) indicating a very poor model fitness.

Table 3. Model validation statistics for the different interpolation methods used when modelling particle size distribution at the three studied soil depths. In bold is the optimal method.

Variable	Method	0–5 cm			5–10 cm			>10 cm		
		ME	R^2	*RMSE*	ME	R^2	*RMSE*	ME	R^2	*RMSE*
Clay (%)	IDW	−0.249	0.723	3.554	0.265	0.689	5.131	1.393	0.648	5.205
	CRS	0.239	0.725	3.505	1.119	0.663	5.313	1.870	0.639	5.800
	SWT	0.223	0.731	3.456	1.123	0.638	5.321	1.822	0.642	5.771
	M-Q	0.200	0.721	4.002	0.147	0.575	6.357	0.740	0.580	6.503
	IM-Q	3.964	0.650	3.964	0.690	0.641	5.569	2.524	0.562	6.590
	LPI	0.223	0.663	4.046	0.970	0.584	5.799	2.077	0.607	6.153
	OK	0.220	0.728	3.523	0.570	0.625	5.480	1.946	0.578	6.376
	SK	0.390	0.674	4.157	0.388	0.603	5.865	2.367	0.591	6.101
	EBK	**0.198**	**0.767**	**3.318**	0.238	0.639	5.525	1.807	0.613	5.972
Silt (%)	IDW	**0.267**	**0.831**	**6.222**	−0.486	**0.897**	**5.103**	1.168	0.905	6.595
	CRS	−0.163	0.823	6.363	−1.365	0.895	5.287	−0.564	0.923	6.438
	SWT	−0.110	0.825	6.331	−1.357	0.895	5.280	−0.418	0.919	6.344
	M-Q	0.489	0.807	6.775	−0.627	0.889	5.494	0.239	0.908	5.804
	IM-Q	−0.416	0.734	7.835	−1.629	0.847	6.375	−2.209	0.761	10.051
	LPI	−0.423	0.793	7.234	−1.415	0.856	6.158	−1.823	0.862	8.939
	OK	−0.264	0.820	6.458	−1.420	0.882	5.626	**−0.609**	**0.921**	**5.605**
	SK	−0.094	0.815	6.563	−1.537	0.884	5.582	0.244	0.923	6.020
	EBK	−0.329	0.820	6.488	−1.040	0.892	5.383	−0.444	0.929	5.725
Sand (%)	IDW	0.847	0.657	5.290	**0.200**	**0.816**	**5.217**	−1.818	0.750	7.039
	CRS	**0.290**	**0.739**	**4.627**	0.730	0.804	5.459	−1.703	0.748	7.072
	SWT	0.204	0.723	4.727	0.832	0.810	9.637	−1.706	0.748	7.069
	M-Q	0.884	0.608	6.406	0.280	0.826	5.729	−1.392	0.732	7.544
	IM-Q	−0.740	0.535	6.233	0.664	0.700	6.680	−2.077	0.514	9.782
	LPI	1.065	0.602	6.279	1.164	0.738	6.881	−0.247	0.615	8.851
	OK	0.471	0.680	5.094	1.460	0.720	6.790	−1.505	0.780	6.586
	SK	0.719	0.677	5.133	0.740	0.693	6.878	**−1.437**	**0.799**	**6.354**
	EBK	1.083	0.637	5.602	0.759	0.816	5.258	−1.794	0.764	6.906

Table 4. Model validation statistics for the different interpolation methods used when modelling the chemical properties selected at the three studied soil depths. In bold is the optimal method.

Variable	Method	0–5 cm ME	R^2	RMSE	5–10 cm ME	R^2	RMSE	>10 cm ME	R^2	RMSE
pH (1:2.5)	IDW	−0.030	0.809	0.350	0.071	0.808	0.349	0.015	0.803	0.413
	CRS	−0.031	**0.834**	**0.333**	0.069	0.826	0.331	0.034	0.803	0.412
	SWT	−0.006	0.794	0.364	0.068	0.827	0.330	0.033	0.804	0.411
	M-Q	−0.033	0.726	0.413	0.030	0.803	0.356	0.007	0.796	0.443
	IM-Q	0.013	0.720	0.424	0.070	0.767	0.381	0.061	0.730	0.487
	LPI	−0.064	0.772	0.399	0.028	0.744	0.837	−0.014	0.761	0.455
	OK	−0.029	0.806	0.352	**0.041**	**0.823**	**0.328**	0.021	0.782	0.441
	SK	−0.043	0.813	0.362	0.054	0.804	0.348	0.030	0.757	0.465
	EBK	−0.018	0.823	0.335	0.064	0.816	0.342	**0.024**	**0.825**	**0.399**
CEC (Cmol kg^{-1})	IDW	**1.396**	**0.854**	**3.778**	1.741	0.683	4.676	0.706	0.690	4.470
	CRS	1.444	0.823	3.990	2.019	0.713	4.555	1.025	0.671	4.625
	SWT	1.438	0.825	3.976	2.018	0.713	4.553	1.021	0.671	4.620
	M-Q	1.402	0.744	5.706	1.893	0.665	5.188	0.418	0.641	4.763
	IM-Q	1.508	0.760	4.436	1.720	0.704	4.538	1.405	0.601	5.239
	LPI	1.158	0.670	5.015	1.819	0.700	4.560	0.575	0.646	4.901
	OK	1.296	0.821	4.108	**1.690**	**0.702**	**4.533**	0.804	0.679	4.583
	SK	1.866	0.770	4.486	1.649	0.703	4.576	1.120	0.704	4.637
	EBK	1.441	0.819	4.199	1.833	0.728	4.567	**0.864**	**0.697**	**4.415**
Ca (Cmol kg^{-1})	IDW	0.006	0.833	2.263	**0.639**	**0.973**	**1.879**	−0.521	0.959	2.326
	CRS	0.254	0.960	1.153	0.451	0.952	2.280	0.040	0.905	2.610
	SWT	0.178	0.964	1.078	0.469	0.956	2.137	0.056	0.904	2.621
	M-Q	0.105	0.963	1.078	0.368	0.897	3.305	−0.812	0.968	2.254
	IM-Q	0.235	0.964	1.093	0.369	0.911	3.208	0.232	0.830	3.664
	LPI	−0.079	0.819	2.509	0.169	0.912	3.217	−0.665	0.961	2.473
	OK	0.053	0.967	1.055	0.325	0.917	2.744	−0.272	0.959	2.241
	SK	0.298	0.970	1.010	0.332	0.901	3.614	0.075	0.905	2.727
	EBK	**0.084**	**0.977**	**0.864**	0.676	0.947	2.233	**−0.400**	**0.964**	**1.955**
Mg (Cmol kg^{-1})	IDW	0.112	0.439	0.907	0.147	0.605	0.539	**0.347**	**0.708**	**0.781**
	CRS	0.268	0.448	0.814	0.246	0.554	0.613	0.417	0.688	0.822
	SWT	0.265	0.422	0.872	0.251	0.551	0.618	0.417	0.694	0.816
	M-Q	0.224	0.291	1.183	−0.052	0.432	0.577	0.370	0.365	1.186
	IM-Q	0.294	0.360	0.995	0.222	0.551	0.661	0.429	0.754	0.782
	LPI	0.303	0.413	0.827	0.229	0.485	0.650	0.362	0.659	0.833
	OK	0.135	0.504	0.789	**0.102**	**0.770**	**0.534**	0.553	0.605	0.988
	SK	**0.301**	**0.569**	**0.741**	0.274	0.333	0.724	0.370	0.535	0.939
	EBK	0.158	0.395	0.955	0.027	0.534	0.561	0.398	0.449	1.105
Na (Cmol kg^{-1})	IDW	0.069	0.828	0.210	0.074	0.505	0.142	**0.105**	**0.044**	**0.196**
	CRS	**0.030**	**0.888**	**0.167**	**0.064**	**0.511**	**0.138**	0.110	0.104	0.198
	SWT	0.040	0.891	0.169	0.067	0.503	0.138	0.116	0.089	0.201
	M-Q	0.008	0.801	0.227	0.040	0.400	0.174	0.085	0.077	0.239
	IM-Q	0.087	0.904	0.202	0.055	0.369	0.175	0.134	0.039	0.227
	LPI	0.048	0.755	0.248	0.102	0.107	0.203	0.139	0.011	0.226
	OK	0.016	0.865	0.178	0.050	0.369	0.137	0.130	0.004	0.234
	SK	0.048	0.855	0.190	0.111	0.303	0.174	0.130	0.103	0.201
	EBK	0.036	0.820	0.213	0.096	0.071	0.200	0.120	0.011	0.218

3.2.3. Selected Elements Content

The statistics obtained as a result of the validation of the selected interpolation methods used for some of the main so-called soil fertilisers are presented in Table 5, and their cartographic representation is shown in Figure A3.

Table 5. Model validation statistics for the different interpolation methods used when modelling nitrogen (N), phosphorus (P), potassium (K) and soil organic matter (SOM) at the three studied soil depths. In bold is the optimal method.

Variable	Method	0–5 cm			5–10 cm			>10 cm		
		ME	R^2	*RMSE*	*ME*	R^2	*RMSE*	*ME*	R^2	*RMSE*
N (%)	IDW	0.010	0.577	0.058	−0.003	0.128	0.039	0.002	0.685	0.024
	CRS	0.006	0.590	0.056	−0.001	0.119	0.039	0.005	0.645	0.026
	SWT	0.006	0.597	0.056	−0.001	0.116	0.039	0.005	0.649	0.026
	M-Q	0.006	0.362	0.076	−0.009	0.064	0.051	0.002	0.541	0.031
	IM-Q	0.010	0.598	0.060	−0.009	0.064	0.051	0.002	0.545	0.029
	LPI	0.010	0.463	0.067	0.003	0.039	0.045	0.009	0.587	0.032
	OK	0.004	0.603	0.056	0.000	0.142	0.038	0.005	0.595	0.028
	SK	**0.006**	**0.634**	**0.053**	**0.004**	**0.161**	**0.033**	0.008	0.442	0.034
	EBK	0.005	0.584	0.060	−0.003	0.038	0.040	0.004	0.495	0.032
P (ppm)	IDW	−1.741	0.836	6.917	−0.780	0.739	5.508	5.518	0.319	14.143
	CRS	**0.840**	**0.840**	**6.726**	1.557	0.519	6.911	7.561	0.590	14.615
	SWT	0.705	0.838	6.710	1.727	0.582	7.024	7.433	0.587	14.374
	M-Q	1.131	0.841	10.297	−0.630	0.673	8.475	12.242	0.584	27.097
	IM-Q	2.255	0.523	10.859	2.975	0.278	9.382	**6.655**	**0.625**	**8.811**
	LPI	−0.731	0.780	7.672	3.034	0.411	8.716	4.663	0.243	12.556
	OK	−0.006	0.811	7.378	**0.616**	**0.745**	**5.366**	7.387	0.151	17.555
	SK	−0.355	0.850	7.336	0.929	0.685	6.385	7.141	0.420	17.683
	EBK	1.145	0.831	8.679	0.929	0.685	6.385	8.544	0.579	16.123
K(Cmol kg^{-1})	IDW	**−0.015**	**0.729**	**0.170**	0.041	0.547	0.147	0.001	0.617	0.146
	CRS	0.000	0.723	0.173	**0.040**	**0.582**	**0.133**	0.032	0.571	0.158
	SWT	0.013	0.686	0.185	0.050	0.522	0.150	0.033	0.575	0.158
	M-Q	−0.032	0.657	0.192	0.002	0.444	0.158	0.004	0.637	0.143
	IM-Q	0.023	0.642	0.203	0.052	0.366	0.194	0.046	0.552	0.185
	LPI	−0.005	0.666	0.188	0.042	0.529	0.159	0.032	0.584	0.140
	OK	−0.020	0.716	0.174	0.027	0.512	0.154	0.021	0.557	0.146
	SK	0.052	0.690	0.187	0.049	0.578	0.141	0.038	0.586	0.145
	EBK	−0.012	0.719	0.172	0.020	0.524	0.139	**0.017**	**0.579**	**0.140**
SOM (%)	IDW	**0.105**	**0.754**	**0.957**	0.31	0.704	0.754	0.109	0.662	0.548
	CRS	−0.041	0.67	1.033	0.268	0.631	0.784	0.083	0.577	0.557
	SWT	−0.078	0.673	1.043	0.267	0.627	0.776	0.084	0.578	0.555
	M-Q	−0.07	0.683	1.072	0.175	0.441	0.881	0.043	0.506	0.630
	IM-Q	−0.052	0.46	1.287	0.264	0.449	0.888	0.022	0.267	0.667
	LPI	0	0.454	1.343	0.318	0.435	0.945	0.131	0.402	0.628
	OK	0.001	0.707	0.97	0.175	0.624	0.79	**0.082**	**0.588**	**0.531**
	SK	−0.07	0.680	1.006	0.199	0.474	0.852	0.091	0.598	0.548
	EBK	−0.008	0.684	1.001	**0.221**	**0.657**	**0.715**	0.046	0.492	0.565

At the 0–5 cm soil layer, the most accurate method predicting nitrogen content was the SK, showing an R^2 of 0.634 and an *RMSE* of 0.053. SK was also optimal for the 5–10 cm soil layer. However, although the *RMSE* value of 0.033 is acceptable, the R^2 value obtained at this depth (0.161) indicates poor model fitness. At the depth >10 cm, IDW provided the best results, with R^2 of 0.685 and *RMSE* of 0.024.

As it is seen in Table 5, when the phosphorus content is interpolated, the optimal methods were the CRS for the uppermost soil layer (R^2 = 0.840 and *RMSE* = 6.726), the OK at 5–10 cm soil depth (R^2 =0.745 and *RMSE* = 5.366) and the IM-Q for the deepest layer (>10 cm) (R^2 = 0.625 and *RMSE* = 8.811).

Regarding the potassium, the IDW was the most accurate method at 0–5 cm (R^2 = 0.729 and *RMSE* = 0.170). In this case, model accuracies were considerably reduced with depth, with the CRS method being better at 5–10 cm with an R^2 of 0.582 and an *RMSE* of 0.133, and at >10 cm soil layer, EBK revealed the most accurate results, very similar to those obtained for the above layer.

For soil organic matter (SOM), IDW was the most accurate method for the 0–5 cm layer, with an R^2 of 0.754 and *RMSE* of 0.957. The model accuracy was also reduced when modelling SOM at deeper layers, with R^2 values of 0.657 at 5–10 cm, where the EBK was the optimal model, and 0.588 at >10 cm layer, with the OK as the best performing method.

3.3. General Assessment of the Interpolation Methods

The general performance of the different interpolation methods considered in this study, attending to the whole set of modelled variables, was evaluated considering their average R^2 and *RMSE* statistics, as reflected in Figure 3.

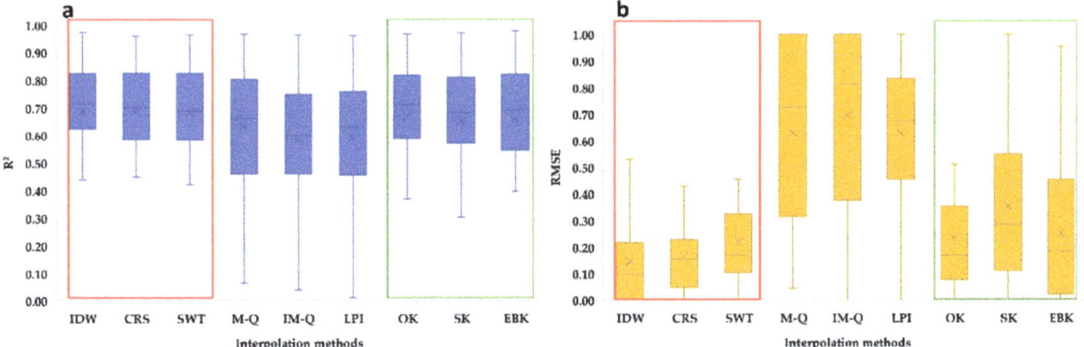

Figure 3. Analysis of the average values of R^2 and *RMSE* for the interpolation methods applied to each study variable. Whiskers show the mean value $\pm$ 0.95 confidence interval. The red rectangle groups the more accurate deterministic interpolation methods and the green rectangle groups the geostatistical interpolation methods. Letter (**a**) refers to R^2 values and letter (**b**) to *RMSE* values.

As expressed in Figure 3a, attending to the model accuracies, as expressed by the R^2, the interpolation methods with the best general performance were IDW, CRS and SWT, reaching, respectively, 0.739, 0.755 and 0.747 on average. The geostatistical methods, on the other hand, showed lower accuracies in general, OK being the one with the highest R^2, as compared to the similar values obtained with EBK and SK methods.

As for the average *RMSE* (Figure 3b), again the IDW, followed by CRS and SWT, were the more accurate methods. In this case, the geostatistical methods yielded similar or slightly higher values, with SK producing the highest mean *RMSE* overall.

When the analysis was performed as a function of the soil depth, the IDW was the method more repeatedly observed as more accurate. It was more reliable at 0–5 cm depth for properties such as silt, cation exchange capacity, potassium, and soil organic matter. At the 5–10 cm layer, it was also preferred when modelling clay, silt, sand, and calcium content. Finally, at depth >10 cm, it was also better when modelling clay, magnesium, sodium, and nitrogen content. Alternatively, CRS was more reliable for variables such as sand, pH, sodium, and phosphorus at the uppermost soil layer, also being good to express sodium and potassium content at 5–10 cm depth. Among the deterministic methods, IM-Q showed the best results when predicting phosphorus content at >10 cm.

As for the geostatistical methods, OK yielded the best results in six cases. At 5–10 cm depth, it was the most reliable method for variables such as pH, cation exchange capacity, and magnesium and phosphorus contents, and also for silt content and soil organic matter at the >10 cm soil layer. SK was the optimal method to model magnesium and nitrogen content at 0–5 cm and for nitrogen content at 5–10 cm and sand content at >10 cm. Finally, the EBK was selected as more accurate in seven cases.

4. Discussion

The lithological and land use characteristics existing in the region under study show a low variability. This highlights that the degree of variation observed was highly dependent on the property itself. The mineral fraction of the soil showed less variability than the selected elements content. In this sense, due to the illuviation process, the clay content showed greater variability as compared with silt and sand content. Those results agree with the findings of Keshavarzi et al. [56] and Addis et al. [57]. In turn, soil nutrients showed a higher variability, since more than 80% of them are concentrated in the upper and shallower soil layers. However, other chemical properties such as pH showed the opposite behaviour, with less variability and higher values at the deepest layers [34].

Currently, geostatistical methods are the most widely used methods for predicting and mapping soil properties [2,58]. However, in this study both deterministic and geostatistical methods have been used to ascertain the more accurate methods. The results of the interpolation analysis in this study showed the deterministic methods as having more potential to interpolate soil particle size distribution as compared to the geostatistical ones. The best results of deterministic methods in the topsoil layers are determined by a sufficiently dense data set. In this way, deterministic functions can capture the extent of local surface variation needed for the analysis and report more accurate results. In contrast to the findings of other authors such as Gozdowski et al. [59] or Radocaj et al. [60], in our study it was the IDW method that provided the best results. However, particularly when interpolating clay content at 0–5 cm or silt and sand content at the >10 cm layer, geostatistical methods were identified as more accurate, as found also by Li et al. [2].

Attending to the chemical variables, it was difficult to identify a unique optimal interpolation technique. The models generated for pH showed adequate fitness and acceptable mean square errors. Although other studies [61] found OK as the best performing method, our analysis showed the CRS as the most accurate for the topsoil, in agreement with Zandi et al. [34]. At the same time, also according with Zandi et al. [34] but contrasting with that observed by Robinson and Metternicht [62], the error and reliability statistics of the subsoil models indicate that kriging techniques were more accurate.

When studying CEC, models showed good performance, even when the average errors obtained were relatively high for two of the three depths. Our results showed that, when interpolating the cation exchange capacity at the 0–5 cm layer, IDW was the best method. Those results are in agreement with the findings of other authors [62,63]. Nevertheless, several works preferred geostatistical techniques, in concordance with the results obtained for the deeper layers, where OK and EBK proved to be more accurate [64–68].

In the case of calcium, models showed good results in general, even though the mean square error increased in depth due to the higher variability of the data. In this case, EBK was the most accurate method for predicting calcium at the shallowest (0–5 cm) and deepest (>10 cm) layers. Our results agree with other authors [69,70], although they disagree with the findings of others [62] that identified IDW as more accurate for interpolating calcium at 5–10 cm depth.

For magnesium, model results improve with depth. When interpolating the shallowest soil layers (0–5 and 5–10 cm), according to John et al. [9], geostatistical methods such as SK and OK were found to be the most accurate. However, IDW was the best method when interpolating Mg at more than 10 cm. This agreed with Schloeder et al. [71], who found no differences between IDW and OK models.

Sodium models provided good results at 0–5 cm depth, with CRS being the most accurate method. Even when at 5–10 cm depth results were less accurate, CRS was also the best method. At >10 cm soil depth, the model showed a weak fitness, indicating the difficulties for mapping Na at this soil depth. In this case, the better accuracy of deterministic vs. geostatistical methods for the first two soil depths coincides with the findings of Fu et al. [72]. However, it differs from Cruz-Cárdenas et al. [70] and Keshavarzi and Sarmadian [73], who found kriging methods as the most accurate.

When interpolating N, P, K and SOM, the results indicate that any of the interpolation techniques dominate in terms of the model's accuracy, i.e., depending on the property and soil depth, the results vary. Good results were observed for nitrogen at 0–5 cm and >10 cm layers, where SK and IDW, respectively, were the most accurate methods. The usefulness of SK coincides with Wang et al. [15]. Nevertheless, the adequacy of IDW at deeper layers in our case differs from the findings of other works [27,74,75]. However, the results of N models at 5–10 cm depth showed poor fitness, which does not allow for accurate predictions for the study area.

When interpolating phosphorus, deterministic techniques such as CRS and IM-Q were the most accurate at 0–5 cm and >10 cm soil depth. These results are in agreement with those obtained by Wollenhaupt et al. [31]. However, coinciding with Shen et al. [12] and Duan et al. [76], OK was the best method for predicting P at 5–10 cm. Furthermore, in our case, OK and IDW showed the same behaviour at 5–10 cm depth, as also found by Schloeder et al. [71].

In the case of potassium models, good results were obtained at 0–5 cm depth. However, ability to ascertain the predictive capacity of the models at 5–10 cm and >10 cm was poor. For the first depth, the deterministic IDW method was the most accurate. In this sense, the results are in agreement with the findings of Bogunovic et al. [77] and Lei et al. [78] in which radial basis functions were observed as more accurate for the shallowest soil layers. These findings agree with our results in the sense that the deterministic CRS was also better to interpolate potassium at 5–10 cm. Conversely, EBK was the best method for predicting values >10 cm. This was also observed by Lingling et al. [79] and Karwariya et al. [5], who found geostatistical methods as more accurate than deterministic ones.

The models developed for the prediction of soil organic matter were good in the case of the first two soil layers. However, at >10 cm, model results were relatively worse, although still keeping acceptable errors. IDW was the most reliable method for predicting SOM at 0–5 cm. Nevertheless, geostatistical methods were the most accurate in the case of Long et al. [3] or Bouasria et al. [80]. At 5–10 and >10 cm, EBK and OK were better. These results could partially agree with Durdevic et al. [7], who found EBK as the most accurate method to interpolate SOM at 0–30 cm, followed by OK and IDW.

In all the case studies analysed in this work, the model's performance varies according to the different properties. This highlights the spatial dependence of the data on intrinsic and soil formation factors. One of the main parameters determining model performance is soil depth. This, in turn, limits the number of samples in many cases due to the shallow soils in this study area. However, the sampling point's density does not imply that the model's performance is reduced but depends on other issues such as the variability of the data at depth, so much so that models with a lower number of samples provide better results than others with more samples, as is the case for silt. In the surface soil layers, the properties show a high variability. However, this spatial variability is considerably reduced at depth.

Another important issue in modelling soil properties is the date of the data used. In some environments (e.g., tropical climates) the use of soil samples exceeding 10–15 years may lead to unrealistic results due to the faster pattern rates of change in soil properties. However, in our environment (Mediterranean climate), soil properties do not usually undergo significant changes. Properties remain stable for 10–15 years, unless there are changes in land use that are very different from the previous ones, as pointed out by Pulido Fernández [81], who compared data from more than 30 years in similar environments. Nevertheless, it is important to add that soil organic matter content changes faster than nutrients and cations, as found by Llorente et al. [82], which shows how in this type of environment only organic matter increases, especially in farms without excessive stocking rates. It is also true that, logically, with proper management, levels increase, but slowly.

When the performance of all the studied interpolation techniques were jointly assessed by means of the analysis described in Section 3.3, it was clearly ascertained that three of the deterministic methods (IDW, CRS and SWT) provided better results than the geostatistical ones. Although the difference is minor in some cases, our results differed from the general consideration by which the geostatistical methods are more commonly used to spatially predict soil properties values.

Working on the search for the most accurate mapping technique is one of the main tasks of this group. However, the main idea is to clarify which parameters have the greatest influence on the model's performance. To this end, the use of deep learning techniques is intended, in which a multitude of environmental covariates can be incorporated. Moreover, new trends in soil mapping are along these lines, as is the use of hybrid techniques for soil mapping.

5. Conclusions

In this study, nine interpolation methods were used to predict 12 soil variables that were measured at three different soil depth intervals. Statistics such as mean error, coefficient of determination and mean square error were used to evaluate the accuracy of the methods. Although a general preference to use geostatistical methods is observed in general, we conclude that deterministic methods provide better results than geostatistical ones. Our results show that geostatistical methods were more accurate in 19 of the 36 case studies. However, the observed difference between the interpolation techniques is negligible in some cases, allowing different ones to be used interchangeably. In this regard, our results indicate that the accuracy of the methods varies depending on the case study. Results also varied, in general, when the different depths are considered, identifying deterministic methods as more accurate for the topsoil and geostatistical ones for the deeper layer. Therefore, we also conclude the necessity to use a variety of soil mapping methods and techniques to achieve the best results.

Author Contributions: This article was written by J.B.-G. and J.F.L.C., and reviewed and edited by M.P.F. The methodology was proposed by J.B.-G. and the data analysis was carried out by J.B.-G., J.F.L.C. and M.P.F. Fieldwork and data acquisition was carried out by J.B.-G. and M.P.F. All authors have read and agreed to the published version of the manuscript.

Funding: This publication has been made possible thanks to funding granted by the Consejería de Economía, Ciencia y Agenda Digital de la Junta de Extremadura and by the European Regional Development Fund of the European Union through the reference grant IB16052.

Acknowledgments: Thanks to the European Social Fund and the Junta de Extremadura for the funding granted to Jesús Barrena González (PD18016).

Conflicts of Interest: The authors declare no conflict of interest.

Appendix A

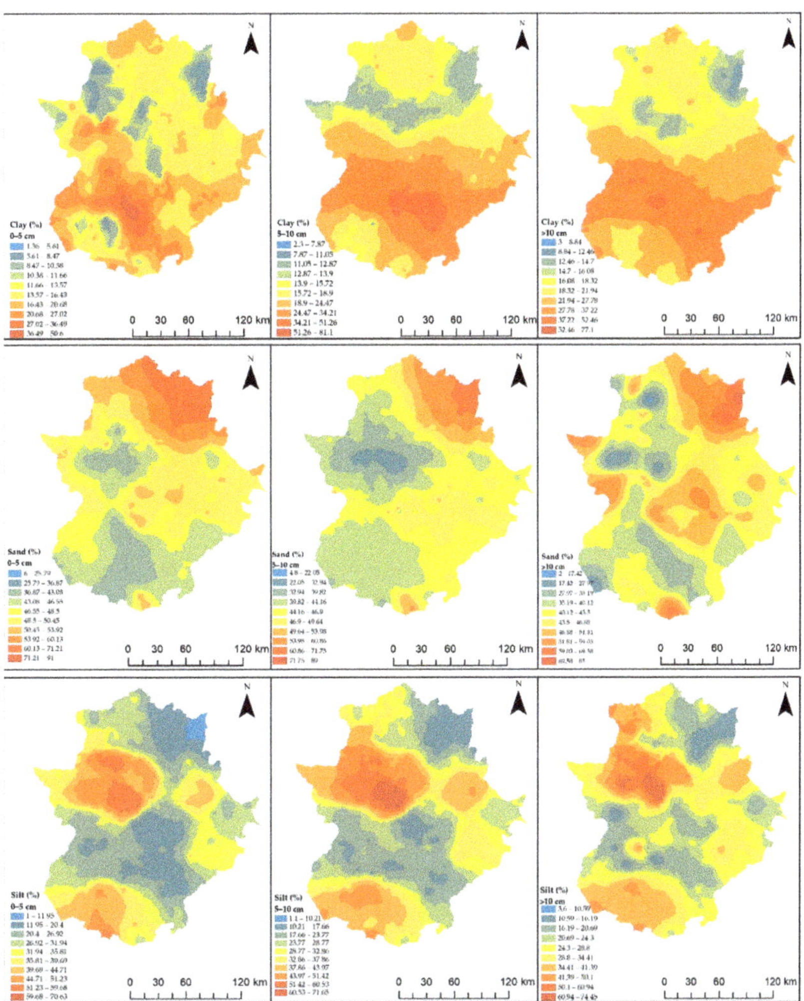

Figure A1. Cartographic representation of the different soil particle sizes and at each of the study depths using the most accurate interpolation method in each case.

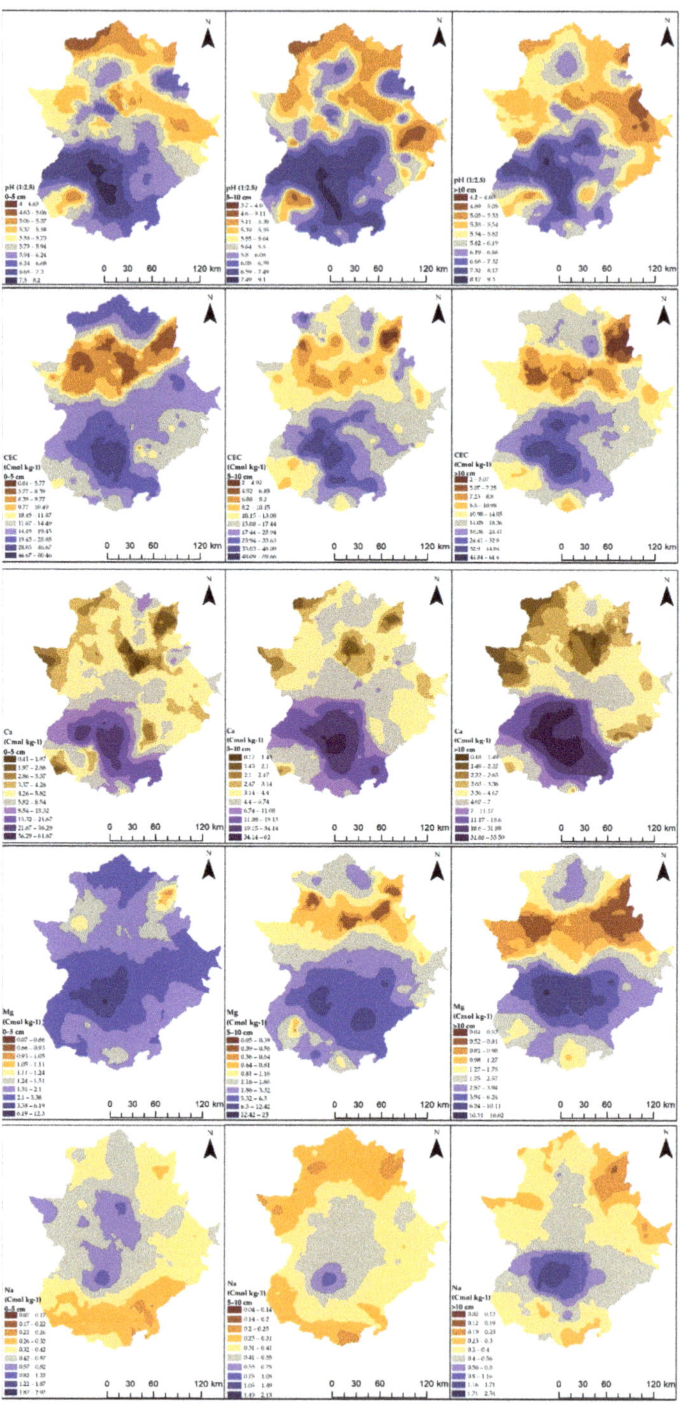

Figure A2. Cartographic representation of the different chemical properties of the soil and at each of the study depths using the most accurate interpolation method in each case.

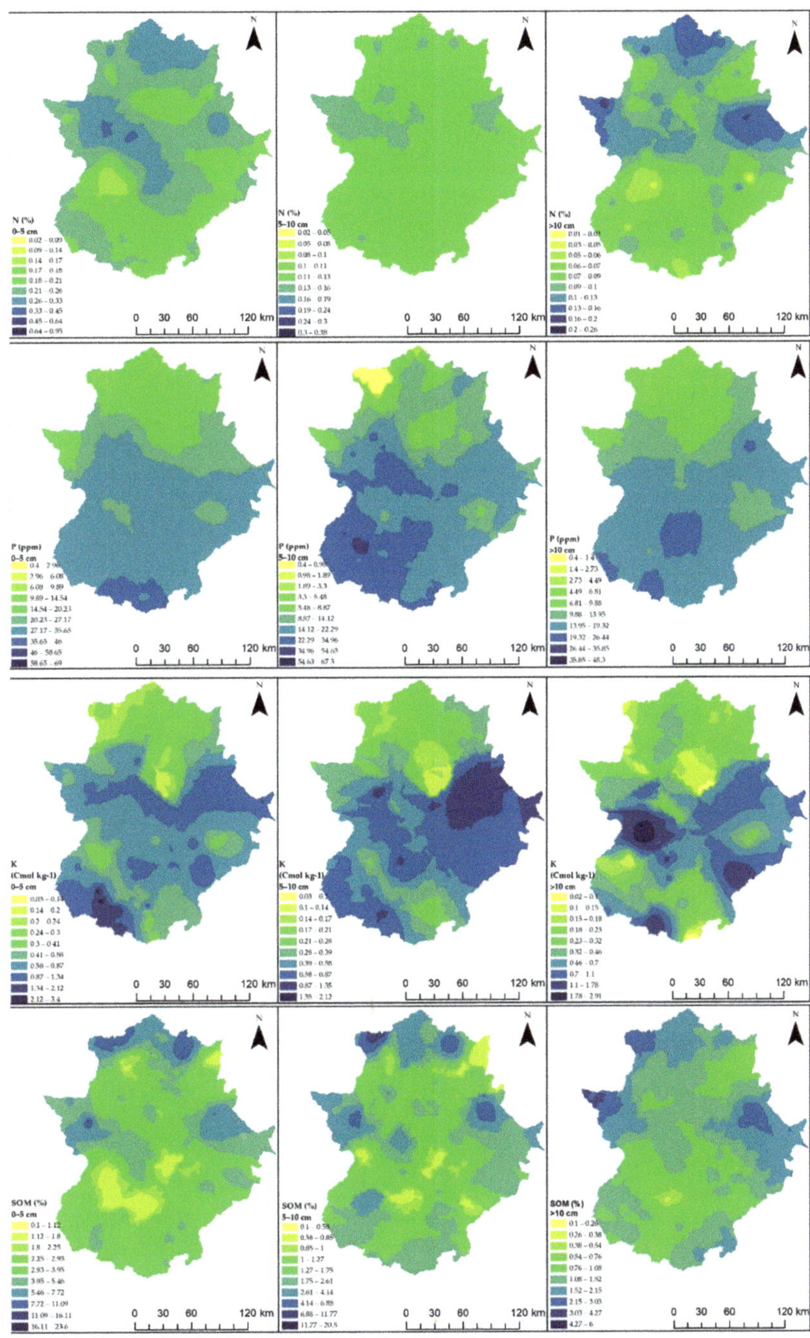

Figure A3. Cartographic representation of the different fertilisers plus soil organic matter and at each of the study depths using the interpolation method with the highest accuracy in each case.

References

1. Igaz, D.; Šinka, K.; Varga, P.; Vrbičanová, G.; Aydın, E.; Tárník, A. The Evaluation of the Accuracy of Interpolation Methods in Crafting Maps of Physical and Hydro-Physical Soil Properties. *Water* **2021**, *13*, 212. [CrossRef]
2. Li, J.; Wan, H.; Shang, S. Comparison of interpolation methods for mapping layered soil particle-size fractions and texture in an arid oasis. *Catena* **2020**, *190*, 104514. [CrossRef]
3. Long, J.; Liu, Y.; Xing, S.; Zhang, L.; Qu, M.; Qiu, L.; Huang, Q.; Zhou, B.; Shen, J. Optimal interpolation methods for farmland soil organic matter in various landforms of a complex topography. *Ecol. Indic.* **2020**, *110*, 105926. [CrossRef]
4. Xie, B.; Jia, X.; Qin, Z.; Zhao, C.; Shao, M. Comparison of interpolation methods for soil moisture prediction on China's Loess Plateau. *Vadose Zone J.* **2020**, *19*, e20025. [CrossRef]
5. Karwariya, S.; Dey, P.; Bhogal, N.S.; Kanga, S.; Singh, S.K. A Comparative Study of Interpolation Methods for Mapping Soil Properties: A Case Study of Eastern Part of Madhya Pradesh, India. In *Recent Technologies for Disaster Management and Risk Reduction*; Springer: Cham, Switzerland, 2021; pp. 431–449.
6. Matcham, E.G.; Subburayalu, S.K.; Culman, S.W.; Lindsey, L.E. Implications of choosing different interpolation methods: A case study for soil test phosphorus. *Crop Forage Turfgrass Manag.* **2021**, *7*, e20126. [CrossRef]
7. Durdevic, B.; Jug, I.; Jug, D.; Bogunovic, I.; Vukadinovic, V.; Stipesevic, B.; Brozovic, B. Spatial variability of soil organic matter content in Eastern Croatia assessed using different interpolation methods. *Int. Agrophys.* **2019**, *33*, 31–39. [CrossRef]
8. Ghorbani, A.; Moghaddam, S.M.; Majd, K.H.; Dadgar, N. Spatial variation analysis of soil properties using spatial statistics: A case study in the region of Sabalan Mountain, Iran. *J. Prot. Mt. Areas Res. Manag.* **2018**, *10*, 70–80. [CrossRef]
9. John, K.; Afu, S.; Isong, I.; Aki, E.; Kebonye, N.; Ayito, E.; Chapman, P.; Eyong, M.; Penížek, V. Mapping soil properties with soil-environmental covariates using geostatistics and multivariate statistics. *Int. J. Environ. Sci. Technol.* **2021**, *18*, 3327–3342. [CrossRef]
10. Abdulmanov, R.; Miftakhov, I.; Ishbulatov, M.; Galeev, E.; Shafeeva, E. Comparison of the effectiveness of GIS-based interpolation methods for estimating the spatial distribution of agrochemical soil properties. *Environ. Technol. Innov.* **2021**, *24*, 101970. [CrossRef]
11. Shi, W.Z.; Tian, Y. A hybrid interpolation method for the refinement of a regular grid digital elevation model. *Int. J. Geogr. Inf. Sci.* **2006**, *20*, 53–67. [CrossRef]
12. Shen, Q.; Wang, Y.; Wang, X.; Liu, X.; Zhang, X.; Zhang, S. Comparing interpolation methods to predict soil total phosphorus in the Mollisol area of Northeast China. *Catena* **2019**, *174*, 59–72. [CrossRef]
13. ZHU, Q.; Lin, H. Comparing ordinary kriging and regression kriging for soil properties in contrasting landscapes. *Pedosphere* **2010**, *20*, 594–606. [CrossRef]
14. Li, W.; Xu, B.; Song, Q.; Liu, X.; Xu, J.; Brookes, P.C. The identification of 'hotspots' of heavy metal pollution in soil–rice systems at a regional scale in eastern China. *Sci. Total Environ.* **2014**, *472*, 407–420. [CrossRef] [PubMed]
15. Wang, K.; Zhang, C.; Li, W. Predictive mapping of soil total nitrogen at a regional scale: A comparison between geographically weighted regression and cokriging. *Appl. Geogr.* **2013**, *42*, 73–85. [CrossRef]
16. Peng, Y.; Xiong, X.; Adhikari, K.; Knadel, M.; Grunwald, S.; Greve, M.H. Modeling soil organic carbon at regional scale by combining multi-spectral images with laboratory spectra. *PLoS ONE* **2015**, *10*, e0142295. [CrossRef] [PubMed]
17. Hou, D.; O'Connor, D.; Nathanail, P.; Tian, L.; Ma, Y. Integrated GIS and multivariate statistical analysis for regional scale assessment of heavy metal soil contamination: A critical review. *Environ. Pollut.* **2017**, *231*, 1188–1200. [CrossRef] [PubMed]
18. Ma, L.; Ma, F.; Li, J.; Gu, Q.; Yang, S.; Wu, D.; Feng, J.; Ding, J. Characterizing and modeling regional-scale variations in soil salinity in the arid oasis of Tarim Basin, China. *Geoderma* **2017**, *305*, 1–11. [CrossRef]
19. Shahriari, M.; Delbari, M.; Afrasiab, P.; Pahlavan-Rad, M.R. Predicting regional spatial distribution of soil texture in floodplains using remote sensing data: A case of southeastern Iran. *Catena* **2019**, *182*, 104149. [CrossRef]
20. Chen, H.; Fan, L.; Wu, W.; Liu, H.-B. Comparison of spatial interpolation methods for soil moisture and its application for monitoring drought. *Environ. Monit. Assess.* **2017**, *189*, 525. [CrossRef] [PubMed]
21. Mishra, U.; Lal, R.; Liu, D.; Van Meirvenne, M. Predicting the spatial variation of the soil organic carbon pool at a regional scale. *Soil Sci. Soc. Am. J.* **2010**, *74*, 906–914. [CrossRef]
22. Yao, X.; Fu, B.; Lü, Y.; Sun, F.; Wang, S.; Liu, M. Comparison of four spatial interpolation methods for estimating soil moisture in a complex terrain catchment. *PLoS ONE* **2013**, *8*, e54660. [CrossRef]
23. Göl, C.; Bulut, S.; Bolat, F. Comparison of different interpolation methods for spatial distribution of soil organic carbon and some soil properties in the Black Sea backward region of Turkey. *J. Afr. Earth Sci.* **2017**, *134*, 85–91. [CrossRef]
24. Hinge, G.; Surampalli, R.Y.; Goyal, M.K. Prediction of soil organic carbon stock using digital mapping approach in humid India. *Environ. Earth Sci.* **2018**, *77*, 172. [CrossRef]
25. Srivastava, P.K.; Pandey, P.C.; Petropoulos, G.P.; Kourgialas, N.N.; Pandey, V.; Singh, U. GIS and remote sensing aided information for soil moisture estimation: A comparative study of interpolation techniques. *Resources* **2019**, *8*, 70. [CrossRef]
26. Denton, O.; Aduramigba-Modupe, V.; Ojo, A.; Adeoyolanu, O.; Are, K.; Adelana, A.; Oyedele, A.; Adetayo, A.; Oke, A. Assessment of spatial variability and mapping of soil properties for sustainable agricultural production using geographic information system techniques (GIS). *Cogent Food Agric.* **2017**, *3*, 1279366. [CrossRef]
27. Yao, X.; Yu, K.; Deng, Y.; Liu, J.; Lai, Z. Spatial variability of soil organic carbon and total nitrogen in the hilly red soil region of Southern China. *J. For. Res.* **2020**, *31*, 2385–2394. [CrossRef]

28. Li, J.; Heap, A.D. *A Review of Spatial Interpolation Methods for Environmental Scientists*; Geoscience Australia: Canberra, Australia, 2008.
29. Kuriakose, S.L.; Devkota, S.; Rossiter, D.; Jetten, V. Prediction of soil depth using environmental variables in an anthropogenic landscape, a case study in the Western Ghats of Kerala, India. *Catena* **2009**, *79*, 27–38. [CrossRef]
30. Sahu, B.; Ghosh, A.K. Deterministic and geostatistical models for predicting soil organic carbon in a 60 ha farm on Inceptisol in Varanasi, India. *Geoderma Reg.* **2021**, *26*, e00413. [CrossRef]
31. Wollenhaupt, N.; Wolkowski, R.; Clayton, M. Mapping soil test phosphorus and potassium for variable-rate fertilizer application. *J. Prod. Agric.* **1994**, *7*, 441–448. [CrossRef]
32. Gotway, C.A.; Ferguson, R.B.; Hergert, G.W.; Peterson, T.A. Comparison of kriging and inverse-distance methods for mapping soil parameters. *Soil Sci. Soc. Am. J.* **1996**, *60*, 1237–1247. [CrossRef]
33. Robinson, T.; Metternicht, G. Comparing the performance of techniques to improve the quality of yield maps. *Agric. Syst.* **2005**, *85*, 19–41. [CrossRef]
34. Zandi, S.; Ghobakhlou, A.; Sallis, P. Evaluation of spatial interpolation techniques for mapping soil pH. In Proceedings of the 19th International Congress on Modelling and Simulation, Perth, Australia, 12–16 December 2011; pp. 1153–1159.
35. IUSS Working Group WRB. *World Reference Base for Soil Resources 2014, Update 2015. International Soil Classification System for Naming Soils and Creating Legends for Soil Maps*; World Soil Resources Reports No. 106; FAO: Rome, Italy, 2015; p. 203.
36. Hernando, V.; Jimeno, L.; Rodríguez, J.; González, R.; Guerra, A.; Monturiol, F.; Gallardo, J.; Labrandero, J.L.; García-Vaquero, J. *Estudio de los suelos de la Tierra de Barros*; Instituto Nacional de Edafología y Biología Vegetal: Madrid, Spain, 1980.
37. International Soil Reference and Information Centre. SoilGrids. Available online: https://soilgrids.org/ (accessed on 25 May 2022).
38. Soil Survey Laboratory Methods Manual. *Soil Survey Investigations Report No. 42*; Version 4.0; USDA-NCRS: Lincoln, NE, USA, 2004; p. 1031.
39. MAPA. *Métodos Oficiales De Análisis: Suelos Y Aguas*; Ministerio de Agricultura, Pesca y Alimentación, Dirección General de Política Alimentaria: Madrid, Spain, 1982; p. 182.
40. Dumas, J.B.A. Procédés de l'analyse organique. *Ann. De Chim. Et De Phys.* **1831**, *247*, 198–213.
41. Olsen, S.R.; Cole, C.V.; Wastanabe, F.S.; Dean, L.A. Estimation of available phosphorus in soils by extraction with sodium bicarbonate. *USDA Circ.* **1954**, *939*, 1–19.
42. USDA. Soil Quality Test Kit Guide; United States Department of Agriculture, Agricultural Research Service and Natural Resources Conservation Service-Soil Quality Institute. 1999. Available online: http://soils.usda.gov/sqi/assessment/files/test_kit_complete.pdf (accessed on 20 May 2022).
43. Walkley, A.; Black, L.A. An examination of Degtjareff method for determining soil organic matter and a proposed modification of the chromic acid titration method. *Soil Sci.* **1934**, *37*, 29–38. [CrossRef]
44. Anselin, L. Local indicators of spatial association—LISA. *Geogr. Anal.* **1995**, *27*, 93–115. [CrossRef]
45. ESRI. *ArcGIS desktop: Release 10*; Environmental Systems Research Institute: Redlands, CA, USA, 2011.
46. Statsoft. *STATISTICA (Data Analysis Software System), Version 6*; Statsoft: Tulsa, OK, USA, 2001.
47. Johnston, K.; Ver Hoef, J.M.; Krivoruchko, K.; Lucas, N. *Using ArcGIS Geostatistical Analyst*; Esri Redlands: Redlands, CA, USA, 2001; Volume 380.
48. Shepard, D. A two-dimensional interpolation function for irregularly-spaced data. In Proceedings of the 1968 23rd ACM National Conference, New York, NY, USA, 27–29 August 1968; pp. 517–524.
49. Hardyr, L. Multi-quadric Equations of Topography and Other Irregular Surface. *J. Geophys. Res.* **1971**, *76*, 1905–1915. [CrossRef]
50. Schaum, A. Principles of local polynomial interpolation. In Proceedings of the 2008 37th IEEE Applied Imagery Pattern Recognition Workshop, Washington, DC, USA, 15–17 October 2008; pp. 1–6.
51. Wackernagel, H. Ordinary Kriging. In *Multivariate Geostatistics*; Springer: Cham, Switzerland, 2003; pp. 79–88.
52. Olea, R.A. Geostatistics for Engineers and Earth Scientists. *Technometrics* **2000**, *42*, 444–445. [CrossRef]
53. Njeban, H.S. Comparison and evaluation of GIS-based spatial interpolation methods for estimation groundwater level in AL-Salman District—Southwest Iraq. *J. Geogr. Inf. Syst.* **2018**, *10*, 362. [CrossRef]
54. Varouchakis, E.; Hristopulos, D. Comparison of stochastic and deterministic methods for mapping groundwater level spatial variability in sparsely monitored basins. *Environ. Monit. Assess.* **2013**, *185*, 1–19. [CrossRef]
55. Khazaz, L.; Oulidi, H.J.; El Moutaki, S.; Ghafiri, A. Comparing and Evaluating Probabilistic and Deterministic Spatial Interpolation Methods for Groundwater Level of Haouz in Morocco. *J. Geogr. Inf. Syst.* **2015**, *7*, 631. [CrossRef]
56. Keshavarzi, A.; Tuffour, H.O.; Brevik, E.C.; Ertunç, G. Spatial variability of soil mineral fractions and bulk density in Northern Ireland: Assessing the influence of topography using different interpolation methods and fractal analysis. *Catena* **2021**, *207*, 105646. [CrossRef]
57. Addis, H.K.; Klik, A.; Strohmeier, S. Performance of frequently used interpolation methods to predict spatial distribution of selected soil properties in an agricultural watershed in Ethiopia. *Appl. Eng. Agric.* **2016**, *32*, 617–626.
58. Zhang, S.-W.; Shen, C.-Y.; Chen, X.-Y.; Ye, H.-C.; Huang, Y.-F.; Shuang, L. Spatial interpolation of soil texture using compositional kriging and regression kriging with consideration of the characteristics of compositional data and environment variables. *J. Integr. Agric.* **2013**, *12*, 1673–1683. [CrossRef]
59. Gozdowski, D.; Stępień, M.; Samborski, S.; Dobers, E.; Szatyłowicz, J.; Chormański, J. Prediction accuracy of selected spatial interpolation methods for soil texture at farm field scale. *J. Soil Sci. Plant Nutr.* **2015**, *15*, 639–650. [CrossRef]

60. Radočaj, D.; Jurišić, M.; Zebec, V.; Plaščak, I. Delineation of soil texture suitability zones for soybean cultivation: A case study in continental Croatia. *Agronomy* **2020**, *10*, 823. [CrossRef]
61. Robinson, T.; Metternicht, G. Testing the performance of spatial interpolation techniques for mapping soil properties. *Comput. Electron. Agric.* **2006**, *50*, 97–108. [CrossRef]
62. AbdelRahman, M.A.; Zakarya, Y.M.; Metwaly, M.M.; Koubouris, G. Deciphering soil spatial variability through geostatistics and interpolation techniques. *Sustainability* **2020**, *13*, 194. [CrossRef]
63. Paz Ferreiro, J.; Pereira De Almeida, V.; Cristina Alves, M.; Aparecida De Abreu, C.; Vieira, S.R.; Vidal Vázquez, E. Spatial variability of soil organic matter and cation exchange capacity in an oxisol under different land uses. *Commun. Soil Sci. Plant Anal.* **2016**, *47*, 75–89. [CrossRef]
64. Houlong, J.; Daibin, W.; Xu, C.; Shuduan, L.; Hongfeng, W.; Yang, C.; Najia, L.; Yiyin, C.; Geng, L. Comparison of kriging interpolation precision between grid sampling scheme and simple random sampling scheme for precision agriculture. *Eurasian J. Soil Sci.* **2016**, *5*, 62–73. [CrossRef]
65. Huang, J.; Bishop, T.; Triantafilis, J. An error budget for digital soil mapping of cation exchange capacity using proximally sensed electromagnetic induction and remotely sensed γ-ray spectrometer data. *Soil Use Manag.* **2017**, *33*, 397–412. [CrossRef]
66. Jung, W.; Kitchen, N.; Sudduth, K.; Anderson, S. Spatial characteristics of claypan soil properties in an agricultural field. *Soil Sci. Soc. Am. J.* **2006**, *70*, 1387–1397. [CrossRef]
67. Bishop, T.; McBratney, A. A comparison of prediction methods for the creation of field-extent soil property maps. *Geoderma* **2001**, *103*, 149–160. [CrossRef]
68. López-Castañeda, A.; Zavala-Cruz, J.; Palma-López, D.J.; Rincón-Ramírez, J.A.; Bautista, F. Digital Mapping of Soil Profile Properties for Precision Agriculture in Developing Countries. *Agronomy* **2022**, *12*, 353. [CrossRef]
69. Medina, H.; van Lier, Q.d.J.; García, J.; Ruiz, M.E. Regional-scale variability of soil properties in Western Cuba. *Soil Tillage Res.* **2017**, *166*, 84–99. [CrossRef]
70. Cruz-Cárdenas, G.; López-Mata, L.; Ortiz-Solorio, C.A.; Villaseñor, J.L.; Ortiz, E.; Silva, J.T.; Estrada-Godoy, F. Interpolation of Mexican soil properties at a scale of 1: 1,000,000. *Geoderma* **2014**, *213*, 29–35. [CrossRef]
71. Schloeder, C.; Zimmerman, N.; Jacobs, M. Comparison of methods for interpolating soil properties using limited data. *Soil Sci. Soc. Am. J.* **2001**, *65*, 470–479. [CrossRef]
72. Fu, T.; Gao, H.; Liu, J. Comparison of Different Interpolation Methods for Prediction of Soil Salinity in Arid Irrigation Region in Northern China. *Agronomy* **2021**, *11*, 1535. [CrossRef]
73. Keshavarzi, A.; Sarmadian, F. Mapping of spatial distribution of soil salinity and alkalinity in a semi-arid region. *Ann. Wars. Univ. Life Sci. SGGW. Land Reclam.* **2012**, *44*, 3–14. [CrossRef]
74. Bangroo, S.; Najar, G.; Achin, E.; Truong, P.N. Application of predictor variables in spatial quantification of soil organic carbon and total nitrogen using regression kriging in the North Kashmir forest Himalayas. *Catena* **2020**, *193*, 104632. [CrossRef]
75. Gia Pham, T.; Kappas, M.; Van Huynh, C.; Hoang Khanh Nguyen, L. Application of ordinary kriging and regression kriging method for soil properties mapping in hilly region of Central Vietnam. *ISPRS Int. J. Geo-Inf.* **2019**, *8*, 147. [CrossRef]
76. Duan, L.; Li, Z.; Xie, H.; Li, Z.; Zhang, L.; Zhou, Q. Large-scale spatial variability of eight soil chemical properties within paddy fields. *Catena* **2020**, *188*, 104350. [CrossRef]
77. Bogunovic, I.; Mesic, M.; Zgorelec, Z.; Jurisic, A.; Bilandzija, D. Spatial variation of soil nutrients on sandy-loam soil. *Soil Tillage Res.* **2014**, *144*, 174–183. [CrossRef]
78. Lei, Z.; Yanbing, Q.; Qingrui, C.J.J.o.N.A. Comparison of spatial interpolation methods of available potassium in Hantai County. *J. Northwest Agric. For. Univ.* **2013**, *40*, 193–199.
79. Liu, L.; Jin, Y.; Wang, J.; Hong, Q.; Pan, Z.; Zhao, J. Comparation of Spatial Interpolation Methods on Slowly Available Potassium in Soils. *IOP Conf. Ser. Earth Environ. Sci.* **2019**, *234*, 012018. [CrossRef]
80. Bouasria, A.; Ibno Namr, K.; Rahimi, A.; Ettachfini, E.M. Geospatial Assessment of Soil Organic Matter Variability at Sidi Bennour District in Doukkala Plain in Morocco. *J. Ecol. Eng.* **2021**, *22*, 120–130. [CrossRef]
81. Pulido Fernández, M. Indicadores de Calidad del Suelo en Áreas de Pastoreo. Ph.D. Thesis, Universidad de Extremadura, Cáceres, Spain, 2014; p. 284.
82. Llorente, M.; Moreno, G. Factors determining distribution and temporal changes in soil carbon in Iberian Dehesa. In Proceedings of the Geophysical Research Abstracts, Vienna, Austria, 7–12 April 2019.

 *sustainability*

Article

Impact of Agricultural Land Use Types on Soil Moisture Retention of Loamy Soils

Szabolcs Czigány [1], Noémi Sarkadi [1,*], Dénes Lóczy [1], Anikó Cséplő [2], Richárd Balogh [2], Szabolcs Ákos Fábián [1], Rok Ciglič [3], Mateja Ferk [3], Gábor Pirisi [1], Marcell Imre [1], Gábor Nagy [4] and Ervin Pirkhoffer [1]

[1] Institute of Geography and Earth Sciences, University of Pécs, Ifjúság útja 6., 7624 Pécs, Hungary
[2] Doctoral School of Earth Sciences, University of Pécs, Ifjúság útja 6., 7624 Pécs, Hungary
[3] Research Centre of the Slovenian Academy of Sciences and Arts, Anton Melik Geographical Institute, Novi trg 2., 1000 Ljubljana, Slovenia
[4] South Transdanubian Water Management Directorate, Köztársaság tér 7, 7623 Pécs, Hungary
* Correspondence: sarkadin@gamma.ttk.pte.hu

Abstract: Increasingly severe hydrological extremes are predicted for the Pannonian Basin as one of the consequences of climate change. The challenges of extreme droughts require the adaptation of agriculture especially during the intense growth phase of crops. For dryland farming, the selections of the optimal land use type and sustainable agricultural land management are potential adaptation tools for facing the challenges posed by increased aridity. To this end, it is indispensable to understand soil moisture (SM) dynamics under different land use types over drought-affected periods. Within the framework of a Slovenian–Hungarian project, soil moisture, matric potential and rainfall time series have been collected at three pilot sites of different land use types (pasture, orchards and a ploughland) in SW Hungary since September 2018. Experiments were carried out in soils of silt, silt loam and clay loam texture. In the summers (June 1 to August 31) of 2019 and 2022, we identified normal and dry conditions, respectively, with regard to differences in water balance. Our results demonstrated that soil moisture is closely controlled by land use. Marked differences of the moisture regime were revealed among the three land use types based on statistical analyses. Soils under pasture had the most balanced regime, whereas ploughland soils indicated the highest amplitude of moisture dynamics. The orchard, however, showed responses to weather conditions in sharp contrast with the other two sites. Our results are applicable for loamy soils under humid and subhumid temperate climates and for periods of extreme droughts, a condition which is expected to be the norm for the future.

Keywords: drought; ecosystem services; land use; soil moisture dynamics; water stress

Citation: Czigány, S.; Sarkadi, N.; Lóczy, D.; Cséplő, A.; Balogh, R.; Fábián, S.Á.; Ciglič, R.; Ferk, M.; Pirisi, G.; Imre, M.; et al. Impact of Agricultural Land Use Types on Soil Moisture Retention of Loamy Soils. *Sustainability* 2023, *15*, 4925. https://doi.org/10.3390/su15064925

Academic Editors: Antonio Ganga, Blaž Repe and Mario Elia

Received: 18 January 2023
Revised: 20 February 2023
Accepted: 3 March 2023
Published: 9 March 2023

1. Introduction

Reports by the Intergovernmental Panel on Climate Change (IPCC) predict that the increasing frequency of both extreme precipitation and prolonged drought periods is very likely in the near future [1]. This pattern was exemplified in Hungary in 2010, with record high annual precipitation totals, followed by the record low annual rainfall total of 2011. However, thanks to the storage of surplus moisture from the previous year in soils, the 2011 drought did not cause remarkable losses in crop yields. Negative water balances, water stress and drought will likely be manifested in diverse ways geographically in the future [2]. Nonetheless, due to the rain shadow effect of the Alps and the Carpathians, the Pannonian Basin will likely be affected by water shortages in the near future [3] and alternating inundations by flash floods, inland excess waters [4–6] and soil erosion [7].

A novel element of sustainable adaptation to climatic conditions of negative water balances could be integrated water management, equally directed to the prevention of excess runoff and prolonged droughts [8,9]. In basin locations and areas of transit waters, such as the Pannonian Basin, water retention is of increasing importance [9,10]. This has been crucial since the river regulation works in the early- and mid-1800s when water conveyance had been accelerated artificially. Over the 1800s and 1900s, flood control measures meant the cost-intensive construction of hydrologic structures (levees, dykes and embankments). Nevertheless, recently, the negative consequences of the rapid conveyance and limited storage of water through the Pannonian Basin have been recognized. The main hydrological constraints are limited land availability and water retention and the reduced storage capacity of the soil [11].

By recognizing the beneficial role of ecosystem services [12–14], a paradigm change occurred in water management policies in many countries in Europe and North America. Water conservation in a sustainable way has priority, especially in floodplains and low-lying areas. Water should be retained in floodplains or various stormwater-mitigation facilities (e.g., raingardens and flood retention pools), in natural and manmade reservoirs and other water bodies instead of increasing the intensity of water conveyance [15].

Adaptation to the increased extremities of hydrologic phenomena (droughts and floods) and the retention of water are indispensable in light of climate change [16]. Therefore, analyzing the suitability of soil textural types, fertility, topography and crop varieties may increase the profitability of the given land use type if managed site-specifically [17–20].

Hence, for sustainable site-specific best management practices, the re-evaluations of landscape diversity and efficiency of increased water retention are essential [21]. As opposed to costly hydrologic structures that are often undesired for natural or seminatural environments [22,23], greener investments and eco-friendly solutions are needed [10], which may comply with the EU Water Framework Directive or other similar frameworks [24–26]. To maintain a more balanced water budget in the long run, the following specific goals should be stated [10]:

- decreasing hydrologic extremities;
- infiltration and subsurface recharge should be intensified over excess runoff;
- increased replenishment and recharge into the vadose zone as well as aquifers;
- canopy density and leaf area index (e.g., employing intercropping) shall be increased to reduce throughfall and decrease evaporation loss.

Land use and management can significantly affect both atmospheric (e.g., greenhouse gas emission rates and vapor content) and soil physical properties, including porosity hydraulic conductivity [16,27], rate of infiltration, and volume of plant-available water [28–31]. Land use changes may influence the intensity of certain elements of the water cycle, especially the magnitude and time of evapotranspiration [32]. Fu et al. revealed the beneficial impacts of intercropping and terraced agriculture on soil moisture [21]. Niu et al. demonstrated that grasslands had the highest mean soil moisture contents among five different land use types (grassland, cropland, poplar land, interdunes and shrubland) in north-eastern China [33]. The degradation of physical soil properties can directly affect moisture dynamics in the vadose zone. For example, soil productivity decreased by converting natural pastures to farmlands in Iran [28].

The present paper reveals the findings of a Hungarian–Slovenian joint research project titled "Possible ecological control of flood hazard in the hill regions of Hungary and Slovenia". The key objective of the project is providing data on moisture dynamics of silty and loamy soils found on surfaces of high relief. The goal of this study is to present the influence of land use type coupled with periods of different water balances ('normal' or dry summer periods) on moisture dynamics. According to our hypothesis, soils of ploughlands should have a lower water retention capacity, whereas balanced moisture dynamics characterize soils under closer-to-natural land use types. The novelty of our paper is to provide data on the effect of agricultural land use types on soil moisture dynamics. The selected area is markedly affected by a changing climate of increasing aridity. Sustainable adaptation to changing climates at local scales helps in maximizing the site-specific efficiency of ecosystem services. The present study complements previous research carried out in Central Hungary [16].

2. Materials and Methods

2.1. Location of Study Sites

For the analyses, three sites were selected in the Transdanubian Hills (SW Hungary) in the vicinity of the city of Pécs: at the villages of Boda (ploughland), Palkonya (orchard) and Almamellék (pasture) (Figure 1). It is a region of a subhumid continental climate influenced by the air masses (whether continental, Atlantic or Mediterranean) and the orographic effect. Although in the long-term precipitation shows an increasing gradient towards the western part of the country, in many years, field rainfall totals show a rather mosaic pattern.

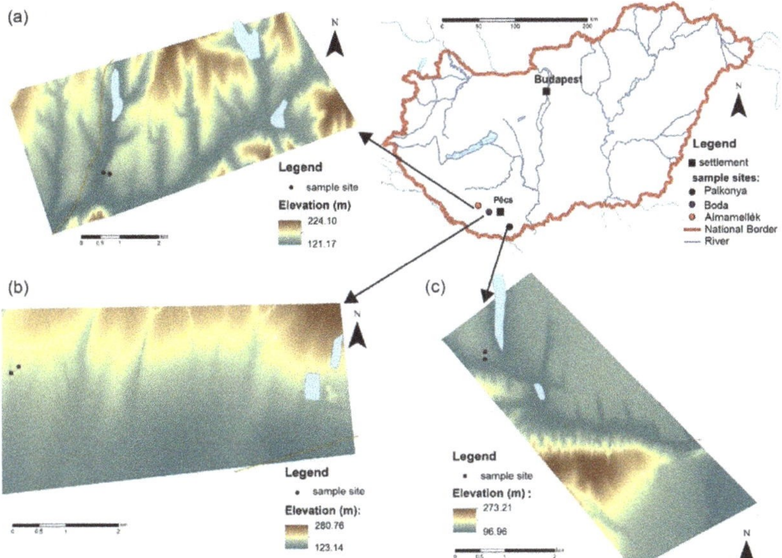

Figure 1. Location of the study sites on the digital elevation model (DEM) generated from a LiDAR survey. (**a**) Almamellék (pasture); (**b**) Boda (ploughland); (**c**) Palkonya-Villánykövesd (orchard).

In all study sites, slightly eroded brown forest soils with clay illuviation (WRB: Endocalcic Luvisol) are found. All soils are formed on loess. The three sites are similar in terms of textural type (silt and silt loam) and diagnostic soil type (Calcaric Phaeozem, WRB).

2.1.1. Ploughland Site (Foothills of the Mecsek Mountains)

This study site is located at a distance of 10 km west of city of Pécs, in the southern footslopes of the Mecsek Mountains, gently sloping to the direction of the Pécs half-basin, west of the village of Boda. The elevation of the lower and upper station is 172 and 182 m, respectively (Table 1). The average slope for this site is 2.63°, whereas the maximum is 6.48°. The land use type is large-scale farming of conventional tillage, with sugar-beet, cereals, sunflower, soybean and rape seed as the most common crops. In both study years, this site was cropped with soybean. The distance between the two monitoring stations of this site is 190 m. A derasional valley and an erosional gully are found at this site, uphill and downhill of the lower station, which is located at the southern margin of a small grove. These landforms and the grove likely influence the soil moisture budget around the station.

2.1.2. Orchard Site

The second study site lies at the southern edge of the village of Palkonya in the northwestern foreland of the Villány Hills. Land utilization is a cherry orchard. The parent material is Pleistocene loess overlying a Mesozoic limestone. Elevation of the lower and upper stations is 175 and 182.7 m. The slope of the uphill station is steeper, with an average slope of 18°, whereas the slope is gradually decreasing to a footslope position closer to the foothill station and the reservoir located in the valley bottom. The two stations of this site were installed at a distance of 156 m.

Table 1. General geographical parameters of the three study sites.

Village Name	Site Area (ha)	Position	EOV X (m)	EOV Y (m)	Elevation (m)	Min	Slope (°) Max	Mean
Almamellék	7.4	Foothill	556,430.4	90,123.7	126.1	0.85	17.4	6.19
		Uphill	556,590.7	90,108.8	141.2			
Boda	16.32	Foothill	571,380.0	81,182.9	175	0.02	6.48	2.63
		Uphill	571,518.2	81,322.1	182.7			
Palkonya	6.85	Foothill	599,400.2	61,098.8	112.6	3.09	9.7	5.92
		Uphill	599,407.4	61,201.5	124.6			

2.1.3. Pasture Site

This study site is situated in the Zselic Hills, where the elevation of the lower and upper is 112.6 and 124.6 m, respectively. The average slope in the Almamellék site is 5.92° uphill from the uphill station, whereas it reaches a maximum of 9.7° immediately downhill from the upper station. Furthermore, the land is utilized here as a natural pasture and meadow. The monitoring stations are located at a distance of 167 m from each other.

2.2. Field Monitoring Setup

To track local moisture dynamics, SM monitoring was installed for each study site in December 2018. At each site, two monitoring stations were deployed. Rainfall was measured using tipping-bucket rain gauges (ECRN-100, Meter Group Inc., Pullman, WA, USA) of 0.2 mm resolution. Rain gauges as well as WP4 temperature and relative humidity sensors were installed only at the uphill stations. At both stations of each site, TDR-type soil moisture sensors (Meter Group Inc., Teros 12) and tensiometers (Teros 21) were used to measure volumetric water contents and matric potential, respectively. At each station, 4 sensors were deployed at depths of 10 and 30 cm (one soil moisture sensor and one tensiometer at each depth). The depths were selected based on soil type (loamy soil) and the typical crops grown in SW Hungary. Soils to a depth of 30 cm experienced the largest fluctuation in soil moisture. TDR sensors had been laboratory-calibrated prior to their installation in the field. Data were logged and stored with EM-60 data loggers at a time interval of 15 min.

2.3. Particle Size Analysis of the Soil Samples

Soil samples were taken from the depths of the sensors. Organic matter and $CaCO_3$ were removed from the samples using H_2O_2 and 10% HCl, respectively. The grain size distribution of the soil samples was determined with static light scattering using a Malvern MasterSizer 3000 (Malvern Inc., Malvern, England, United Kingdom) particle size analyzer. The textural type was determined using an MS Excel macro (https://www.nrcs.usda.gov/resources/education-and-teaching-materials/soil-texture-calculator, accessed on 10 January 2023) and clay–silt and silt–sand boundaries at 2 and 63 μm, respectively.

2.4. Calculation of the Pálfai Drought Index and the Aridity Indices

The Pálfai Drought Index (hereafter PaDI, [34]) was calculated for the three study sites using mean monthly temperatures and weighted monthly precipitation totals. Potential evapotranspiration was calculated using the Thornthwaite equation [35,36], widely applied for the estimation of PET under humid and subhumid temperate climates.

2.5. Analysis of Field Data

The field data were statistically analyzed using MATLAB R2020b and MS Excel programs. The statistics focused on descriptive statistical parameter calculations (mean, median, standard deviation, minimum, maximum and range of the data). Boxplots were generated by the MATLAB program from raw data (excluding missing or inappropriate values). The general description of the boxplots was the following: the boxes' sizes show the interquartile range (IQR, data fall into 25–75% percentile), the median was plotted on the boxplots (red lines) and the outliers were measured (all variables located at a distance from the median 1.5 times larger than the IQR were outliers). The whiskers showed the 5–95% percentiles.

3. Results

3.1. Soil Textural Types

Soil texture at the three sites was dominated by the silt fraction, hence the soils were classified as either silt or silt loam types (Table 2). In general, all samples of the ploughland and the foothill pasture had higher sand content (24–32%) than the other sites.

Table 2. Fine earth fractions and textural types of the soils of the monitoring sites.

Land Use	Depth [cm]	Slope Position	Clay [%]	Silt [%]	Sand [%]	Textural Type
Pasture	10	Uphill	4.88	95.02	0.10	Silt
Pasture	30	Uphill	4.56	87.79	7.65	Silt
Pasture	10	Foothill	2.48	72.56	24.96	Silt loam
Pasture	30	Foothill	2.87	64.53	32.60	Silt loam
Orchard	10	Uphill	4.43	86.56	9.01	Silt
Orchard	30	Uphill	6.63	93.37	0.00	Silt
Orchard	10	Foothill	3.68	86.97	9.35	Silt
Orchard	30	Foothill	4.08	86.56	9.36	Silt
Ploughland	10	Uphill	1.23	73.95	26.05	Silt loam
Ploughland	30	Uphill	0.75	69.24	30.76	Silt loam
Ploughland	10	Foothill	0.80	72.10	27.90	Silt loam
Ploughland	30	Foothill	1.21	75.67	24.33	Silt loam

3.2. Water Balance

Rainfall distribution showed a rather contrasting picture among the three sites. The highest rainfall for both the summer and the period of January to August was measured for the orchard site in 2019 and for the pasture in 2022, whereas the lowest rainfall total in 2022 was observed in the orchard (Table 3).

Table 3. Precipitation totals [mm] for the periods of January to August and June to August, 2019 and 2022.

Land Use	2019		2022	
	1–8	6–8	1–8	6–8
Pasture	469.7	187.4	390	180
Ploughland	466.7	185.3	310.4	163
Orchard	516.7	265.3	231.7	115

Both the antecedent (January to May) and summer precipitation of all three study sites indicated marked contrasts between the two studied years (Figure 2). January and May's monthly precipitation totals demonstrated the largest variation between 2019 and 2022.

Mean monthly temperatures showed a less diverse pattern than rainfall among the three sites. Both the highest mean annual and the highest summer mean temperatures were recorded in the orchard in 2019 and 2022 (Table 4). Consistently, the lowest temperatures were registered at the pasture in both studied years. Mean summer and mean annual temperatures were about 1 °C and 0.2 °C higher in 2022 than in 2019, respectively. The differences between spring season average temperatures were minor; however, the growing season temperature at all sites was about 0.7 °C higher in 2022 than in 2019. The greatest variation was measured in autumn and differed by 0.8 to 1.1 °C from site to site (higher in 2019).

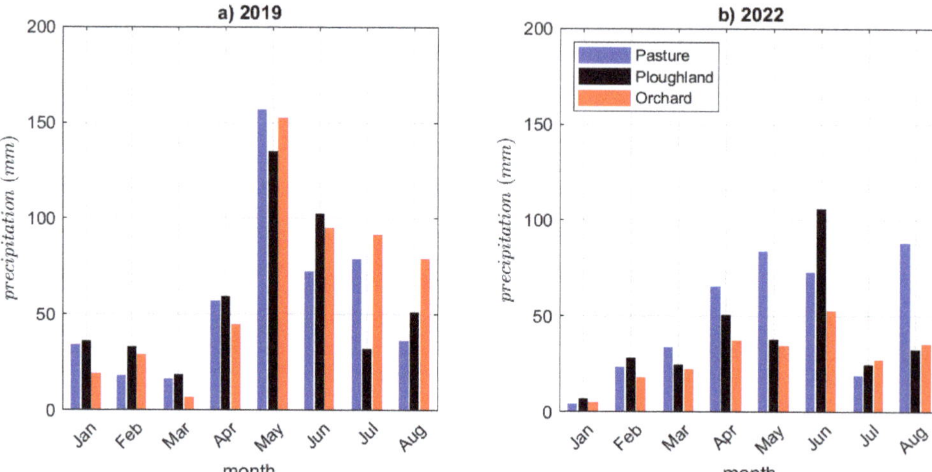

Figure 2. Monthly precipitation totals from January to August in (**a**) 2019 and (**b**) 2022.

Table 4. Mean annual, growing season (April–October), spring (March to May), summer (June to August) and autumn (September to November) temperatures [°C] in 2019 and 2022.

Land Use	Annual		Growing Season		Spring		Summer		Autumn	
	2019	2022	2019	2022	2019	2022	2019	2022	2019	2022
Ploughland	12.17	12.29	17.4	18.1	11.2	11.4	22.28	23.31	12.7	11.5
Orchard	12.66	12.81	17.8	18.5	11.7	11.8	22.69	23.60	13.3	12.2
Pasture	12.02	12.18	17.1	17.8	11	11.3	21.91	22.89	12.5	11.5

The higher summer temperatures and lower precipitations of 2022 compared to 2019 generated significant differences in PET, aridity index and PaDI both seasonally and annually. Monthly potential evapotranspiration revealed the greatest variations between 2019 and 2022 in the summer months. The close-to-record temperatures in July (T_{max} = 39.2 °C in Palkonya) generated a monthly PET of almost 160 mm at all three sites. The largest differences in PET were registered in May.

PaDI and the aridity index demonstrated great variations between the two studied years. Due to the high rainfall totals of 2019 at the orchard site, its PaDI did not indicate any water stress in the first study year, whereas the other two sites had a PaDI of 4.2 and 4.6 which referred to mild drought conditions. Although drought conditions remained in the class of mild drought in 2022 at the pasture and the ploughland, all three sites experienced water stress with the highest increase in PaDI at the orchard site, which entered the class of moderate drought (Figure 3a).

Aridity indices were around 1 in 2019 (common for Hungary since the onset of meteorological measurements), whereas they turned into a negative water balance (AI > 1) for the ploughland and the orchard by 2022 (Figure 3b).

Figure 4 shows the aridity index, total monthly precipitation and monthly evapotranspiration in 2019 and in 2022 for the ploughland, pasture and orchard locations. All sites in 2022 experienced semiarid or arid conditions especially during the growing season (from April to October), except for September. However, in 2019 aridity was less severe, but still reached the semiarid category during this most critical season. The orchard site remained under subhumid conditions. Compared to Figure 4 (PaDI), the orchard site showed more intense drought in 2022, since over the growing season it experienced a lower amount of plant-available water (less precipitation) against high evapotranspiration.

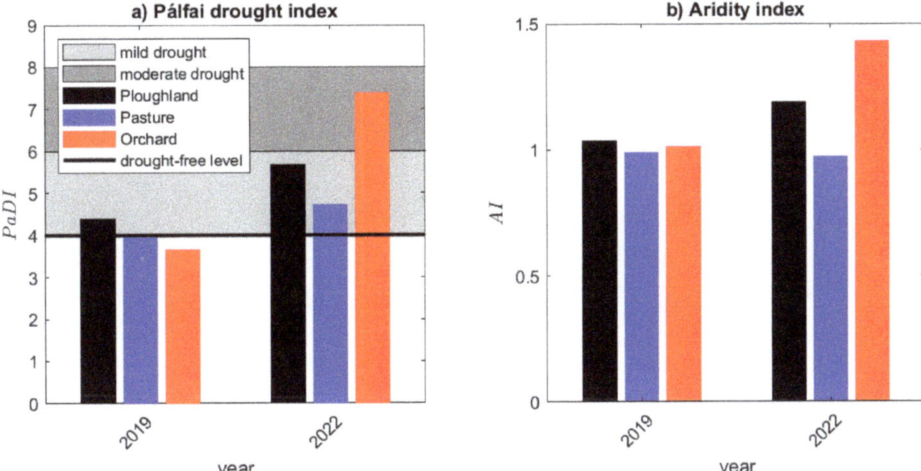

Figure 3. (**a**) PaDI and (**b**) aridity indices of 2019 and 2022.

In terms of the average aridity index, precipitation and evapotranspiration, summer periods had intensively critical values for almost all sites, especially in 2022. The main difference was found in the growing season when higher evapotranspiration and lower precipitation in 2022 resulted in more arid conditions (Table 5). Nevertheless, on an annual basis and in autumn the differences were negligible. During the winter of 2021/2022, all sites received a substantial precipitation amount, but this was not able to compensate for the low rainfall during the growing season in 2022. It is to be pointed out that in 2021 the summer, and, in fact, the entire growing season were also moderately dry (mild drought, with around PaDI = 5, not shown here).

Table 5. Summary statistics of the annual, growing season (April–October) and seasonal (spring, summer, autumn and winter. Seasons are the same as in Table 4, and the winter periods have been selected as the following: 2018/2019: December 2018–February 2019 and 2021/2022: December 2021–February 2022 mean of aridity index (AI), precipitation (P) and evapotranspiration (ET_0).

	Annual		Growing Season		Spring		Summer		Autumn		Winter	
	2019	2022	2019	2022	2019	2022	2019	2022	2019	2022	2018/2019	2021/2022
Pasture												
P	59.5	62.2	67.7	79.6	76.7	73.9	62.5	60	59.3	90.8	22.7	44.5
ET_0	58.9	60.4	90.8	95.9	50.6	54.8	130.1	137.2	49.6	43.7	-	-
AI	1.2	1.4	1.7	2.1	1.1	0.8	2.3	3.8	1.2	0.6	-	-
Ploughland												
P	57.5	51.4	66.8	54.8	70.9	64.8	61.8	54.3	54	69.1	29.4	41.3
ET_0	59.6	61.1	92.3	97.6	50.9	54.9	132.5	140.2	50.1	43.7	-	-
AI	1.2	1.7	1.8	2.7	1.0	1.4	2.7	3.8	1.1	1.2	-	-
Orchard												
P	60.1	43.8	79.2	47.3	67.9	41.3	88.4	38.3	52	72.6	18.4	33.0
ET_0	60.9	62.5	93.9	99.0	52.3	56.2	134.8	141.7	51.2	45.4	-	-
AI	1.2	1.8	1.3	2.8	2.2	1.7	1.5	4.0	1.1	1.3	-	-

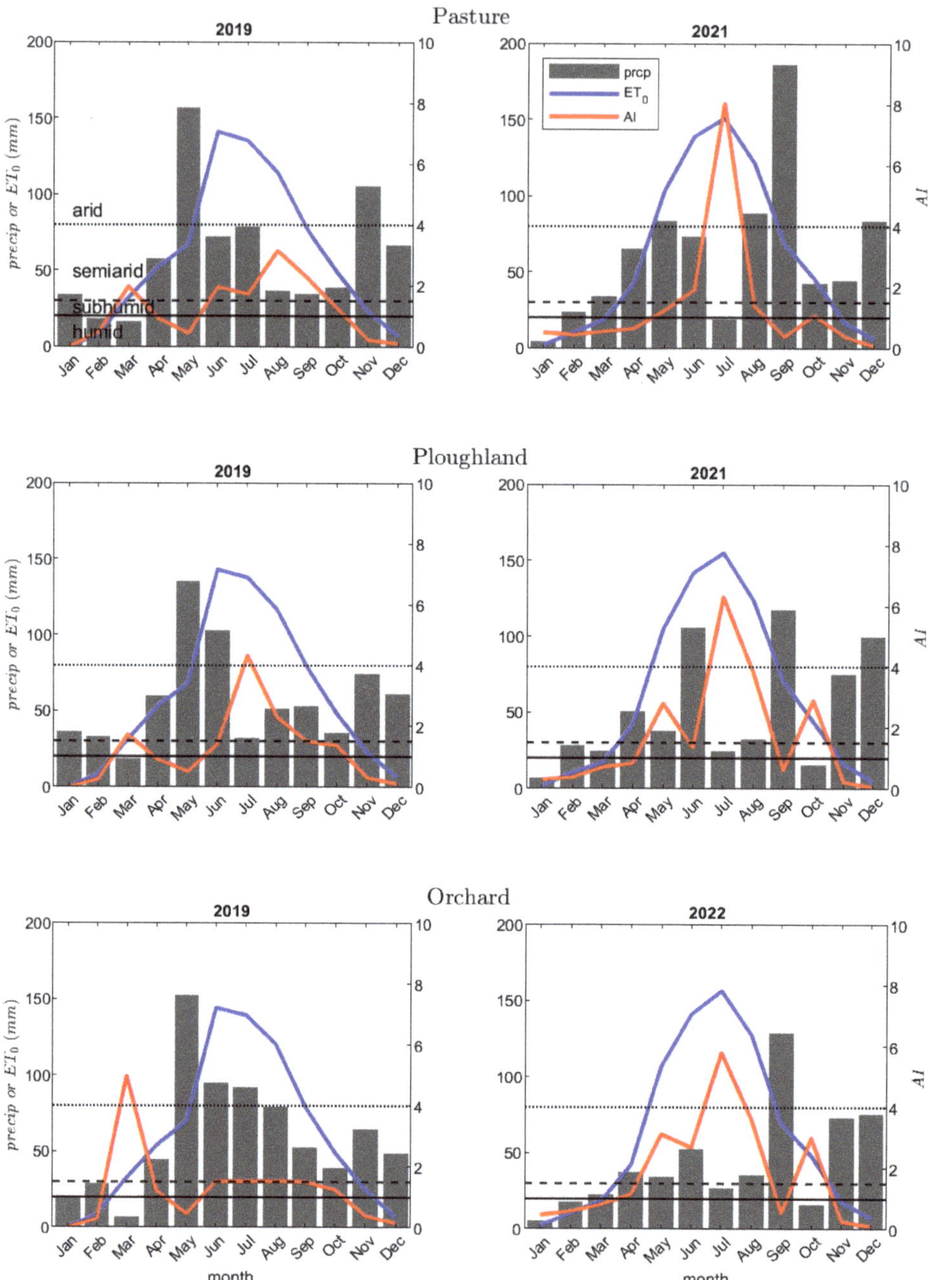

Figure 4. Monthly precipitation totals (prcp), evapotranspiration (ET$_0$) and aridity indices (AI) for the different sites (pasture, ploughland and orchard) of 2019 and 2022. Solid lines depict the value of AI = 1, dashed lines AI = 1.5 and dotted lines AI = 4 as a reference for humid (AI < 1), subhumid (1 ≤ AI ≤ 1.5), semiarid (1.5 ≤ AI ≤ 4) and arid (AI ≥ 4) conditions, respectively.

3.3. Soil Moisture Regime

In general, the SM regime showed a rather variable picture for the three study sites. On average, the natural pasture had the highest volumetric water content and soils and that site had the most responsive behaviour to rainfall events. The foothill site of the ploughland, however, had the lowest SM contents and the least variability of SM among all sites (Figure 5). The orchard site revealed a contrasting picture between the two studied years due to (i) necrosis, tree removal above the upper monitoring station and (ii) the markedly less precipitation in 2022 compared to 2019. While the moisture regime at this site demonstrated a great variability in 2019, the SM content showed a monotonously decreasing trend over the summer of 2022.

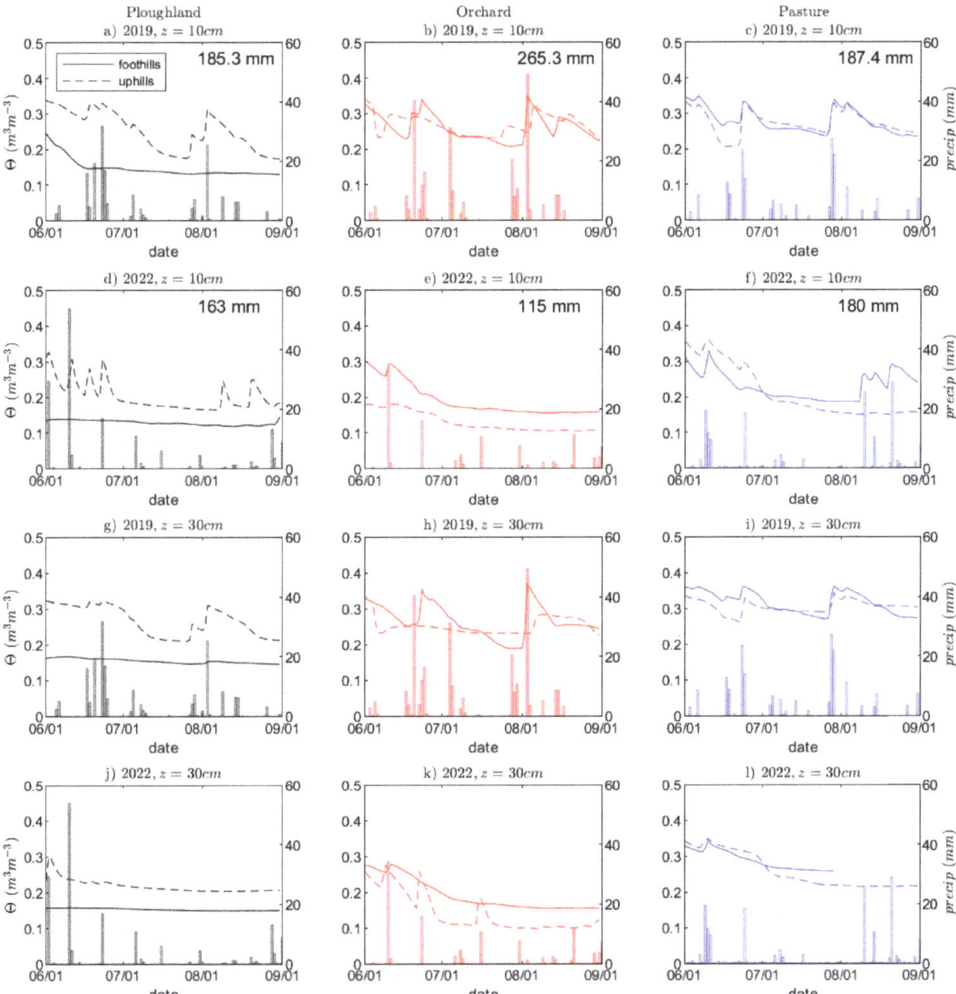

Figure 5. Soil moisture dynamics of the study sites (first column ploughland, middle column pasture and right column orchard) in the summers of 2019 (1st and 3rd rows) and 2022 (2nd and 4th rows), for (**a**–**f**) 10 cm and (**g**–**l**) 30 cm. The values on the right upper corners in subplots from (**a**) to (**f**) represent the precipitation amount during the summer (June to August).

The statistics of SM indicated marked variations among the three land use types. Commonly, the pasture showed the highest SM content, and the ploughland had the lowest median SM values. The footslope station of the ploughland revealed the lowest SM content and the lowest SM range, likely influenced by the grove uphill. On the other hand, the orchard showed the greatest range of SM over the two studied summers due to its extreme water balance and the removal of tree canopy at the upper station (Figures 6 and 7). The natural pasture presented the greatest water stress tolerance and the most homogeneous water dynamics among the three land use types. At all stations, mean and median SM contents in 2022 were either equal to (at the ploughland footslope, 30 cm) or lower than those in 2019 (at all other stations).

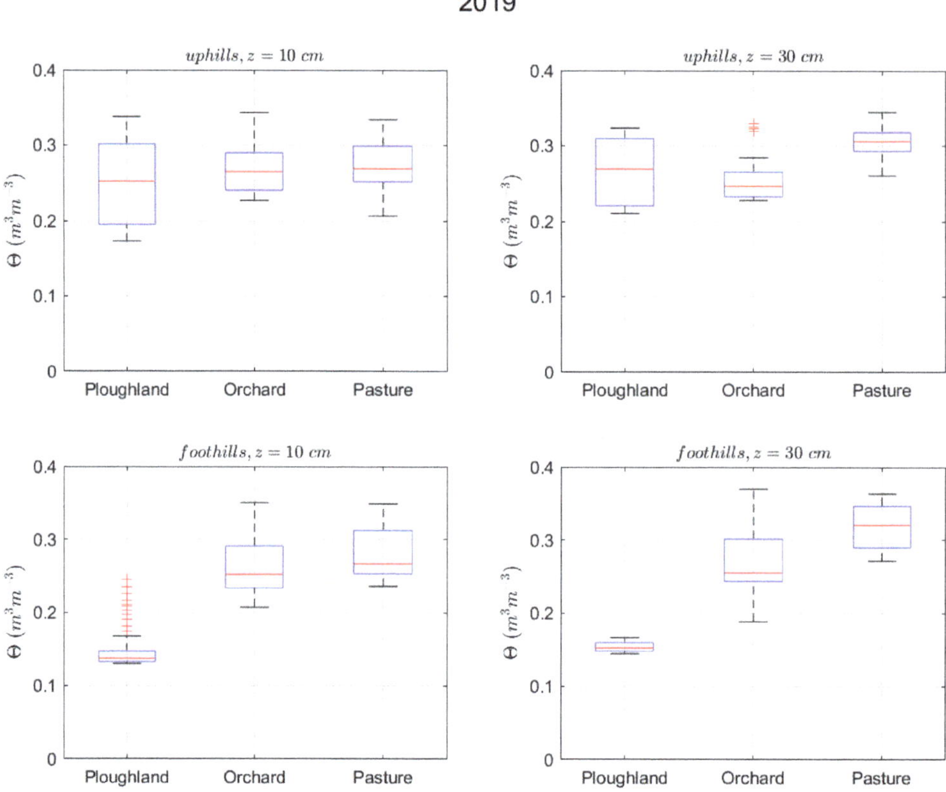

Figure 6. Mean soil moisture values of the study sites in the summer of 2019.

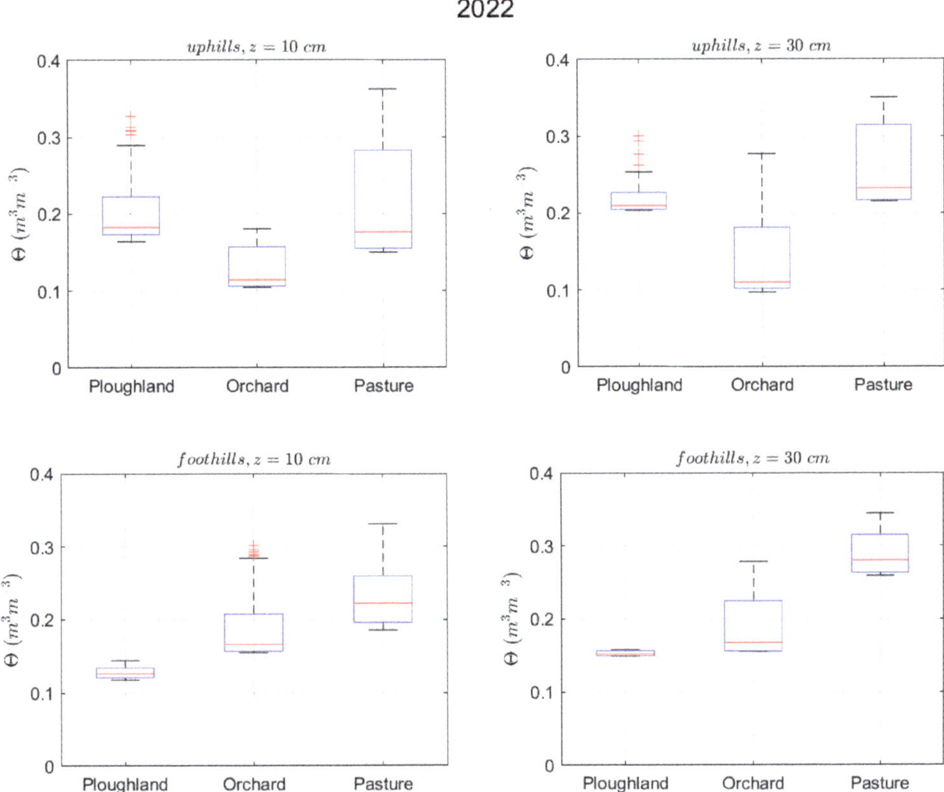

Figure 7. Mean soil moisture values of the study sites in the summer of 2022.

4. Discussion

Our research confirmed former results [16,21,27,30,37–45], i.e., moisture content of the vadose zone is markedly influenced by land use type, distance from landscape and morphological features, local water and moisture balance and soil texture.

In both summers, the most optimal moisture content (i.e., the highest median with the lowest variability) was revealed in the pasture. The highest median SM content was found here over both studied summers. A higher drought risk during the two studied summers was found both at the ploughland and the orchard. SM contents were low in the ploughland on many occasions and for prolonged periods in 2019, during which the matric potential fell below the permanent wilting point (data are not shown here). The longest of such periods lasted for 73 and 101 days in 2019 and 2022, respectively, at the upper station of the ploughland. According to our a priori hypothesis, the ploughland should have had a large evaporation loss and hence low mean SM content. Yet, due to the dense canopy cover of soybean until late September, evaporation loss caused by direct solar radiation was limited. The orchard performed well when the canopy was present at the site (in 2019); however, when the land use type was changed by 2022, its water retention capacity deteriorated. As a consequence, the orchard showed a markedly more negative water balance in 2022 compared to 2019, further exacerbated by the effect of the absence of canopy. The negative water balance of 2022 in the orchard produced the lowest median SM content at both monitored depths of the upper stations.

Our results have been partly confirmed by the findings of Wang et al. [46], who found the second highest SM content in grasslands in eastern China. Nonetheless, in their study, corn had the highest SM content due to the reduced evaporation from the soil. However, Shi et al. [47] pointed out that intense transpiration created extreme water stress in orchards in the Loess Plateau of China, underpinning our result concerning the high variations of SM at the orchard site.

Opposed to the findings of Tölgyesi et al. [48], we did not find compelling evidence of the drying effect of trees in the top 30 cm of the soil. This may be explained by (i) the removal of trees, (ii) the finer soil textural types of our site and (iii) the extreme water balance of the orchard site in 2022 compared to the ploughland and the pasture. Our results, however, indicated the enhanced water retention and water storage capacity of the soil due to the canopy cover and shading of the orchard (cherry) trees during the summer of 2019, and hence contributed to the overall roles of natural ecosystem services. This finding is corroborated by the results of Syrbe and Grunewald [41] and Ribeiro and Šmid Hribar [49]. In a good agreement with the findings of the present study, previous studies also demonstrated the benefit of low-impact agricultural practices for the reduction or possibly the termination of the decline of plant-available water [50,51].

Depending on the water balance and the proximity of the groundwater table, the direction of water motion may differ temporally or seasonally. During the summer, according to matric potential data, capillary rise was common at all monitoring stations [30]. Such capillary rise-dominated periods were intermittently interrupted by intense infiltration events such as surface runoff and probably also through flow rates also intensified during heavy thunderstorms (e.g., 2 August 2019: 51.5 mm and 9 June 2022: 55.2 mm). Leitinger et al. revealed the marked influence of the slope gradient on the distribution of SM along the hillslope in the Eastern Alps [40]. They also found a significant influence of water balance and land management type on infiltration and surface runoff [40]. Their findings revealed the impact by cattle trampling and treading; however, at our pasture site, no grazing animals were kept.

5. Conclusions

The marked variations among the three study sites can be partially explained by the difference in land use types, whereas the contrast between 2019 and 2022 can be explained by the influence of water balance. The most stable behaviour was found for the natural grazing land in both years. Our hypothesis, however, according to which the ploughland should have demonstrated the worst moisture dynamics, was not proven, especially in 2022. This is likely attributed to the (i) spatial variability of water balance and (ii) crop type, as soybean forms a relatively dense canopy early in the growing season and harvesting is commonly timed to late September and early October in SW Hungary. Hence, evaporation loss from the soil due to direct irradiation is limited. A third possible reason for the observed variation among the three sites is the insufficient spatial resolution of field-monitored SM data.

Our result may be utilized by stakeholders in the field of agricultural and farming businesses especially under subhumid climates of the temperate zone.

Although tackling the challenges posed by drought is not a novel phenomenon in the Pannonian Basin, adaptation to altered climates is not just an option presently: it is crucial. Therefore, a long-term analysis of climatic trends and water budget is indispensable, similar to the site-specific differentiation and optimization of agriculture. The present study aimed at providing data for studies of this type and delivering a more accurate understanding of these processes with the aim to adjust ecosystem services at a local scale. The present study could be improved with analyses performed (i) under more controlled conditions, (ii) within a more restricted geographical area, (iii) using hydrologic models and (iv) estimating local water balances at a higher spatial resolution.

Author Contributions: Conceptualization, S.C.; methodology, S.Á.F.; formal analysis, N.S., E.P., A.C. and M.I.; investigation, S.Á.F.; data curation, G.N., S.C. and R.B.; writing—original draft preparation, S.C.; writing—review and editing, N.S., D.L., G.P., R.C., M.F. and S.Á.F.; visualization, N.S. and

G.N.; project administration, D.L.; funding acquisition, D.L., R.C. and M.F. All authors have read and agreed to the published version of the manuscript.

Funding: The authors are grateful for the financial support from the National Research, Development and Innovation Office (NKFIH) within the framework of the Hungarian–Slovenian collaborative project "Possible ecological control of flood hazard in the hill regions of Hungary and Slovenia" (contract no SNN 125727) and within the framework of the program Excellence in Higher Education, Theme II. 3. ("Innovation for sustainable life and environment"). The study was financed by the project "Possible ecological control of flood hazard in the hill regions of Hungary and Slovenia" (SNN 125727; Slovenian Research Agency ARRS, N6-0070) and a research programme Geography of Slovenia (Slovenian Research Agency ARRS, P6-0101).

Institutional Review Board Statement: Not applicable.

Informed Consent Statement: Not applicable.

Data Availability Statement: Data are located on the University of Pécs data storage system (One Drive–Pécsi Tudományegyetem), which due to system safety issues are available upon request.

Conflicts of Interest: The authors declare no conflict of interest.

References

1. IPCC 2013. *Climate Change 2013: The Physical Science Basis. Summary for Policymakers. Intergovernmental Panel of Climate Change*; Cambridge University Press: Cambridge, UK, 2013; 222p, Available online: https://www.ipcc.ch/site/assets/uploads/2018/03/WG1AR5_SummaryVolume_FINAL.pdf (accessed on 18 January 2023).
2. Li, M.; Chu, R.; Islam, A.R.M.T.; Jiang, Y.; Shen, S. Attribution Analysis of Long-Term Trends of Aridity Index in the Huai River Basin, Eastern China. *Sustainability* **2020**, *12*, 1743. [CrossRef]
3. Didovets, I.; Krysanova, V.; Bürger, G.; Snizhko, S.; Balabukh, V.; Bronstert, A. Climate change impact on regional floods in the Carpathian region. *J. Hydrol. Reg. Stud.* **2019**, *22*, 100590. [CrossRef]
4. Fábián, S.Á.; Kovács, J.; Lóczy, D.; Schweitzer, F.; Varga, G.; Babák, K.; Lampért, K.; Nagy, A. Geomorphologic hazards in the Carpathian Foreland, Tolna County (Hungary). *Stud. Geomorphol. Carpatho Balc.* **2006**, *11*, 107–118. Available online: https://www.igipz.pan.pl/tl_files/igipz/ZGiHGiW/sgcb/sgcb_40/sgcb_40_07.pdf (accessed on 18 January 2023).
5. Czigány, S.; Pirkhoffer, E.; Balassa, B.; Bugya, T.; Bötkös, T.; Gyenizse, P.; Nagyváradi, L.; Lóczy, D.; Geresdi, I. Villámárvíz mint természeti veszélyforrás a Dél-Dunántúlon. *Földrajzi Közlemények* **2010**, *134*, 281–298. Available online: https://www.foldrajzitarsasag.hu/downloads/foldrajzi_kozlemenyek_2010_134_evf_3_szam.pdf (accessed on 18 January 2023).
6. Lóczy, D.; Czigány, S.; Pirkhoffer, E. Flash Flood Hazards. In *Studies on Water Management Issues*; Kumarasamy, M., Ed.; InTech: Rijeka, Croatia, 2012; pp. 27–52. [CrossRef]
7. Lukić, T.; Micić Ponjiger, T.; Basarin, B.; Sakulski, D.; Gavrilov, M.; Marković, S.; Zorn, M.; Komac, B.; Milanović, M.; Pavić, D.; et al. Application of Angot precipitation index in the assessment of rainfall erosivity: Vojvodina Region case study (North Serbia). *Acta Geogr. Slov.* **2021**, *61*, 123–153. [CrossRef]
8. Grobicki, A.; MacLeod, F.; Pischke, F. Integrated policies and practices for flood and drought risk management. *Water Policy* **2015**, *17*, 180–194. [CrossRef]
9. Ferk, M.; Ciglič, R.; Komac, B.; Lóczy, D. Management of small retention ponds and their impact on flood hazard prevention in the Slovenske Gorice Hills. *Acta Geogr. Slov.* **2020**, *60*, 107–125. [CrossRef]
10. European Commission. *A Guide to Support the Selection, Design and Implementation of Natural Water Retention Measures in Europe: Capturing the Multiple Benefits of Nature-Based Solutions*; European Commission: Brussels, Belgium, 2014; 98p, Available online: http://www.ecrr.org/Portals/27/Publications/NWRMpublication.pdf (accessed on 18 January 2023).
11. Castellini, M.; Iovino, M. Pedotransfer functions for estimating soil water retention curve of Sicilian soils. *Arch. Agron. Soil Sci.* **2019**, *65*, 1401–1416. [CrossRef]
12. Fisher, B.; Turner, K. Ecosystem services: Classification for valuation. *Biol. Conserv.* **2008**, *141*, 1167–1169. [CrossRef]
13. Haines-Young, R.; Potschin, M. *Common International Classification of Ecosystem Services (CICES): 2011 Update*; European Environment Agency: London, UK, 2011; 17p, Available online: https://cices.eu/content/uploads/sites/8/2009/11/CICES_Update_Nov2011.pdf (accessed on 18 January 2023).
14. Dezső, J.; Lóczy, D.; Salem, A.M.; Nagy, G. Floodplain Connectivity. In *The Drava River: Environmental Problems and Solutions*; Lóczy, D., Ed.; Springer International Publishing: Cham, Switzerland, 2019; pp. 215–230. [CrossRef]
15. Lóczy, D. *Hydromorphological-Geoecological Foundations of Floodplain Management: CaseS from Hungary*; Lambert Academic Publishing: Saarbrücken, Germany, 2013; 382p.
16. Horel, Á.; Zsigmond, T.; Farkas, C.; Gelybó, G.; Tóth, E.; Kern, A.; Bakacsi, Z. Climate Change Alters Soil Water Dynamics under Different Land Use Types. *Sustainability* **2022**, *14*, 3908. [CrossRef]
17. Abdelkadir, A.; Yimer, F. Soil water property variations in three adjacent land use types in the Rift Valley area of Ethiopia. *J. Arid Environ.* **2011**, *75*, 1067–1071. [CrossRef]

18. Jakab, G.; Németh, T.; Csepinszky, B.; Madarász, B.; Szalai, Z.; Kertész, Á. The influence of short term soil sealing and crusting on hydrology and erosion at Balaton Uplands, Hungary. *Carpathian J. Earth Environ. Sci.* **2013**, *8*, 147–155.
19. Saeidi, S.; Grósz, J.; Sebők, A.; Barros, V.D.d.; Waltner, I. Analysis results of land-use and land cover changes of Szilas catchment from 1990. *Tájökológiai Lapok* **2019**, *17*, 265–275. [CrossRef]
20. Hota, S.; Mishra, V.; Mourya, K.K.; Giri, K.; Kumar, D.; Jha, P.K.; Saikia, U.S.; Prasad, P.V.V.; Ray, S.K. Land use, landform, and soil management as determinants of soil physicochemical properties and microbial abundance of Lower Brahmaputra Valley, India. *Sustainability* **2022**, *14*, 2241. [CrossRef]
21. Fu, B.; Wang, J.; Chen, L.; Qui, Y. The effects of land use on soil moisture variation in the Danangou catchment of the Loess Plateau, China. *Catena* **2003**, *54*, 197–213. [CrossRef]
22. Jørgensen, D.; Jørgensen, F.A. Aesthetics of energy landscapes. *Environ. Space Place* **2018**, *10*, 1–14. [CrossRef]
23. Peng, S.H.; Han, K.T. Assessment of aesthetic quality on soil and water conservation engineering using the scenic beauty estimation method. *Water* **2018**, *10*, 407–422. [CrossRef]
24. European Commission. Directive 2000/60/EEC. Establishing a framework for community action in the field of water policy. *Off. J. Eur. Communities Luxembg.* **2000**, *L327*, 1–71.
25. European Commission. *Ecoflood Guidelines: How to Use Floodplains for Flood Risk Reduction*; Office for Official Publications of the European Communities: Luxembourg, 2006; 144p.
26. European Commission. Directive 2007/60/EC of the European Parliament and of the Council on the assessment and management of flood risks. *Off. J. Eur. Union L* **2007**, *288*, 27–34.
27. Horel, Á.; Tóth, E.; Gelybó, G.; Kása, I.; Bakacsi, Z.; Farkas, C. Effects of Land Use and Management on Soil Hydraulic Properties. *Open Geosci.* **2015**, *1*, 742–754. [CrossRef]
28. Haghighi, F.; Gorji, M.; Shorafa, M. A study of the effects of land use changes on soil physical properties and organic matter. *Land Degrad. Dev.* **2010**, *21*, 496–502. [CrossRef]
29. Sławiński, C.; Cymerman, J.; Witkowska-Walczak, B.; Lamorski, K. Impact of diverse tillage on soil moisture dynamics. *Int. Agrophys.* **2012**, *26*, 301–309. [CrossRef]
30. Nagy, G.; Lóczy, D.; Czigány, S.; Pirkhoffer, E.; Fábián, S.Á.; Ciglič, R.; Ferk, M. Soil moisture retention on slopes under different agricultural land uses in hilly regions of Southern Transdanubia. *Hung. Geogr. Bull.* **2020**, *69*, 263–280. [CrossRef]
31. Tang, M.; Li, W.; Gao, X.; Wu, P.; Li, H.; Ling, Q.; Zhang, C. Land use affects the response of soil moisture and soil temperature to environmental factors in the loess hilly region of China. *PeerJ* **2022**, *10*, e13736. [CrossRef] [PubMed]
32. Wang-Erlandsson, L.; Ent, R.J.; Gordon, L.J.; Savenije, H.H.G. Contrasting roles of interception and transpiration in the hydrological cycle—Part 1: Temporal characteristics over land. *Earth Syst. Dynam.* **2014**, *5*, 441–469. [CrossRef]
33. Niu, C.Y.; Musa, A.; Liu, Y. Analysis of soil moisture condition under different land uses in the arid region of Horqin sandy land, northern China. *Solid Earth* **2015**, *6*, 1157–1167. [CrossRef]
34. Pálfai, I. Az aszály definíciói, befolyásoló tényezői és mérőszámai. (Definition, influencing factors and indices of drought). In *Belvizek és Aszályok Magyarországon (Inland Excess Water and Drought in Hungary)*; Pálfai, I., Ed.; Hidrológiai Tanulmányok: Budapest, Hungary, 2004; pp. 255–263.
35. Thornthwaite, C.W. A contribution to the report of the committee on transpiration and evaporation. *Trans. Am. Geophys. Union* **1944**, *25*, 686–693.
36. Thornthwaite, C.W. A re-examination of the concept and measurement of potential evapotranspiration. *John Hopkins Univ. Publ. Climatol.* **1954**, *7*, 200–209.
37. Kocsis, T.; Anda, A. Microclimate simulation of climate change impacts in a maize canopy. *Időjárás* **2006**, *116*, 109–122.
38. Emerson, C.H.; Welty, C.; Traver, R.G. Watershed-scale evaluation of a system of storm water detention basins. *J. Hydrol. Eng.* **2005**, *10*, 237–242. [CrossRef]
39. Davis, A.P.; Hunt, W.F.; Traver, R.G.; Clar, M. Bioretention technology: Overview of current practice and future needs. *J. Environ. Eng. ASCE* **2009**, *135*, 109–117. [CrossRef]
40. Leitinger, G.; Tasser, E.; Newesely, C.; Obojes, N.; Tappeineret, U. Seasonal dynamics of surface runoff in mountain grassland ecosystems differing in land use. *J. Hydrol.* **2010**, *385*, 95–104. [CrossRef]
41. Syrbe, R.-U.; Grunewald, K. Restrukturierungsbedarf für regionaltypische Landschaftselemente und Biotopstrukturen am Beispiel Sachsens. *Nat. Und Landsch.* **2013**, *88*, 103–111. [CrossRef]
42. Wang, S.; Fu, B.; Gao, G.; Liu, Y.; Zhou, J. Responses of soil moisture in different land cover types to rainfall events in a re-vegetaion catchment area of the Loess Plateau, China. *Catena* **2013**, *101*, 122–128. [CrossRef]
43. Zucco, G.; Brocca, L.; Moramarco, T.; Morbidelli, R. Influence of land use on soil moisture spatial–temporal variability and monitoring. *J. Hydrol.* **2014**, *516*, 193–199. [CrossRef]
44. Výleta, R.; Danáčová, M.; Škrinár, A.; Fencík, R.; Hlavčová, K. Monitoring and assessment of water retention measures in agricultural land. *Earth Environ. Sci.* **2017**, *95*, 022008. [CrossRef]
45. Juhos, K.; Czigány, S.; Madarász, B.; Ladányi, M. Interpretation of soil quality indicators for land suitability assessment—A multivariate approach for Central European arable soils. *Ecol. Indic.* **2019**, *99*, 261–272. [CrossRef]
46. Wang, H.; Gao, J.E.; Zhang, S.L.; Zhang, M.J.; Li, X.H. Modeling the impact of soil and water conservation on surface and ground water based on the SCS and Visual MODFLOW. *PLoS ONE* **2013**, *8*, e79103. [CrossRef]

47. Shi, D.; Tan, H.; Rao, W.; Liu, Z.; Elenga, H.I. Variations in water content of soil in apricot orchards in thewestern hilly regions of the Chinese Loess Plateau. *Vadose Zone J.* **2020**, *19*, e20034. [CrossRef]
48. Tölgyesi, C.; Török, P.; Hábenczyus, A.A.; Bátori, Z.; Valkó, O.; Deák, B.; Kelemen, A. Underground deserts below fertility islands? Woody species desiccate lower soil layers in sandy drylands. *Ecography* **2020**, *43*, 848–859. [CrossRef]
49. Ribeiro, D.; Šmid Hribar, M. Assessment of land-use changes and their impacts on ecosystem services in two Slovenian rural landscapes. *Acta Geogr. Slov.* **2019**, *59*, 143–160. [CrossRef]
50. Guerrero, B.; Amosson, S.; Nair, S.; Marek, T. The importance of regional analysis in evaluating agricultural water conservation strategies. *J. Reg. Anal. Policy* **2017**, *47*, 188–198. [CrossRef]
51. Fuentes, J.P.; Flury, M.; Bezdicek, F.D. Hydraulic Properties in a Silt Loam Soil under Natural Prairie, Conventional Till, and No-Till. *Soil Sci. Soc. Am.* **2004**, *68*, 1679–1688. [CrossRef]

 *sustainability*

Article

Assessing Soil Erosion by Monitoring Hilly Lakes Silting

Yamuna Giambastiani [1,*], Riccardo Giusti [1], Lorenzo Gardin [1], Stefano Cecchi [1], Maurizio Iannuccilli [1], Stefano Romanelli [2], Lorenzo Bottai [2], Alberto Ortolani [1,2] and Bernardo Gozzini [1,2]

[1] CNR-IBE, National Research Council, Institute of Bioeconomy, 50019 Florence, Italy; giusti@lamma.toscana.it (R.G.); gardin@lamma.toscana.it (L.G.); cecchi@lamma.toscana.it (S.C.); iannuccilli@lamma.toscana.it (M.I.); ortolani@lamma.toscana.it (A.O.); gozzini@lamma.toscana.it (B.G.)

[2] Environmental Modelling and Monitoring Laboratory for Sustainable Development, LaMMA Consortium, 50019 Florence, Italy; romanelli@lamma.toscana.it (S.R.); bottai@lamma.toscana.it (L.B.)

* Correspondence: giambastiani@lamma.toscana.it

Abstract: Soil erosion continues to be a threat to soil quality, impacting crop production and ecosystem services delivery. The quantitative assessment of soil erosion, both by water and by wind, is mostly carried out by modeling the phenomenon via remote sensing approaches. Several empirical and process-based physical models are used for erosion estimation worldwide, including USLE (or RUSLE), MMF, WEPP, PESERA, SWAT, etc. Furthermore, the amount of sediment produced by erosion phenomena is obtained by direct measurements carried out in experimental sites. Data collection for this purpose is very complex and expensive; in fact, we have few cases of measures distributed at the basin scale to monitor this phenomenon. In this work, we propose a methodology based on an expeditious way to monitor the volume of hilly lakes with GPS, sonar sensor and aquatic drone. The volume is obtained by means of an automatic GIS procedure based on the measurements of lake depth and surface area. Hilly lakes can be considered as sediment containers. Time-lapse measurements make it possible to estimate the silting rate of the lake. The volume of 12 hilly lakes in Tuscany was measured in 2010 and 2018, and the results in terms of silting rate were compared with the estimates of soil loss obtained by RUSLE and MMF. The analyses show that all the lakes measured are subject to silting phenomena. The sediment estimated by the measurements corresponds well to the amount of soil loss estimated with the models used. The relationships found are significant and promising for a distributed application of the methodology, which allows rapid estimation of erosion phenomena. Substantial differences in the proposed comparison (mainly found in two cases) can be justified by particular conditions found on site, which are difficult to predict from the models. The proposed approach allows for a monitoring of basin-scale erosion, which can be extended to larger domains which have hilly lakes, such as, for example, the Tuscany region, where there are more than 10,000 lakes.

Keywords: sediment monitoring; remote sensing; lakes; water capacity; sonar; aquatic drone; USLE; MMF

Citation: Giambastiani, Y.; Giusti, R.; Gardin, L.; Cecchi, S.; Iannuccilli, M.; Romanelli, S.; Bottai, L.; Ortolani, A.; Gozzini, B. Assessing Soil Erosion by Monitoring Hilly Lakes Silting. *Sustainability* **2022**, *14*, 5649. https://doi.org/10.3390/su14095649

Academic Editors: Blaž Repe, Mario Elia and Antonio Ganga

Received: 30 March 2022
Accepted: 3 May 2022
Published: 7 May 2022

Publisher's Note: MDPI stays neutral with regard to jurisdictional claims in published maps and institutional affiliations.

1. Introduction

1.1. Erosion: A Worldwide Threat

After almost a century of research and studies on the territory, soil erosion caused by water, wind and tillage is known to be the greatest threat to soil health, and to the ecosystem services it provides, in many regions of the world [1–3]. Its impact on global crop production has been estimated at a reduction of 0.4% per year [4]. Some authors argue that nearly a third of the world's arable land has been lost due to erosion over the past 40 years and continues to decrease at a rate greater than ten million hectares per year [5]. Erosion is a natural phenomenon which consists of the loss of the most superficial layer of the soil due to the action of precipitation or wind. With the advent of modern agriculture and, above all, with (I) the introduction of extensive mechanization, (II) the leveling of the slopes, (III) the abandonment of traditional hydraulic-agricultural solutions and (IV) the specialization of crops, erosion has assumed worrying proportions [6–8]. Erosion is now a worldwide threat, especially in hilly areas with significant economic impacts, particularly in areas with valuable crops [9,10]. Water erosion represents one of the main threats to the correct functionality of the soil, through (I) the removal of the fertile surface soil horizon, (II) the denser subsoil incorporation in the surface layer and (III) the possible decrease in the root zone [11].

A reliable assessment of this phenomenon is therefore particularly useful as a decision-support tool for planning soil conservation interventions [12–14]. To address these issues, the Community Agricultural Policy has ensured that agriculture is in line with the EU soil protection policies. Effective management of these issues is considered essential for many strategies and priorities of the European Green Deal, as defined primarily in the (I) thematic strategy and the sustainable management of soil [15], (II) the fight against erosion, and (III) the fight against the loss of organic carbon and biodiversity in soil. The quantitative assessment of soil erosion, due to both water and wind, is generally carried out through modeling the phenomenon or with experimental tests (plots, rain simulators, etc.) carried out directly on the field [16]. In recent decades, researchers in Italy have also conducted several direct studies on the phenomenon of erosion [17–22].

1.2. Models for Erosion Estimation

The most commonly used erosion estimation model is the universal equation of soil loss (USLE) [23], and its revised version (RUSLE) [24], which estimates the annual mean long-term loss of soil due to sheet (interrill) and rill erosion. It should be noted that soil loss caused by (ephemeral) gully erosion is not predicted by RUSLE [25]. Despite its shortcomings, RUSLE is still the most widely used model on a large scale [26,27]. It can process data input for large regions and provides a basis for scenario analysis and taking actions against erosion [28]. A recent work [29] estimated erosion on a European scale using the most in-depth processing of the single factors [29–32]. USLE has been applied in comparative studies between various analysis methods, and the authors have shown that it does not lead to greater errors than process-based physical models (WEPP and PESERA), although it has some limitations due to the simple empirical nature of the model [33–35]. Soil loss is also analyzed worldwide through the revised MMF—Morgan–Morgan–Finney model [36,37], in order to evaluate the land degradation and ecological status of specific catchment areas or wider territories [38–40]. This model allows an erosion simulation to be developed in relation to the characteristics of the vegetation cover. The comparison between these empirical models shows similar results [41,42]. The Soil and Water Assessment Tool (SWAT) is a semi-empirical model used for the assessment of erosion phenomena at the basin scale [43–45]. It also allows analyses related to hydrological processes [46], land management and climate change [47,48]. Other simulations have been performed directly on reconstructions of hydrographic basins in miniature or in experimental sites [49–51].

In Tuscany in 2009, soil erosion estimates were made at a regional scale with the USLE model through the elaboration of single climatic, pedological, land use and morphometric factors at high resolution [52,53]. Direct measurements carried out over the years in experimental fields [54–56] have allowed the model to be properly constrained and tested.

1.3. Scope of Work

With this work, we propose a new approach to monitor erosion phenomena at the basin scale, based on an expeditious estimate of the hilly lakes silting rate, through remote sensing techniques. The basic assumption is that a relationship exists between the soil loss (or sediment production) from the basin with the volume loss of the reservoir. The silting rate, estimated by sonar and aquatic drone [57], is compared with the soil loss obtained from two models (RUSLE and MMF) in order to evaluate erosion through direct measurements of the sediment produced. The hilly lakes distributed throughout the territory can be considered the containers of the sediment coming from the erosion phenomena of the slope, and therefore constitute a net of distributed monitoring. This expeditious method of estimating the reservoir capacity for the study of erosion phenomena is an innovative tool that enables the estimation of the sediment produced by a slope. The aim is to demonstrate that through the repetition over time of a simple procedure of silting estimation, it is possible to observe the evolution of the erosive phenomena and better understand the impact of anthropogenic actions or climate change on the quality of soils. In Tuscany, there are about 5000 lakes with a surface greater than 1000 square meters, which can become the mean for the distributed monitoring of erosive phenomena.

2. Materials and Methods

2.1. Lakes Analyzed and Volume Changes

The study took into consideration 12 hilly lakes in Tuscany, shown in Figure 1, of which the volume calculated in 2010 is known, thanks to a past survey by the former Agency for Development and Innovation in Agriculture (ARSIA) of Regione Toscana (RT— regional administration of Tuscany). The volume estimate was carried out by a private company (Aquaterra, Florence) using a boat, a depth sounder and GPS, thus carrying out a bathymetric survey [58]. LaMMA Consortium, applying the methodology described by Giambastiani et al. 2020 [57], measured the lakes again in 2018 thanks to a monitoring project. We assume that the two methodologies are comparable, as the same types of tools and procedures are used. Furthermore, the measurements were carried out for both years in early spring, when the reservoir tends to be full from winter rains and agricultural use is limited. Each lake and its basin was evaluated and investigated in order to carry out a modeling analysis of the surface erosion with the common models (RUSLE, MMF). These lakes are mainly used for irrigation of agricultural crops; however, some are used for sport fishing, forest-fire-fighting and other objectives. Table 1 shows the main characteristics of the basins corresponding to such lakes. Table 2 reports historical data regarding the lakes volume at the time of construction. The comparison between 2010 and 2018 is summarized in Table 3.

Figure 1. Lakes geographic distribution in Tuscany, Italy.

In order to implement the erosion estimation models (RUSLE and MMF), data for the hydrographic basins were collected relative to the precipitation (Figure 2) of the meteorological stations closest to the lakes; the hydrological network was elaborated, for each basin, from a DTM (Digital Terrain Model) with resolution 10 × 10 m (Figure 3); and land cover was processed via photointerpretation (Figure 4, Table 4) in 9 main classes.

Table 1. Main characteristics of the basins corresponding to the studied lakes (Appendix A): Altitude and slope are obtained from a DTM with 10 m resolution. Hydrographic networks are taken from a database of the regional administration of Tuscany (https://www.regione.toscana.it/-/geoscopio, accessed on 1 April 2018); viability is taken from the OpenStreetMap database.

GID	Lake Name	Area (ha)	Altitude Max (m agl)	Altitude Lake (m agl)	Slope Mean (%)	Hydrographic Network (m)	Road Network (m)
1156	Romena	9.55	293	154	15.1	618.8	331.5
2629	Cavalcanti	61.64	212	156	10.9	2594.6	751.4
3036	Galliano	67.25	409	281	8.3	3515.7	2883.1
5171	Fabbrica	218.99	413	229	14.1	11,268.2	11,812.8
7438	Pavone	50.83	202	135	14.4	1998.9	1029.9
7719	Schifanoia	87.04	281	242	4.6	4549.5	2297.4
8454	Castelfalfi 1	52.70	261	158	14.8	2459.3	4034.4
8477	Castelfalfi 3	127.19	177	99	10.3	7852.2	6069.4
8967	Potenti 2	65.34	183	48	11.5	4135.1	1091.1
8969	Potenti 1	43.79	128	39	9.4	2351.5	0.0
11525	Angiola	136.16	195	35	14.6	8183.8	11,074.5
12964	Castelfalfi 2	64.00	338	177	17.5	3507.2	1476.0

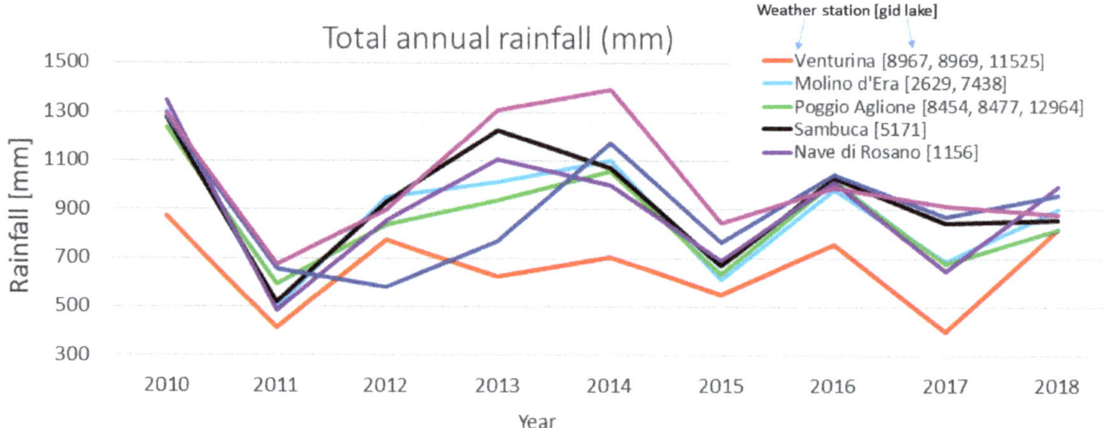

Figure 2. Rainfall trends in the period object of study, for the nearest weather stations. The legend on the right side associates each weather station to the corresponding lake(s).

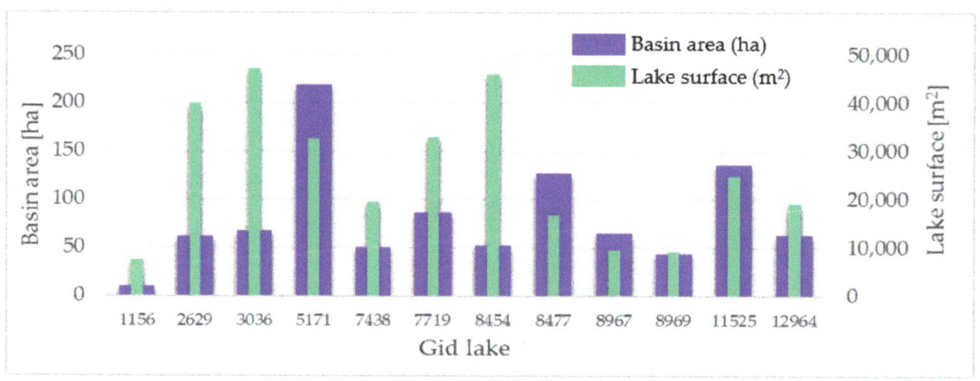

Figure 3. Relationships between lake surface and corresponding drainage basin.

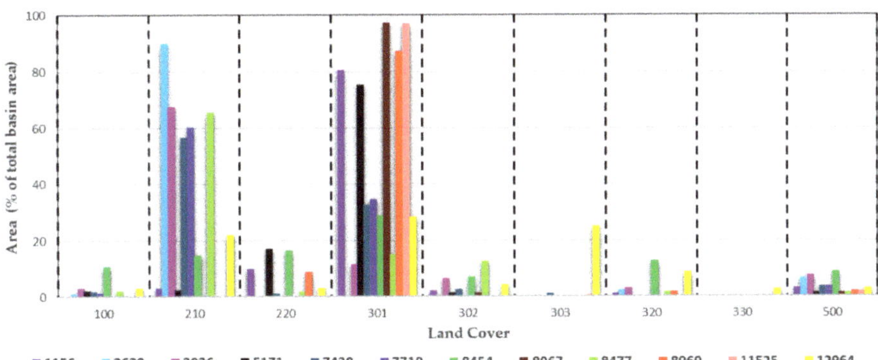

Figure 4. Distribution of land-cover classes for each lake. The land-use classes are shown along the x axis, described in Table 4, while the legend shows the lake GID. In X axis: 100 = artificial surfaces; 210 = lands under a rotation system used for annually harvested plants and fallow lands; 220 = permanent crops (vineyards and olive groves); 301 = forest with a complete canopy closure or a little less; 302 = forest with a sparse canopy closure (40–60%), shrubs and max 10% of soil bare; 303 = degraded forest (canopy closure less of 40%), shrubs cover of 40% and bare soil max 30%; 320 = permanent shrub and/or herbaceous vegetation associations; 330 = degraded soil or bare rock; 500 = water bodies.

From research conducted in the archives of the body in charge of the regulation and authorization of artificial lakes, it was also possible to recover the lake volume on the project deposited during the authorizations phase. From what has been learned, however, unregistered changes have often occurred regarding the morphology of the reservoir, so the related dimensional parameters are to be considered just indicative (Table 2, Figure 5).

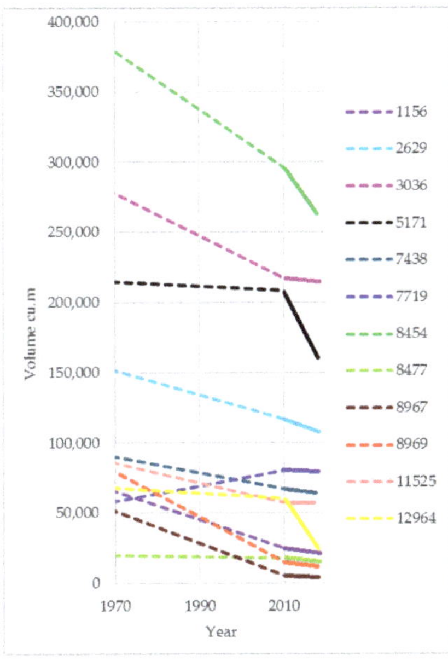

Figure 5. Trend of the reservoir capacity from construction to the analysis period. Dashed lines indicate the uncertain range, while solid lines indicate the trend found with the actual analysis.

Table 2. Construction year and project volumes.

GID	Construction Year	Volume of Design Phase (mc)
1156	1964	76,000
2629	1958	160,000
3036	1958	293,150
5171	1956	216,420
7438	1959	96,000
7719	1970	52,500
8454	1970	400,000
8477	1963	20,507
8967	1970	63,000
8969	1970	96,000
11525	1970	93,000
12964	1967	69,803

Table 3. Lake parameters for the years 2010 and 2018: surface area, volume, variation (in volume, percentage and percentage per year), and silting rate, the latter normalized to the lake surface area. For 2010, we show harmonized volumes, indicated as 2010 h.

GID	Surface (m^2)		Volume (m^3)		Variation			Silting	
	2010	2018	2010-h	2018	m^3	%	%/y	Mg	Mg/y
1156	7570	7599	24,855	21,597	−3258	−13.1	−1.6	2821	353
2629	38,875	39,942	116,826	108,254	−8572	−7.3	−0.9	7423	928
3036	49,986	47,241	217,520	215,144	−2376	−1.1	−0.1	2057	257
5171	35,293	32,713	208,807	159,454	−49,353	−23.6	−3.0	42,740	5342
7438	20,729	19,561	67,659	64,228	−3431	−5.1	−0.6	2971	371
7719	35,080	33,029	80,651	79,407	−1244	−1.5	−0.2	1077	135
8454	48,412	46,044	296,296	261,833	−34,463	−11.6	−1.5	29,845	3731
8477	13,389	16,747	18,018	15,315	−2703	−15	−1.9	2341	293
8967	8059	9654	5792	4141	−1651	−28.5	−3.6	1430	179
8969	7744	9180	15,220	12,070	−3150	−20.7	−2.6	2728	341
11525	21,246	24,886	57,625	57,238	−387	−0.7	−0.1	335	42
12964	22,135	19,204	60,549	23,953	−36,596	−60.4	−7.6	31,692	3961

Table 4. Land cover classes corresponding to C and P factors.

UCS Code	Description	USLE_C	USLE_P
100	Artificial surfaces	0	1
210	Lands under a rotation system used for annually harvested plants and fallow lands	0.15	1
220	Permanent crops (vineyards and olive groves)	0.4	1
301	Forest with a complete canopy closure or a little less	0.01	1
302	Forest with a sparse canopy closure (40–60%), shrubs and max 10% of soil bare	0.08	1
303	Degraded forest (canopy closure less of 40%), shrubs cover of 40% and bare soil max 30%	0.20	1
320	Permanent shrub and/or herbaceous vegetation associations	0.1	1
330	Degraded soil or bare rock	0.75	1
500	Water bodies	0	1

Harmonization

In order to compare the two volume measures, a lake-surface-based harmonization was applied, as this parameter (surface area) was easily obtained from the Tuscany Region orthophotos at 20 cm resolution (https://www.regione.toscana.it/-/geoscopio, accessed on 1 April 2018). Harmonization is necessary because in two different years we could have a different reservoir capacity due to the water level, according to different previous rainfall. It is based on the surface variation between 2010 and 2018 (Figure 6), according to the following equations.

Figure 6. Orthophotos of lake "gid 8477" where it is possible to check the difference in water level.

We can write a generic expression for the lake volume V, according to Giambastiani [57], as:

$$V = \iint h(x,y)\cdot dx\,dy = \int_S h\cdot d\sigma \tag{1}$$

where h is the height that depends on the coordinate positions (x,y), which we have omitted in the second step, rewriting the integral as a surface one (σ). Using the integral mean value theorem, we can write V as:

$$V = \langle h \rangle_S \cdot S \tag{2}$$

$\langle h \rangle_S$ being the lake height average (over the surface S). If we assume that its variation is negligible for limited variation of S, we can write the variation ΔV of the volume as a linear function of ΔS, the latter being the variation of the surface with time (depending for instance to rainfall, evaporation, etc.).

$$\Delta V = \langle h \rangle_S \cdot \Delta S = \underline{h} \cdot \Delta S \tag{3}$$

The last step is just to rename $\langle h \rangle_S$ with $\underline{h}$, both for simplicity and to highlight the assumption that it is no more dependent on the surface extension S.

In practice, we have measured $\underline{h}$ for the reference year y_0, which for us was the year 2010, for which we had the ARSIA measurements of the surface areas with the corresponding volumes.

Where the "Surface measured 2010" is the ARSIA surface, measured simultaneously with the volume

$$\underline{h} = \frac{V_{y_0}}{S_{y_0}} \tag{4}$$

For a generic year y, the volume to be compared with the one at the reference year y_0 becomes:

$$V_y = V_{y_0} + \Delta V = V_{y_0} + \underline{h} \cdot \Delta S \tag{5}$$

with ΔS as the surface difference between 2018 and 2010; surfaces were obtained through photointerpretation of orthophotos.

2.2. Erosion Simulation by the Morgan–Morgan–Finney Model

The MMF model divides the soil erosion process into two phases: the phase related to the water component, which determines the energy of the rainfall, and the phase related to the production of sediments, based on the characteristics of the soil. Soil loss in relation to erosion is determined on the basis of precipitation and transport capacity, influenced by soil cover and slope [41,59]. The MMF model is implemented within the open-source SAGA GIS software. Input data come from various sources. Starting from the Digital Terrain Model (DTM), with a resolution of 10 m, the slope map and the channel network were elaborated, while the "plant height" map was obtained through the Crown Height Model (CHM—10 × 10 m). Canopy cover, permanent interception and ground cover derive from Sentinel-2 image processing, in particular based on the NDVI calculation [60], with 10 m resolution. The characteristics of the soils (bulk density, effective hydrological depth, percentages of clay, sand and silt, etc.) are derived from the soil database of RT (http://www502.regione.toscana.it/geoscope/pedologia.html, accessed on 1 April 2018). Other necessary input variables for the model were obtained from direct processing of the land use and land cover map, carried out by photointerpretation. Annual precipitation data were taken from the meteorological stations closest to the lakes in question. In particular, the average distance between the lakes and the rain gauges is 4.5 km, with a standard deviation of 1.8 km.

2.3. Erosion Estimate by RUSLE Model

The Universal Soil Loss Equation (USLE) [23] and subsequent revisions (RUSLE) [24], is an empirical relationship, as it derives from experimental plots carried out in the United States and from the mathematical definition of the results found from these plots, which models soil erosion as a process resulting from a set of six main factors: the energy and intensity of precipitation (R factor), the erodibility of the soil (K factor), the length and slope of the plot (LS factor), vegetation cover (C factor) and conservation practices (P factor). The employed R factor is derived from the SIAS project (ISPRA 2016), through which a simplified relationship is elaborated between the amount of rainfall and the erosion value [61]. This simplification does not critically affect the accuracy, as it emerges from comparison work [31]. In fact, the mean value of R factor for Tuscany is 1765 (standard deviation = 710) against 1748 (standard deviation 365), as found by Panagos 2015 [29]. Larger fieldwork, developed in 2006 for punctual soil data, allowed the calculation of K factor carried out following the original methodology [23]. The LS factor was calculated for the basins of each lake, using a digital elevation model (DEM) of 10 × 10 m, with the method developed by Desmet and Govers 1996 [62]. In order to avoid overestimation of the LS factor in heterogeneous landscapes, the lengths of long slopes were limited to a value of 333 m [24,62]. The length exponent (m) is based on the original USLE method [63]. The C factor was completely updated, carrying out a detailed photo interpretation on orthophotos, using functional classes for the purpose. For each soil cover class identified, the values C and P were attributed as in Table 4, consistent with the values reported in the bibliography [24,31,64,65].

From the photo interpretation, it was possible to verify that in the agricultural landscapes of the study areas, there were no particular conservation techniques similar to those already codified in the RUSLE model: for this reason, we decided to always adopt a factor P = 1.

Sediment Delivery Ratio

The Sediment delivery ratio (SDR) is applied to estimate the amount of sediments produced by the erosion phenomena that reaches the lake [66]. SDR is the erosion fraction, generated by each single source cell, which reaches the nearest permanent drainage line. As a first approximation, the SDR can be considered constant for the whole basin or sub-basin [67], but in recent works, it is calculated pixel by pixel as a function of the length and slope of the path in the downstream direction [68]. The most complete algorithm for its modeling is proposed by [65], which accounts for the connectivity index (IC) for each pixel, considering the morphological and hydrological characteristics of both the hydrological upstream and downstream portion of the pixel. It describes the hydrological link between sediment sources and collection and transfer points such as streams [69]. For SDR calculation in the study area, we have used the InVEST model (https://naturalcapitalproject.stanford.edu/, accessed on 3 September 2021), which implemented the algorithm of Borselli [65], making some minor simplifications.

3. Results

The silting rate (SR) (Figure 7) is very variable among the lakes under study and no significant relationships appear between the lake volume or the surface of the basin, or other main dimensional parameters. Some lakes have a high volume variation (e.g., 12964, 5171) and this is not related to either the size of the lake or the basin. The annual silting rate is obtained by dividing the volume variation by 8. We consider this operation significant as the surveys were carried out, in both cases, during the months of March and April, with a difference of a few weeks. The lakes' surfaces did not vary much during the 8 years of analysis. Figure 8a shows the relationship between the area of 2010 and 2018. The strong correlation indicates that the variation in the surface responds to ordinary dynamics. The surface change is present in almost all lakes; from this, we can confirm the need to harmonize volumes. While this process may be the source of further processing errors, we believe it is robust as the volume variations are small. A greater variation may exist in the relationship between the lakes' mean depths (Figure 8b). For all cases, we found that the mean depth has decreased. This is further confirmation of the soundness of the harmonization process. The correlation matrix highlights significant relationships between several parameters considered (Figure 9). Among these, we find a good correlation between soil loss and the erodibility of the lithology (K_lito_Sl, r = −0.72), the quantity of specialized (olive grove, vineyard) agricultural land cover (r = 0.85) and with the presence of roads (r = 0.68).

In Figure 9, we summarize the many relationships developed between the silting and the characteristics of the lake or basin. Among these are the physical characteristics of the basin, such as the altimetry and the difference in height (alti-max and altit_lake, disl_basin), the length of the network present in the basin (hydro_network), the presence of roads (road_net), the various soil classes (sup_ucs: Table 4) and rainfall (rain_acc, num_event).

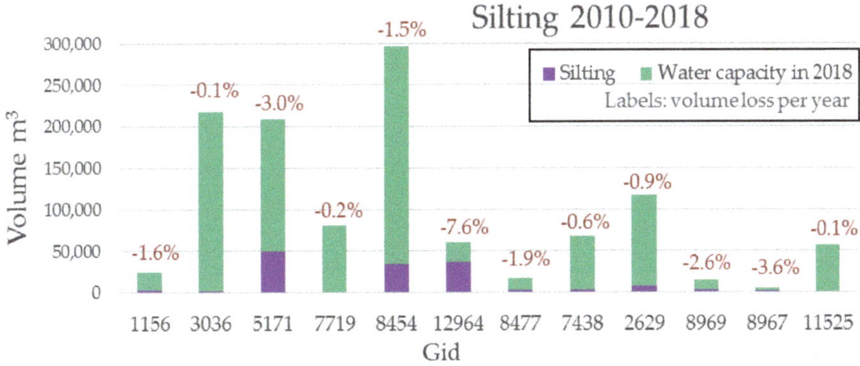

Figure 7. Silting rate and water capacity of the lakes under study.

Comparing the silting rate (SR) of each lake, obtained with the proposed methodology [57], with the annual soil-loss values obtained from the described empirical models, the link between these methodologies is evident (Figure 10). In absolute terms, the sediment values produced (SL) were almost always lower than the SR values. SL by RUSLE was similar to SR for three lakes. In seven cases, however, it was different, even if it was consistent with the MMF results. For example, Lake 5171 had the highest silting rate with 3701 Mg y^{-1}, which was similar to the MMF estimate (3384), while with RUSLE, we found a third of the soil was lost. It should be noted that this lake has a much larger basin than all the others. Lake 8477 showed an inverse and very different trend compared to the models: 202 (SR), 630 (SL-RUSLE), 675 (SL-MMF). For other lakes, we found a close similarity. For example, lake 2629 had SR = 643; RUSLE = 499; MMF = 558 Mg y^{-1}. Additionally, in statistical terms, SR showed higher absolute values (SR/ha mean = 15.1; SL MMF/ha mean = 7.9; SL RUSLE/ha mean = 6.3).

Using the RUSLE and SAGA MMF models, it was possible to estimate the soil-loss rate of each catchment area [70], which was compared with the average silting rate of the reservoirs (considered as the closure section of the basin), in the period 2010–2018. The average sediment produced per hectare for the entire analysis sample was in line with the work of Angeli et al. 2004 [71]. The average annual soil loss per hectare obtained by RUSLE was 6.35 Mg, while MMF returned 7.96 Mg. The average silting per hectare of catchment area was 15.13 Mg y^{-1} (Table 5). Both models showed a good correlation with the silting rate measured for each lake (Figure 11), but with a stronger relationship with the RUSLE model. The good significance of such relationships suggests a close link between the loss of soil and the silting rate, as visible in Figure 10.

Table 5. Summary of sediment volumes and average values per hectare, obtained from field surveys (silting) and models (soil loss for RUSLE and MMF).

GID Lake	Basin Area (ha)	Silting (Mg/y)	SL RUSLE (Mg/y)	SL MMF (Mg/y)	Silting (Mg y^{-1} ha^{-1})	RUSLE (Mg y^{-1} ha^{-1})	MMF (Mg y^{-1} ha^{-1})
1156	10	244	49	4	24.393	4.928	0.422
2629	60	643	499	559	10.703	8.306	9.299
3036	81	178	344	609	2.197	4.242	7.511
5171	213	3701	1248	3384	17.390	5.862	15.900
7438	50	257	290	871	5.183	5.842	17.543
7719	101	93	192	7	0.925	1.899	0.071
8454	41	2585	655	721	63.529	16.092	17.733
8477	106	203	630	675	1.906	5.925	6.349
8967	53	124	11	3	2.341	0.209	0.049
8969	32	236	17	3	7.329	0.523	0.088
11525	135	29	31	1	0.215	0.229	0.004
12964	60	2745	1342	1247	45.475	22.231	20.653

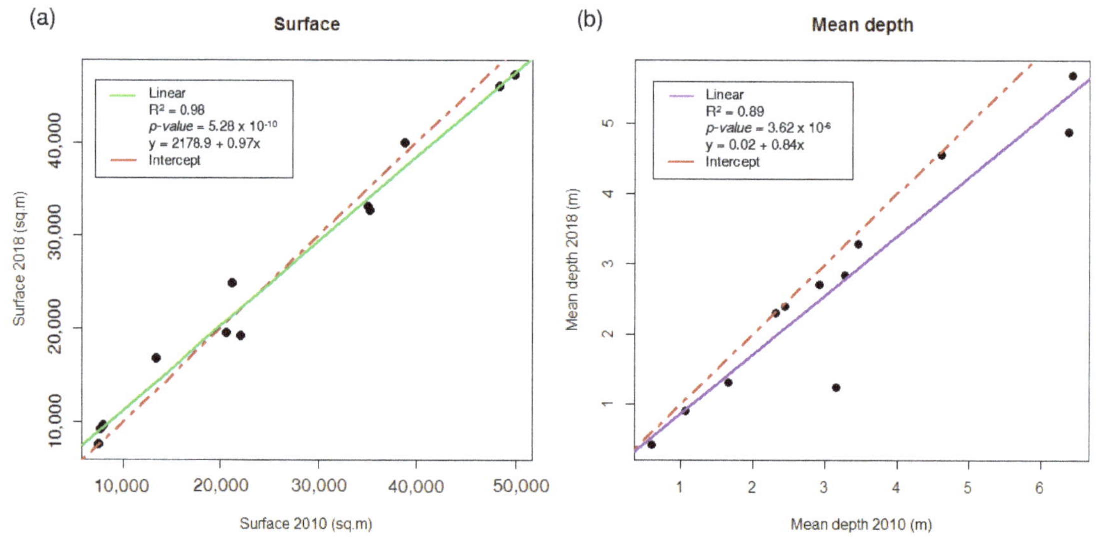

Figure 8. Relationships between variations in surface area (**a**) and mean depth (**b**).

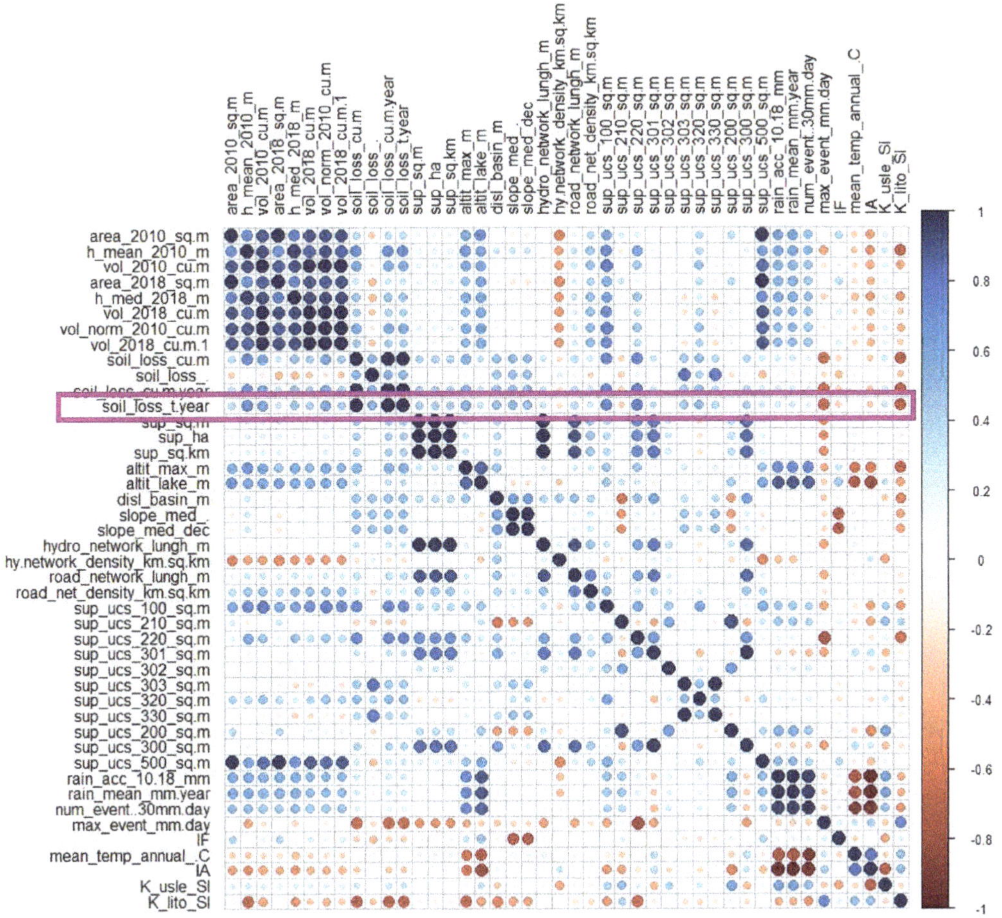

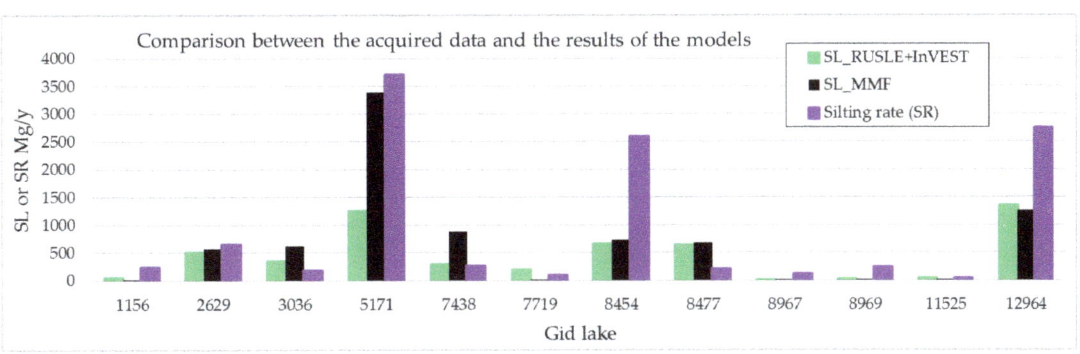

Figure 9. Correlation matrix, based on Pearson's analysis.

Figure 10. Comparison between the soil loss (SL) calculated with the models (RUSLE and MMF) and the silting rate for each individual lake.

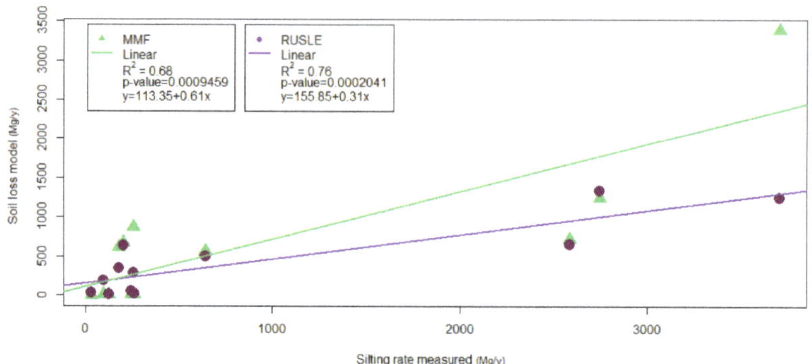

Figure 11. Relationship between the measured silting rate and the soil loss obtained from the MMF and RUSLE models.

4. Discussion

The methods that allow the monitoring of erosive phenomena are often very expensive to apply: the models require several input variables and therefore (I) challenging surveying campaigns, (II) data collection from different sources to be harmonized and (III) aerial images and high-resolution digital terrain models for GIS analysis and photointerpretation. All this involves high costs, low reaction time in carrying out analyses at critical moments, and possible relevant evaluation errors due to the high number of interactions and variables to take into account [4]. The proposed approach for the erosion evaluation is based on monitoring the water capacity of hilly lakes, which are considered as real sediment containers. The variation in lake volume, following silting, can be linked to the production of sediments from the afferent basin. Given the high distribution of artificial hilly reservoirs in the Tuscan territory, the proposed approach could create conditions for a periodic assessment of the degradation phenomena of the soil and the territory in general. The use of input data with greater accuracy and precision, such as using LiDAR Dems [70] and an in situ meteorological station for precipitation measurements, could lead to improvements. This approach is also easily applicable to different contexts, as long as they have a good distribution of hilly reservoirs. The methodology, however, has some limitations for calculating the volume of the lake, in particular due to the simplification of the shape of the lake profile [57]. Anyway, the application of this methodology, from a monitoring point of view, can be valuable for reducing major evaluation errors arising from too-long sampling-time intervals that we have when using classical approaches. In fact, field measurements for medium-sized reservoirs (up to 4–5 hectares of surface area) can be carried out in a short time (e.g., 30 min), and processing can be carried out automatically within a few minutes. This allows a wide and rapid application, analyzing wide domains with relatively low effort and costs. The results obtained show significant measures of the lake volume variation, well correlated with other physical lake variables. The comparison of the mean depth between 2010 and 2018 (Figure 8) shows a very strong relationship (R^2 = 0.89 and p-value = 3.619×10^{-6}), which indicates a good significance of the methodology. Furthermore, the values are located below the bisector, as in no case do we find negative silting (greater lake depth). This leads us to think further that the methodology is correct. Given the small variation in terms of volume in most cases (Figure 7), calculation errors, however accepted, could lead to mathematical anomalies. Mathematical models verify the physical process of sediment accumulation in the lake. The 12 basins analyzed are very different from each other (average surface = 78 ± 54 hectares); they mainly have soil covers relating to arable land and forest, so with high variability. We also found evident differences in terms of precipitation. The lakes examined are well distributed and are representative enough of the study area, Tuscany. Some lakes exhibit very high-volume variations (Figure 7). For example, Lake 12964 has a silting rate of 45 tons/year per hectare of basin.

Close to the lake, we found a pig farm on an area of about seven hectares, a practice that has probably greatly affected erosion. Lake 5171 has the largest basin of all the lakes under study, with a greater total length of variability, another important source of sediment when maintenance is limited. Lake 8454 also has a strong silting; in this case, the data available do not show any trend and the lack of knowledge of the specific territory does not allow us to put forward hypotheses. Lake 11525 has a silting rate of 0.21 ton/year per hectare of basin, and has the basin completely covered by forest. The models used provided data compatible with the results of other authors [34,40]. The soil-loss values obtained from the basins show a significant correlation with the silting rate values (Figure 10), regardless of the intrinsic great variability. In some cases, soil loss is greater than the volume variation (8 out of 24, Table 5); in particular, this concerns large basins (about 100 hectares). One reason can be that the sediment produced by the basin does not reach the lake. Another factor that has not been taken into account concerns the suspended sediment lost by the reservoir spillway. Tauro [72] indirectly estimates suspended solids by means of water turbidity. The concentration of suspended solids typically increases with the flow speed. In a lake, the effect of containment and slowdown of the outflows allows greater sedimentation. In fact, the water flowing in the spillways usually appears clear [73]. The opposite behavior occurs in smaller basins (less than 50 hectares), which are characterized by higher silting rates and lower soil loss values. The actual erosion is caused by phenomena that, in some cases, the models are unable to consider, which concern the slopes closest to the lake (such as the case of the lake with pig breeding). Having such present shortcomings in mind, the applied methodology can be useful for building decision-support systems in spatial planning and monitoring areas that show critical characteristics related to hydrological processes [74].

5. Conclusions

A new approach for the soil erosion analysis was proposed. It consists of the use of an aquatic boat bringing a GPS sonar. In few minutes, it is able to detect data about water depth in the reservoir. Using an automatic GIS process, it is possible to obtain an estimation of the volume. The methodology was applied to 12 hilly lakes mainly for irrigation purposes, for which the volume (or reservoir capacity) was measured and repeated after 8 years, using comparable instruments (sonar with GPS). The volume variation in this period was compared with soil-loss estimates obtained from well-established models widely known in the scientific community (RUSLE and MMF), obtaining a clear relationship between the two variables. This approach allows low-cost monitoring of the soil erosion phenomena in relation to changes in land use or climate change. Being based on lakes, the analysis can refer to specific portions of the territory. The main advantage is the speed of carrying out the survey on the lake; the instrumentation is inexpensive and it is not necessary to acquire other parameters relating to the basin. The processing procedure can be automated in the GIS environment. The method can be applied to land management issues as a tool for a decision-support system. The harvesting of more data could permit the development of some estimate models about the erosion phenomena based on the silting rate of lakes.

Author Contributions: Conceptualization, S.C. and R.G.; methodology, L.G. and S.C.; software, L.B. and S.R.; validation, R.G. and M.I.; formal analysis, Y.G. and L.G.; investigation, Y.G., R.G., S.C.; resources, B.G. and A.O.; data curation, Y.G.; writing—original draft preparation, Y.G.; writing—review and editing, A.O., S.C. and R.G.; visualization, M.I.; supervision, L.B. and A.O.; project administration, B.G.; funding acquisition, B.G. and L.B. All authors have read and agreed to the published version of the manuscript.

Funding: This research was carried out with the ordinary funds of the LaMMA Consortium and CNR-IBE and with external funding from the Tuscany Region: Decrees 100/2018 and 1345/2018 "Mappatura della totalità dei laghi in Regione Toscana e costituzione del catasto informatizzato".

Institutional Review Board Statement: Not applicable.

Informed Consent Statement: Informed consent was obtained from all subjects involved in the study.

Data Availability Statement: Not applicable.

Acknowledgments: The data used for this article derive from the hilly lakes census of the Tuscany Region.

Conflicts of Interest: The authors declare no conflict of interest.

Appendix A

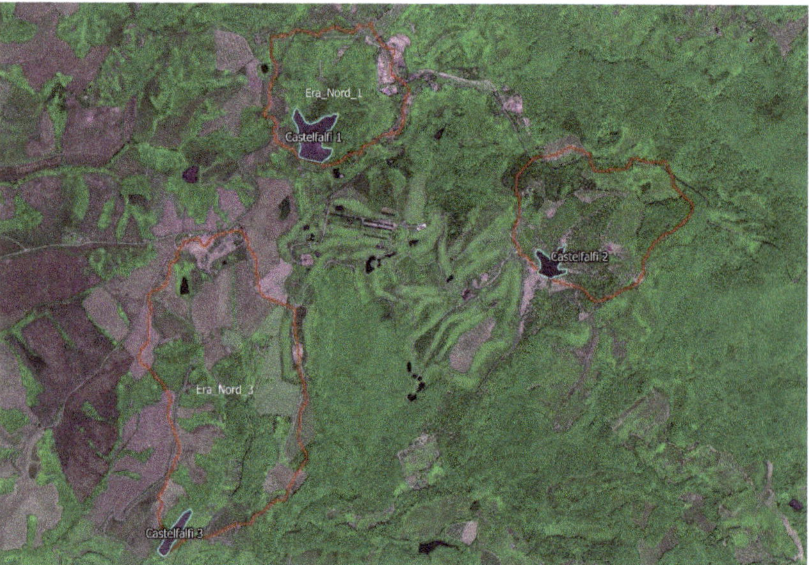

Figure A1. Castelfalfi lakes and basins. GID: 8454 = Castelfalfi 1; 12964 = Castelfalfi 2; 8477 = Castelfalfi 3.

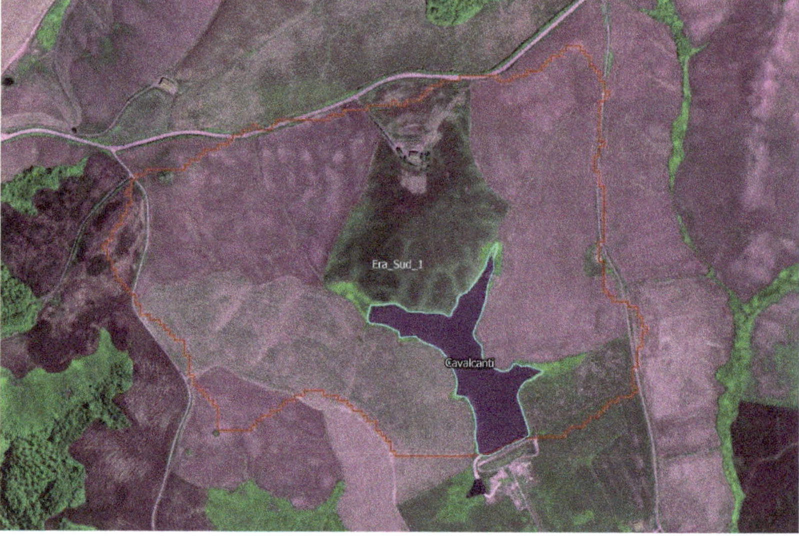

Figure A2. Cavalcanti lake and basin. GID: 2629.

Figure A3. Fabbrica lake and basin. GID: 5171.

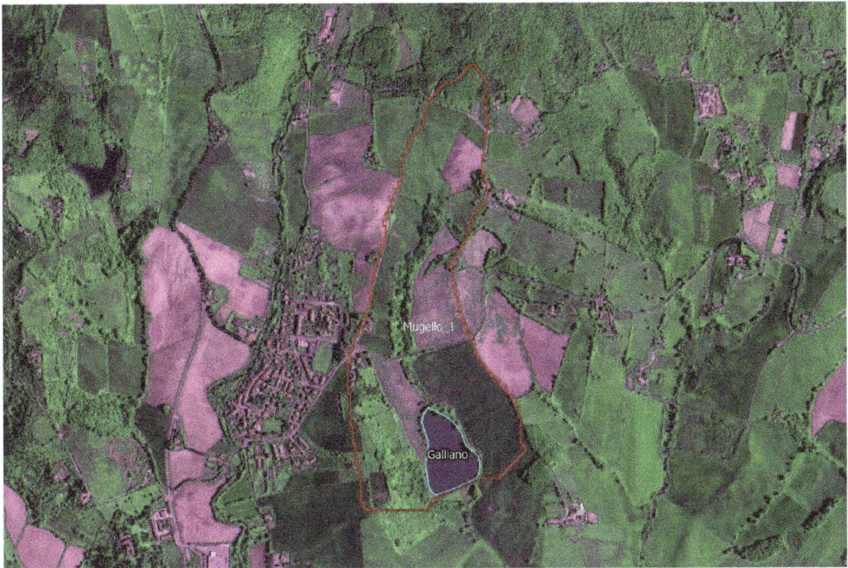

Figure A4. Galliano lake and basin. GID: 3036.

Figure A5. Pavone lake and basin. GID: 7438.

Figure A6. Romena lake and basin. GID: 1156.

Figure A7. Schifanoia lake and basin. GID: 7719.

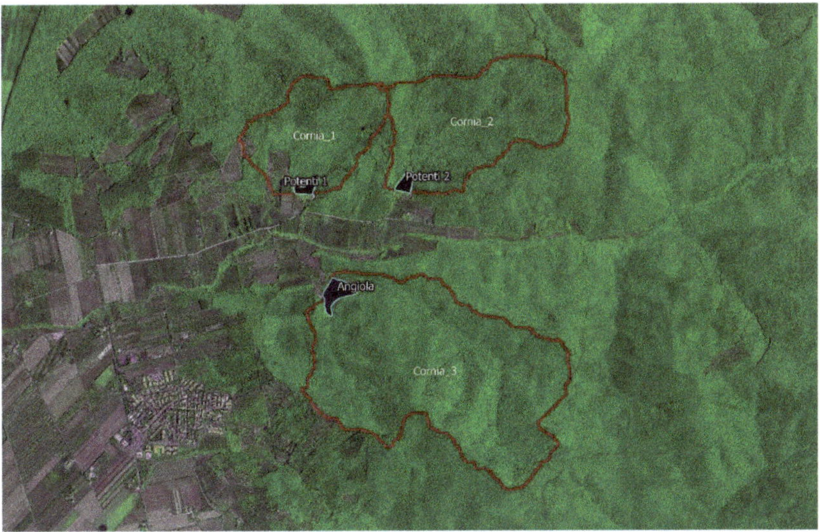

Figure A8. Cornia lakes and basins. GID: 8969 = Potenti 1; 8967 = Potenti 2; 11525 = Angiola.

References

1. Borrelli, P.; Robinson, D.A.; Panagos, P.; Lugato, E.; Yang, J.E.; Alewell, C.; Wuepper, D.; Montanarella, L.; Ballabio, C. Land use and climate change impacts on global soil erosion by water (2015–2070). *Proc. Natl. Acad. Sci. USA* **2020**, *117*, 21994–22001. [CrossRef]
2. Pimentel, D. Soil erosion: A food and environmental threat. *Environ. Dev. Sustain.* **2006**, *8*, 119–137. [CrossRef]
3. Pimentel, D.; Burgess, M. Soil erosion threatens food production. *Agriculture* **2013**, *3*, 443–463. [CrossRef]
4. Pennock, D.; Pennock, L.; Sala, M. *Soil Erosion: The Greatest Challenge for Soustainable Soil Management*; Food & Agriculture Organization: Rome, Italy, 2019; ISBN 9789251314265.
5. Pimentel, D.; Harvey, C.; Resosudarmo, P.; Sinclair, K.; Kurz, D.; McNair, M.; Crist, S.; Shpritz, L.; Fitton, L.; Saffouri, R.; et al. Environmental and economic costs of soil erosion and conservation benefits. *Science* **1995**, *267*, 1117–1123. [CrossRef] [PubMed]

6. Kirkby, M.; Jones, R.J.; Irvine, B.; Gobin, A.G.G.; Cerdan, O.; van Rompaey, J.J.; Huting, J.R.M. *Pan-European Soil Erosion Risk Assessment for Europe: The PESERA Map, version 1 October 2003*; Office for Official Publications of the European Communities: Luxembourg, Germany, 2004.

7. Nearing, M.A.; Foster, G.R.; Lane, L.J.; Finkner, S.C. A process-based soil erosion model for USDA-Water Erosion Prediction Project technology. *Trans. ASAE* **1989**, *32*, 1587–1593. [CrossRef]

8. Laflen, J.; Lane, L.; Foster, G.R. WEPP: A new generation of erosion prediction technology. *J. Soil Water Conserv.* **1991**, *46*, 34–38.

9. Assennato, F.; Antona, M.D.; Di Leginio, M.; Strollo, A. Assessing land consumption impact on ecosystem services provision: An insight on biophysical and economic dimension of loss of erosion control in Italy. *Authorea* **2020**, 1–12. [CrossRef]

10. Mastrorosa, S.; Crosetto, M.; Congedo, L.; Munafò, M. Land consumption monitoring: An innovative method integrating SAR and optical data. *Environ. Monit. Assess.* **2018**, *190*, 588. [CrossRef]

11. Van Oost, K.; Bakker, M. Soil Productivity and Erosion. *B Soil Ecol. Ecosyst. Serv.* **2012**, *2012*, 301–314. [CrossRef]

12. Stokes, A.; Douglas, G.B.; Fourcaud, T.; Giadrossich, F.; Gillies, C.; Hubble, T.; Kim, J.H.; Loades, K.W.; Mao, Z.; Mcivor, I.R.; et al. Ecological mitigation of hillslope instability: Ten key issues facing researchers and practitioners. *Plant Soil* **2014**, *377*, 1–23. [CrossRef]

13. Fox, D.M.; Laaroussi, Y.; Malkinson, L.D.; Maselli, F.; Andrieu, J.; Bottai, L.; Wittenberg, L. POSTFIRE: A model to map forest fire burn scar and estimate runoff and soil erosion risks. *Remote Sens. Appl. Soc. Environ.* **2016**, *4*, 83–91. [CrossRef]

14. Capra, G.F.; Ganga, A.; Grilli, E.; Vacca, S.; Buondonno, A. A review on anthropogenic soils from a worldwide perspective. *J. Soils Sediments* **2015**, *15*, 1602–1618. [CrossRef]

15. SEC. Communication from the Commission to the Council, the European Parliament, the European Economic and Social Committee and the Committee of the Regions. In *Thematic Strategy for Soil Protection*; [SEC(2006)620], [SEC(2006)1165]; Communication from the Commission to the Council: Brussels, Belgium, 2006.

16. Borrelli, P.; Robinson, D.A.; Fleischer, L.R.; Lugato, E.; Ballabio, C.; Alewell, C.; Meusburger, K.; Modugno, S.; Schütt, B.; Ferro, V.; et al. An assessment of the global impact of 21st century land use change on soil erosion. *Nat. Commun.* **2017**, *8*, 2013. [CrossRef]

17. Caracciolo, C.; Napoli, M.; Porcù, F.; Prodi, F.; Dietrich, S.; Zanchi, C.; Orlandini, S. Raindrop Size Distribution and Soil Erosion. *J. Irrig. Drain. Eng.* **2012**, *138*, 461–469. [CrossRef]

18. Napoli, M.; Marta, A.D.; Zanchi, C.A.; Orlandini, S. Assessment of soil and nutrient losses by runoff under different soil management practices in an Italian hilly vineyard. *Soil Tillage Res.* **2017**, *168*, 71–80. [CrossRef]

19. Napoli, M.; Cecchi, S.; Orlandini, S.; Mugnai, G.; Zanchi, C.A. Simulation of field-measured soil loss in Mediterranean hilly areas (Chianti, Italy) with RUSLE. *Catena* **2016**, *145*, 246–256. [CrossRef]

20. Canuti, P.; Moretti, S.; Bazzoffi, P.; Rodolfi, G.; Zanchi, C. Soil erosion as a result of the man activity-geological environment relationship an example of quantitative evaluation in the mugello valley (Tuscany, Italy). *Bull. Int. Assoc. Eng. Geol. l'Assoc. Int. Géol. l'Ingénieur* **1986**, *33*, 109–114. [CrossRef]

21. Bagarello, V.; Iovino, M.; Elrick, D. A Simplified Falling-Head Technique for Rapid Determination of Field-Saturated Hydraulic Conductivity. *Soil Sci. Soc. Am. J.* **2004**, *68*, 66. [CrossRef]

22. Vacca, A.; Loddo, S.; Ollesch, G.; Puddu, R.; Serra, G.; Tomasi, D.; Aru, A. Measurement of runoff and soil erosion in three areas under different land use in Sardinia (Italy). *Catena* **2000**, *40*, 69–92. [CrossRef]

23. Wischmeier, W.; Smith, D. *Predicting Rainfall Erosion Losses: A Guide to Conservation Planning*; Agriculture Handbook (No. 537); United States Department of Agriculture: Washington, DC, USA, 1978; pp. 1–67.

24. Renard, K.G. *Predicting Soil Erosion by Water: A Guide to Conservation Planning with the Revised Universal Soil Loss Equation (RUSLE)*; Agriculture Handbook (No. 703); United States Department of Agriculture: Washington, DC, USA, 1997; p. 300.

25. Poesen, J.; Nachtergaele, J.; Verstraeten, G.; Valentin, C. Gully erosion and environmental change: Importance and research needs. *Catena* **2003**, *50*, 91–133. [CrossRef]

26. Renschler, C.S.; Harbor, J. Soil erosion assessment tools from point to regional scales—The role of geomorphologists in land management research and implementation. *Geomorphology* **2002**, *47*, 189–209. [CrossRef]

27. Kinnell, P.I.A. Event soil loss, runoff and the Universal Soil Loss Equation family of models: A review. *J. Hydrol.* **2010**, *385*, 384–397. [CrossRef]

28. Lu, H.; Prosser, I.P.; Moran, C.J.; Gallant, J.C.; Priestley, G.; Stevenson, J.G. Predicting sheetwash and rill erosion over the Australian continent. *Aust. J. Soil Res.* **2003**, *41*, 1037–1062. [CrossRef]

29. Panagos, P.; Borrelli, P.; Meusburger, K. A new European slope length and steepness factor (LS-factor) for modeling soil erosion by water. *Geosciences* **2015**, *5*, 117–126. [CrossRef]

30. Panagos, P.; Meusburger, K.; Ballabio, C.; Borrelli, P.; Alewell, C. Soil erodibility in Europe: A high-resolution dataset based on LUCAS. *Sci. Total Environ.* **2014**, *479–480*, 189–200. [CrossRef] [PubMed]

31. Panagos, P.; Borrelli, P.; Meusburger, K.; Alewell, C.; Lugato, E.; Montanarella, L. Estimating the soil erosion cover-management factor at the European scale. *Land Use Policy* **2015**, *48*, 38–50. [CrossRef]

32. Panagos, P.; Ballabio, C.; Borrelli, P.; Meusburger, K.; Klik, A.; Rousseva, S.; Tadić, M.P.; Michaelides, S.; Hrabalíková, M.; Olsen, P.; et al. Rainfall erosivity in Europe. *Sci. Total Environ.* **2015**, *511*, 801–814. [CrossRef]

33. Benavidez, R.; Jackson, B.; Maxwell, D.; Norton, K. A review of the (Revised) Universal Soil Loss Equation ((R)USLE): With a view to increasing its global applicability and improving soil loss estimates. *Hydrol. Earth Syst. Sci.* **2018**, *22*, 6059–6086. [CrossRef]

34. Alewell, C.; Borrelli, P.; Meusburger, K.; Panagos, P. Using the USLE: Chances, challenges and limitations of soil erosion modelling. *Int. Soil Water Conserv. Res.* **2019**, *7*, 203–225. [CrossRef]
35. Le Roux, J.J.; Newby, T.S.; Sumner, P.D. Monitoring soil erosion in South Africa at a regional scale: Review and recommendations. *S. Afr. J. Sci.* **2007**, *103*, 329–335.
36. Morgan, R.P.C.; Duzant, J.H. Modified MMF (Morgan–Morgan–Finney) model for evaluating effects of crops and vegetation cover on soil erosion. *Earth Surf. Process. Landf.* **2008**, *32*, 90–106. [CrossRef]
37. Sterk, G. A hillslope version of the revised Morgan, Morgan and Finney water erosion model. *Int. Soil Water Conserv. Res.* **2021**, *9*, 319–332. [CrossRef]
38. Li, C.; Qi, J.; Feng, Z.; Yin, R.; Guo, B.; Zhang, F.; Zou, S. Quantifying the effect of ecological restoration on soil erosion in china's loess plateau region: An application of the MMF approach. *Environ. Manag.* **2010**, *45*, 476–487. [CrossRef]
39. Tesfahunegn, G.B.; Tamene, L.; Vlek, P.L.G. Soil erosion prediction using Morgan-Morgan-Finney model in a GIS environment in northern Ethiopia catchment. *Appl. Environ. Soil Sci.* **2014**, *2014*, 468751. [CrossRef]
40. Shrestha, D.P.; Jetten, V.G. Modelling erosion on a daily basis, an adaptation of the MMF approach. *Int. J. Appl. Earth Obs. Geoinf.* **2018**, *64*, 117–131. [CrossRef]
41. Mondal, A.; Khare, D.; Kundu, S. A comparative study of soil erosion modelling by MMF, USLE and RUSLE. *Geocarto Int.* **2018**, *33*, 89–103. [CrossRef]
42. Das, S.; Deb, P.; Bora, P.K.; Katre, P. Comparison of rusle and mmf soil loss models and evaluation of catchment scale best management practices for a mountainous watershed in india. *Sustainability* **2021**, *13*, 232. [CrossRef]
43. Shen, Z.Y.; Gong, Y.W.; Li, Y.H.; Hong, Q.; Xu, L.; Liu, R.M. A comparison of WEPP and SWAT for modeling soil erosion of the Zhangjiachong Watershed in the Three Gorges Reservoir Area. *Agric. Water Manag.* **2009**, *96*, 1435–1442. [CrossRef]
44. Mosbahi, M.; Benabdallah, S.; Boussema, M.R. Assessment of soil erosion risk using SWAT model. *Arab. J. Geosci.* **2013**, *6*, 4011–4019. [CrossRef]
45. Tibebe, D.; Bewket, W. Surface runoff and soil erosion estimation using the SWAT model in the Keleta Watershed, Ethiopia. *Land Degrad. Dev.* **2011**, *22*, 551–564. [CrossRef]
46. Napoli, M.; Orlandini, S.; Grifoni, D.; Zanchi, C.A. Modelling soil and nutrient runoff yields from an Italian vineyards using swat. *Trans. ASABE* **2013**, *56*, 1397–1406.
47. Pacetti, T.; Castelli, G.; Schröder, B.; Bresci, E.; Caporali, E. Water Ecosystem Services Footprint of agricultural production in Central Italy. *Sci. Total Environ.* **2021**, *797*, 149095. [CrossRef] [PubMed]
48. Castelli, G.; Foderi, C.; Guzman, B.H.; Ossoli, L.; Kempff, Y.; Bresci, E.; Salbitano, F. Planting waterscapes: Green infrastructures, landscape and hydrological modeling for the future of Santa Cruz de la Sierra, Bolivia. *Forests* **2017**, *8*, 437. [CrossRef]
49. Bagarello, V.; Di Piazza, G.; Ferro, V.; Giordano, G. Predictiong unit plot soil loss in Sicily, south Italy. *Hydrol. Process.* **2008**, *22*, 586–595. [CrossRef]
50. Bagarello, V.; Ferro, V.; Keesstra, S.; Comino, J.R.; Pulido, M.; Cerdà, A. Testing simple scaling in soil erosion processes at plot scale. *Catena* **2018**, *167*, 171–180. [CrossRef]
51. Wu, S.; Chen, L.; Wang, N.; Zhang, J.; Wang, S.; Bagarello, V.; Ferro, V. Variable scale effects on hillslope soil erosion during rainfall-runoff processes. *Catena* **2021**, *207*, 105606. [CrossRef]
52. Angeli, L.; Bottai, L.; Costantini, R.; Ferrari, R.; Innocenti, L.; Märker, M. Carta della suscettibiltià all'erosione: Analisi e confronto fra modelli di erosione del suolo. *Atti ASITA* **2009**. Available online: http://atti.asita.it/Asita2005/Pdf/0223.pdf (accessed on 29 March 2022).
53. ISPRA. *Stato dell'Ambiente—Annuario dei Dati Ambientali*; Geosfera 13/2009 1-23; ISPRA: Roma, Italy, 2016.
54. Zanchi, C.A. Primi risultati sperimentali sull'influenza di differenti colture nei confronti del ruscellamento superficiale e dell'erosione. In *Annali dell'Istituto Sperimentale per lo Studio e la Difesa del Suolo*; Autorità di Bacino del Fiume Arno: Firenze, Italy, 1983; Volume XIV, pp. 277–289.
55. Zanchi, C.A. Influenza dell'azione battente della pioggia e del ruscellamento nel processo erosivo e variazioni dell'erodibilità del suolo nei diversi periodi stagionali. In *Annali dell'Istituto Sperimentale per lo Studio e la Difesa del Suolo*; Autorità di Bacino del Fiume Arno: Firenze, Italy, 1983; Volume 14, pp. 347–358.
56. Rodolfi, G.; Zanchi, C. Caratteristiche fondamentali e dinamica del paesaggio dell'Appennino Tosco Romagnolo. In *Annali dell'Istituto Sperimentale per lo Studio e la Difesa del Suolo*; Autorità di Bacino del Fiume Arno: Firenze, Italy, 1983; Volume XIV, pp. 289–336.
57. Giambastiani, Y.; Giusti, R.; Cecchi, S.; Palomba, F.; Manetti, F.; Romanelli, S.; Bottai, L. Volume estimation of lakes and reservoirs based on aquatic drone surveys: The case study of Tuscany, Italy. *J. Water Land Dev.* **2020**, *46*, 84–96. [CrossRef]
58. Schweizer, S.; Pini Prato, E. Gestione e tutela (applicazione direttiva nitrati) delle risorse idriche e valutazione degli approvvigionamenti nel settore agricolo ed elaborazione di cartografia tematica (gis) sulle risorse idriche ad uso irriguo. *ARSIA—Reg. Toscana* **2010**, *1*, 1–41.
59. Fernández, C.; Vega, J.A.; Vieira, D.C.S. Assessing soil erosion after fire and rehabilitation treatments in NW Spain: Performance of rusle and revised Morgan-Morgan-Finney models. *Land Degrad. Dev.* **2010**, *21*, 58–67. [CrossRef]
60. Giannetti, F.; Pegna, R.; Francini, S.; McRoberts, R.E.; Travaglini, D.; Marchetti, M.; Mugnozza, G.S.; Chirici, G. A new method for automated clearcut disturbance detection in mediterranean coppice forests using landsat time series. *Remote Sens.* **2020**, *12*, 3720. [CrossRef]

61. Borselli, L.; Cassi, P.; Sanchis, P.S.; Ungaro, F.; Menduni, G.; Brugioni, M.; Sulli, L.; Montini, G. *Studio della Dinamica delle Aree Sorgenti Primarie di Sedimento Nell'area Pilota del Bacino di Bilancino: Progetto (BABI)*; Relazione Attività di Progetto; Autorità di Bacino del Fiume Arno: Firenze, Italy, 2007. [CrossRef]
62. Desmet, P.J.J.; Govers, G. GIS-based simulation of erosion and deposition patterns in an agricultural landscape: A comparison of model results with soil map information. *Catena* **1995**, *25*, 389–401. [CrossRef]
63. Oliveira, A.H.; da Silva, M.A.; Silva, M.L.N.; Curi, N.; Neto, G.K.; de Freitas, D.A.F. *Development of Topographic Factor Modeling for Application in Soil Erosion Models*; InTech: London, UK, 2013; pp. 111–138.
64. Bazzoffi, P. *Erosione del Suolo e Sviluppo Rurale Sostenibile. Fondamenti e Manualistica per la Valutazione Agroambientale*; Il Sole 24 Ore Edagricole: Milan, Italy, 2007; ISBN 8850652283.
65. Borselli, L.; Cassi, P.; Torri, D. Prolegomena to sediment and flow connectivity in the landscape: A GIS and field numerical assessment. *Catena* **2008**, *75*, 268–277. [CrossRef]
66. Rajbanshi, J.; Bhattacharya, S. Assessment of soil erosion, sediment yield and basin specific controlling factors using RUSLE-SDR and PLSR approach in Konar river basin, India. *J. Hydrol.* **2020**, *587*, 124935. [CrossRef]
67. Vanoni, V.A. *Sedimentation Engineering*; American Society of Civil Engineers: New York, NY, USA, 1975; 745p.
68. Ferro, V.; Porto, P. Sediment delivery distributed (sedd) model. *J. Hydrol. Eng.* **2000**, *5*, 411–422. [CrossRef]
69. Saygın, S.D.; Ozcan, A.U.; Basaran, M.; Timur, O.B.; Dolarslan, M.; Yılman, F.E.; Erpul, G. The combined RUSLE/SDR approach integrated with GIS and geostatistics to estimate annual sediment flux rates in the semi-arid catchment, Turkey. *Environ. Earth Sci.* **2014**, *71*, 1605–1618. [CrossRef]
70. Li, P.; Mu, X.; Holden, J.; Wu, Y.; Irvine, B.; Wang, F.; Gao, P.; Zhao, G.; Sun, W. Comparison of soil erosion models used to study the Chinese Loess Plateau. *Earth-Sci. Rev.* **2017**, *170*, 17–30. [CrossRef]
71. Angeli, L.; Costantini, R.; Costanza, L.; Ferrari, R.; Gardin, L.; Innocenti, L. *Rapporto Finale Ufficiale Sulla Realizzazione Della "Carta del Rischio di Erosione Idrica Attuale Della Regione Toscana in Scala 1:250.000"—Region Toscana*; Consorzio LaMMA: Sesto Fiorentino, Italy, 2004.
72. Tauro, F.; Petroselli, A.; Fiori, A.; Romano, N.; Rulli, M.C.; Porfiri, M.; Palladino, M.; Grimaldi, S. "Cape Fear"-A hybrid hillslope plot for monitoring hydrological processes. *Hydrology* **2017**, *4*, 35. [CrossRef]
73. Tazioli, A. Evaluation of erosion in equipped basins: Preliminary results of a comparison between the Gavrilovic model and direct measurements of sediment transport. *Environ. Geol.* **2009**, *56*, 825–831. [CrossRef]
74. Romano, N.; Nasta, P.; Bogena, H.; De Vita, P.; Stellato, L.; Vereecken, H. Monitoring Hydrological Processes for Land and Water Resources Management in a Mediterranean Ecosystem: The Alento River Catchment Observatory. *Vadose Zone J.* **2018**, *17*, 180042. [CrossRef]

 *sustainability*

Article

Mapping of Land Degradation Vulnerability in the Semi-Arid Watershed of Rajasthan, India

Lal Chand Malav [1], Brijesh Yadav [1,*], Bhagwati L. Tailor [1], Sarthak Pattanayak [2], Shruti V. Singh [3], Nirmal Kumar [4], Gangalakunta P. O. Reddy [4], Banshi L. Mina [1], Brahma S. Dwivedi [4] and Prakash Kumar Jha [5]

[1] ICAR—National Bureau of Soil Survey & Land Use Planning, Regional Centre, Udaipur 313001, India
[2] Krishi Vigyan Kendra, Odisha University of Agriculture and Technology, Bolangir 751003, India
[3] Krishi Vigyan Kendra, ICAR—Indian Institute of Vegetable Research, Kushinagar 274406, India
[4] ICAR—National Bureau of Soil Survey & Land Use Planning, Amravati Road, Nagpur 440033, India
[5] Feed the Future Innovation Lab for Collaborative Research on Sustainable Intensification, Kansas State University, Manhattan, KS 66506, USA
* Correspondence: brijesh.yadav@icar.gov.in

Abstract: Global soils are under extreme pressure from various threats due to population expansion, economic development, and climate change. Mapping of land degradation vulnerability (LDV) using geospatial techniques play a significant role and has great importance, especially in semi-arid climates for the management of natural resources in a sustainable manner. The present study was conducted to assess the spatial distribution of land degradation hotspots based on some important parameters such as land use/land cover (LULC), Normalized Difference Vegetation Index (NDVI), terrain characteristics (Topographic Wetness Index and Multi-Resolution Index of Valley Bottom Flatness), climatic parameters (land surface temperature and mean annual rainfall), and pedological attributes (soil texture and soil organic carbon) by using Analytical Hierarchical Process (AHP) and GIS techniques in the semi-arid region of the Bundi district, Rajasthan, India. Land surface temperature (LST) and NDVI products were derived from time-series Moderate-Resolution Imaging Spectroradiometer (MODIS) datasets, rainfall data products from Climate Hazards Group InfraRed Precipitation with Station data (CHIRPS), terrain characteristics from Shuttle Radar Topography Mission (SRTM), LULC from Landsat 9, and pedological variables from legacy soil datasets. Weights derived for thematic layers from the AHP in the studied area were as follows: LULC (0.38) > NDVI (0.23) > ST (0.15) > LST (0.08) > TWI (0.06) > MAR (0.05) > SOC (0.03) > MRVBF (0.02). The consistency ratio (CR) for all studied parameters was <0.10, indicating the high accuracy of the AHP. The results show that about 20.52% and 23.54% of study area was under moderate and high to very high vulnerability of land degradation, respectively. Validation of LDV zones with the help of ultra-high-resolution Google Earth imageries indicates good agreement with the model outputs. The research aids in a better understanding of the influence of land degradation on long-term land management and development at the watershed level.

Keywords: analytical hierarchical process; land degradation vulnerability; NDVI; land surface temperature; soil properties

Citation: Malav, L.C.; Yadav, B.; Tailor, B.L.; Pattanayak, S.; Singh, S.V.; Kumar, N.; Reddy, G.P.O.; Mina, B.L.; Dwivedi, B.S.; Jha, P.K. Mapping of Land Degradation Vulnerability in the Semi-Arid Watershed of Rajasthan, India. *Sustainability* **2022**, *14*, 10198. https://doi.org/10.3390/su141610198

Academic Editors: Antonio Ganga, Blaž Repe and Mario Elia

Received: 19 June 2022
Accepted: 11 August 2022
Published: 17 August 2022

Publisher's Note: MDPI stays neutral with regard to jurisdictional claims in published maps and institutional affiliations.

1. Introduction

Land is a vital and precious resource to produce food, fiber, fuel, and other ecosystem services for the survival of humans and animals [1,2]. However, the constant pace of degradation and deterioration due to persistent human-induced disturbances and climatic irregularities [3] places livelihood and sustainable progress under acute threat [4]. Land degradation is a major environmental problem all around the world and influences human society and its livelihoods. Globally, the life of around 3.2 billion people totally depends on degraded lands, and around one-third of the world's lands are affected by land degradation [5,6]. In recent years, land degradation has been considered a pivotal factor in environmental issues and has attracted the attention of all stakeholders [7]. The United Nations General Assembly adopted Sustainable Development Goal 15.3 in September 2015, which focuses on achieving land degradation neutrality (LDN) by implementing the best management practices that reduce the loss of healthy land and maintain or improve the productivity of the land [8,9]. Land degradation can be defined as a spatio-temporal deterioration of physico-chemical and biological properties of land, making it unsuitable for human society, and a deterioration of the soil ecosystem, influencing agricultural production and ecological instability [10,11].

Around 24% of the world's total geographic area (approximately 3500 Mha) is severely affected by land degradation [11,12]. Around 20% of cropland, 10% of grassland, and 30% of forests are under the process of land degradation throughout the world [13]. In India, around 36.7% of total geographical area (TGA) (120.7 Mha) is under different types of land degradation such as soil erosion, soil acidity, soil salinity and alkalinity, and waterlogging [14], and soil salinity and alkalinity alone affect 6.73 Mha in different arid, semi-arid, and sub-humid areas [15]. According to the Indian Space Research Organization (ISRO), land degradation accounts for around 29.32% of the TGA of India. It covers 96.4 Mha of agricultural, forest, and non-forest land spread across the country [16]. India joined the Bonn Challenge and the United Nations Decade on Ecosystem Restoration 2021–2030 to maximize ecological and economic advantages from the restoration of degraded ecosystems, pledging to rehabilitate 26 Mha of degraded land by 2030 [17].

The problem of land degradation is especially severe in arid and semi-arid areas of the country, such as the state of Rajasthan. Land degradation affects 67% of Rajasthan's land, where wind erosion contributes to the maximum percentage (44.2%), and water erosion (11.2%), vegetal degradation (6.25%), and salinization (1.07%) are the next most common forms of degradation. Chambal ravines in the state of Rajasthan are perhaps among the worst physically degraded lands, as cultivated fertile lands were engulfed by ravines and rendered unsuitable for agricultural activities [18]. The Chambal ravines are very typical as they are deep to very deep (>20 m) and are devoid of any kind of vegetation, with ravines and gullies being the typical forms of degradation [19]. For the development of effective strategies to minimize and lessen the effects of land degradation, it is a prerequisite to understand the process of land degradation, including the causes and its consequences for major functions of the ecosystem and the proper identification of the affected area and the regions at high risk.

Modeling and assessing the vulnerability of land degradation play a pivotal role in land degradation neutrality planning and prioritization processes and in fulfilling targets for restoration. Assessment of land degradation requires various information such as climate, soil properties, topography, land use, etc. Several techniques are being adopted in monitoring and evaluating the area, rate, and type of land degradation. A survey using satellite images overcomes the time-consuming and expensive traditional survey, particularly in areas tough to assess [20]. Geospatial techniques such as remote sensing (RS) and geographic information system (GIS) play an important role in the assessment and monitoring of land degradation vulnerability. Satellite imageries with precise spatial and spectral resolution are excellent resources for detecting, mapping, and monitoring various degradation kinds and issues in a rapid, consistent, reliable, and cost-effective manner [21–25].

The integrated use of geospatial techniques with the multi-criterion decision analysis (MCDA) method is the most feasible option to assess and map land degradation vulnerability. This MCDA technique has numerous applications in multiple areas such as groundwater potential mapping, crop suitability zonation, and land degradation vulnerable mapping. It is mostly used to solve complex problems by breaking them up into sections, then solving and integrating each section to obtain the ultimate results. The AHP, which was first developed by Saaty (1980), is the most widely used multi-criterion decision method for the mapping of vulnerable zones [26,27]. Decisions may be made using this strategy based on judgements, hierarchical structure, and accurate perception, all of which have a dominant influence on the final decision [27,28]. The AHP approach is a widely recognized, basic, and well-structured decision-making technique. Few research findings have been generated by other researchers [12,13,29] with respect to the assessment and mapping of land degradation vulnerability zones (LDVZ) based on AHP and GIS modeling approaches and their validation with Google Earth imageries. Considering the importance of land degradation vulnerability assessment through remote sensing and GIS and AHP approaches, the present study was carried out in the semi-arid region of Rajasthan, western India. In the present study area, water erosion is the most important cause of land degradation due to favorable erosion geology, vegetal degradation, and the perennial Chambal River. Despite this fact, so far, no studies have been carried out in this area to assess and prepare a land degradation susceptibility map. The core objectives of the study are to (i) characterize the terrain, climatic, vegetative, and pedological variables of the watershed and (ii) identify the most vulnerable areas to land degradation using remote sensing and geospatial techniques. Furthermore, the research provides important information for long-term land use management and development.

2. Materials and Methods

2.1. Study Area

The Chanda Kalan Watershed is in the Bundi district of Rajasthan, western India, and it lies between latitude 25°41′ N to 25°46′ N and longitude 76°16′ E to 76°22′ E. Geographically, it covers an area of 2629 hectares (Figure 1). The watershed falls within the Northern Plain (and Central Highlands) including Aravalli, a hot semi-arid eco-region (4.2) denoted as an agro-ecological sub-region (AESR). The climate of the study area is semi-arid with an average annual rainfall of 681 mm, in which the southwest (SW) monsoon contributes roughly 90% of the rainfall. The altitude ranges from 187 to 459 m from the mean sea level (MSL). The watershed is mainly drained by the Chambal River and its tributaries. Major soils are deep brown loamy and brown clayey. The important crops cultivated in the study area are wheat, maize, rapeseed, soybean, paddy, etc. Geologically, the watershed is exposed by rock formations belonging to the Vindhyan Super Group. Vindhyan sedimentary sequences have occupied a major part of the watershed. The Bhander Group of the Vindhyan Super Group and their formations (Upper Bhander shale, Balwan Limestone, Maihar Sandstone) are well exposed in the study area [30]. The watershed has a systematic drainage system, and most of the study area is drained by the southwest to northeast flowing Chambal River and its tributaries. The aquifer area formed in the watershed comes under younger alluvium.

2.2. Dataset Used

In the current study, eight thematic layers were considered to identify the land degradation vulnerable zones, including Moderate Resolution Imaging Spectroradiometer (MODIS) Normalized Difference Vegetation Index (NDVI), MODIS land surface temperature (LST), Climate Hazards Group Infrared Precipitation with Station data (CHIRPS) rainfall, land use/land cover (LULC), Topographical Wetness Index (TWI), Multi-Resolution Index of Valley Bottom Flatness (MRVBF), and soil texture and soil organic carbon. The Landsat 9 images and Shuttle Radar Topography Mission (SRTM) DEM data were collected from the United States Geological Survey (USGS) website (https://earthexplorer.usgs.gov, accessed on 10 February 2022). The CHIRPS rainfall, MODIS NDVI, and MODIS LST products were downloaded for the period of 10 years (2011–2020) using Google Earth Engine. Soil organic carbon data were downloaded from Soil Grids (https://soilgrids.org/, accessed on 11 February 2022). In addition, soil texture data were collected from the ICAR—National Bureau of Soil Survey and Land Use Planning, Nagpur, at 1:250,000 scale. Various datasets and their specifications are summarized in Table 1.

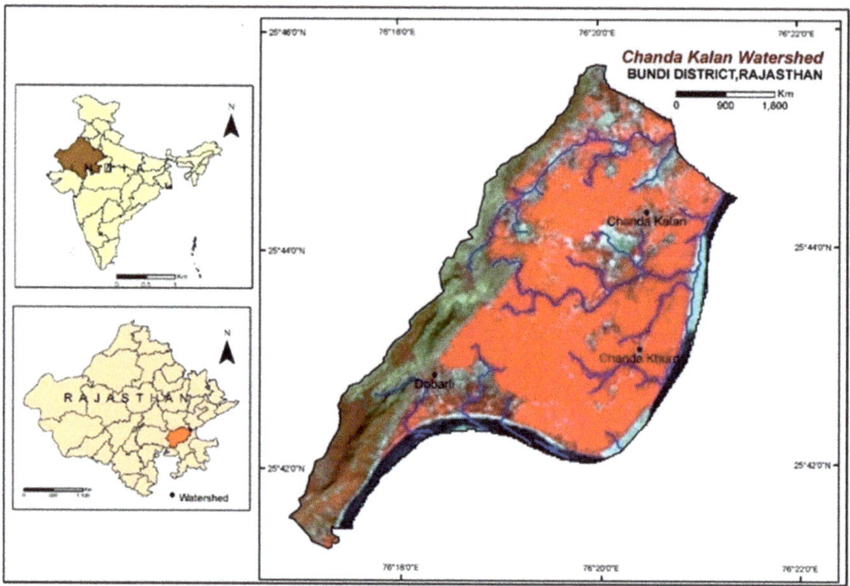

Figure 1. Location map of the study area.

Table 1. Datasets and their specifications.

S. No.	Dataset	Variable	Temporal Resolution	Spatial Resolution	Temporal Coverage
1	MODIS MOD13Q1	NDVI	16 days	250 m	2011–2020
2	MODIS MOD11A2	LST	8 days	1 km	2011–2020
3	SRTM DEM	Elevation	-	30 m	-
4	Soil Grids 250 m	Soil organic carbon	-	250 m	-
5	CHIRPS	Rainfall	-	5 km	2011–2020
6	SRM data, NBSS&LUP	Soil texture	-	2.5 km	-

2.3. Processing of Data

2.3.1. Processing of Terrain Parameters

The SRTM DEM was downloaded and reprojected to Universal Transverse Mercator (UTM), 43 N coordinate system, and filled in QGIS. After that, the filled DEM was used to produce TWI and MRVBF of the watershed. The TWI is widely used to evaluate the impact of topography on different hydrological processes, and it is considered an important indicator of the wetness conditions of a particular region [31]. TWI depicts the water accumulation tendency of a region [32]. Therefore, with respect to land degradation, a higher value of the TWI indicates less vulnerability to degradation or water erosion, and vice versa. In this study, TWI was computed in the SAGA GIS using the following equation:

$$TWI = \ln(As/\tan\beta)$$

(1)

where As is the area of the ascending slope and β is the gradient of the slope.

The flatness and lowness of valley bottoms are measured by an index called the Multi-Resolution Index of Valley Bottom Flatness (MRVBF). A higher value of MRVBF denotes a flatter valley with higher deposition, and vice versa. MRVBF values range from 0 to a positive integer value. In this study, MRVBF was computed in the SAGA GIS.

2.3.2. Processing of Climate Parameters

The combined effects of climate, physical processes, and land use practices are often the cause of land degradation. Rainfall is the most significant factor in land degradation, and it has a direct impact on the detachment of soil particles and migration of eroded sediment [33]. As a result, it is recognized as a major factor in assessing land degradation. In this study, CHRIPS-based rainfall products of 5 km spatial resolution were downloaded for 10 years (2011–2020) using Google Earth Engine and reprojected from the Geographic Coordinate System (GCS) to the UTM 43N coordinate system in QGIS. The downloaded products were resampled to 30 m resolution in QGIS by using the bilinear interpolation technique. The intensity and distribution of land surface temperature are directly linked to the vegetative condition of a region [13]. Therefore, land surface temperature is considered as an important indicator of land degradation. In the present study, MODIS MOD11A2 products for the period of 10 years (2011–2020) were downloaded using Google Earth Engine. The data were converted to degrees Celsius (°C) by using Equation (2).

$$\text{LST} = 0.02 \times \text{DN} - 273.15 \tag{2}$$

Subsequently, the datasets were reprojected from the sinusoidal coordinates system to the geographical coordinate system and resampled to 30 m by using the bilinear interpolation technique in QGIS. Finally, the thematic layer was classified into five subclasses: <32.70 °C, 32.70–33.30 °C, 33.30–33.91 °C, 33.91–34.52 °C, and >34.52 °C.

2.3.3. Processing of Vegetation Parameters

Vegetal degradation is a direct indicator of land degradation. Therefore, LULC and NDVI were taken as important thematic layers for assessing land degradation. The LULC map was prepared from the downloaded Landsat 9 images using supervised classification in the QGIS environment. In the present study, MODISMOD13Q1 NDVI products were downloaded for a period of 10 years (2011–2020) using Google Earth Engine. NDVI, which is a dimensionless index, depicts the difference of reflectance between near-infrared and red bands and can be used to analyze vegetative greenness over an area. It ranges from −1 to +1, where low NDVI values indicate stressed vegetation and higher values indicate healthy vegetation. Temporal smoothing of the NDVI time series data was carried out with the Savitzky–Golay (SG) filter [34]. SG filter fits a polynomial function based on a weighted least squares regression approach. The processing was executed in Google Earth Engine and downloaded. It was then reprojected from GCS to the UTM 43N coordinate system in QGIS. Datasets were resampled to 30 m resolution by using the bilinear interpolation technique in the QGIS. The layer was classified into six subclasses: <0.15, 0.15–0.20, 0.20–0.25, 0.25–0.30, 0.35–0.40, and >0.40.

2.3.4. Processing of Soil Parameters

Soil organic carbon data downloaded from Soil Grids were reprojected to the geographical coordinate system and resampled to 30 m resolution in QGIS. The layer was classified into five subclasses: <122.5, 122.5–176.5, 176.5–230.5, 230.5–284.5 and >284.5 decigram/kg. Soil texture data were taken from NBSS&LUP and resampled to 30 m in the QGIS environment. Soil texture data were classified into three classes, namely, fine loamy, clayey, and rock outcrops. The detailed methodology is given in Figure 2.

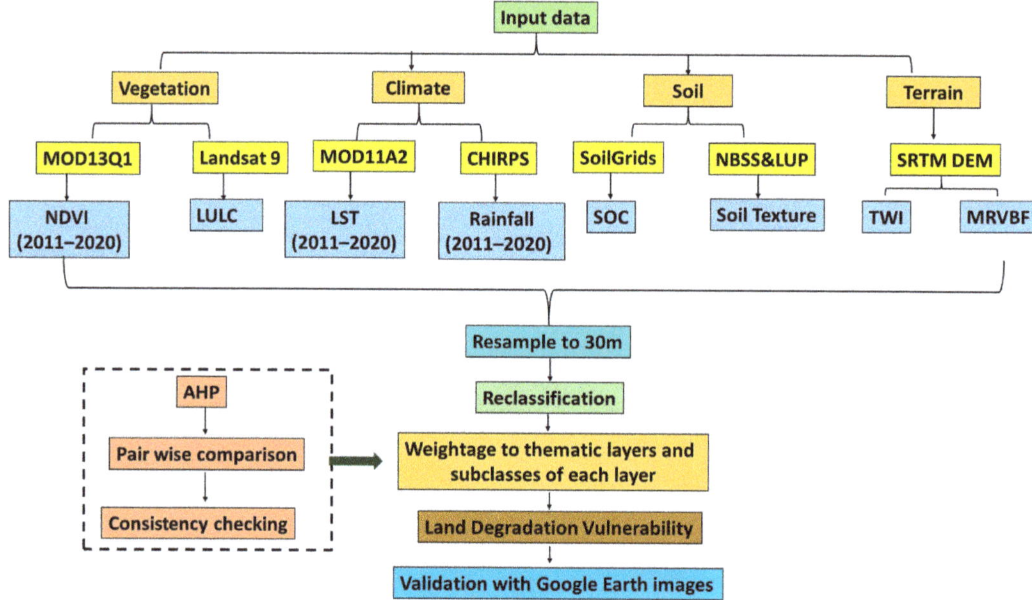

Figure 2. Methodology followed in the present study.

2.4. Analytical Hierarchical Process (AHP)

The most widely used and well-known GIS-based method for demarcating land degradation vulnerability zones is MCDA using the AHP technique. To make an organized decision of priorities, we need to make comparisons and a scale of numbers that show how much more important one parameter is in comparison to another in terms of the criterion being compared. The AHP is a pairwise comparison assessment theory, where parameters are compared with each other using Saaty's scale of relative importance (Table 2) [31,35].

Table 2. Saaty's 1–9 scale of relative importance in AHP.

Scale	Importance
1	Equal significance
2	Intermediate between 1 and 3
3	Moderate significance
4	Intermediate between 3 and 5
5	Strong
6	Intermediate between 5 and 7
7	Very strong
8	Intermediate between 7 and 9
9	Maximum importance

The relative weight of each variable was determined by a knowledge-based spatial decision support system and referring to the literature [13,29]. We selected LULC as the first significant layer since LULC changes are one of the main human-induced activities affecting the land degradation of a region. NDVI was selected as the second most important parameter in the hierarchy as it is the most significant indicator of vegetal degradation. The soil texture was chosen as the third element in the hierarchy because soil erosion is directly controlled by the size and distribution of soil particles. The LST was selected as the fourth layer in the hierarchy, and it was mainly based on the assumption that higher LST zones have low vegetation cover compared to low LST. Other remaining layers were assigned lower order in the hierarchy. Consequently, all layers were compared to one another in a pair-wise comparison matrix.

2.5. Consistency Analysis

To authenticate the decision on the pair-wise comparison of the thematic layers and their sub-classes, the consistency ratio (CR) was utilized [28]. For computing the CR, the following equation was used:

$$CR = \frac{CI}{RCI} \tag{3}$$

where RCI stands for Random Consistency Index, and its values are based on Saaty's standard (Table 3). CI indicates consistency index, which was computed using the following equation:

$$CI = \frac{(\lambda max - n)}{(n - 1)} \tag{4}$$

where λ_{max} is the principal eigenvalue and n is the total number of thematic layers used in the study.

Table 3. Saaty's Random Consistency Index.

N	1	2	3	4	5	6	7	8	9	10	11
RCI	0	0	0.58	0.9	1.12	1.24	1.32	1.41	1.45	1.49	1.51

N, Order of the matrix; RCI, Random Consistency Index.

A CR value ≤ 0.10 is acceptable to conduct a weighted overlay analysis using AHP. If the CR is >0.10, the judgement must be revised to identify the cause of the inconsistency and fix it until the CR ≤ 0.10 is reached.

2.6. Mapping of Land Degradation Vulnerability Zones

In GIS-based modeling, AHP-based weights were given to thematic layers and their sub-classes to demarcate the land degradation vulnerability (LDV) zones. In the present study, the following equation was used to delineate the land degradation vulnerability map:

$$
\begin{aligned}
LDV = & \; LULCCwi \times LULCSCwi + NDVICwi \times NDVISCwi + STCwi \times STSCwi + \\
& LSTCwi \times LSTSCwi + TWICwi \times TWISCwi + MARCwi \times MARSCwi + \\
& SOCCwi \times SOCSCwi0.05 \times MRVBFCwi + MRVBFSCwi
\end{aligned}
\tag{5}
$$

where LULC, NDVI, ST, LST, TWI, MAR, SOC, and MRVBF indicate land use/land cover, Normalized Difference Vegetation Index, soil texture, land surface temperature, Topographic Wetness Index, mean annual rainfall, soil organic carbon, and Multi-Resolution Index of Valley Bottom Flatness, respectively; Cwi is the class weight and SCwi is the sub-class weight. The generated LDV map was classified into five classes, namely, very low, low, moderate, high, and very high. Using the ultra-high resolution Google Earth imagery of 2022, the very high and high LDV classes were validated at five randomly selected sites. Finally, validation of the results was performed using the ROC curve generated from the site selected from the Google Earth image. The area under the curve (AUC) was estimated from the ROC curve and its values range from 0.5 to 1. The AUC value closer to 1 implies great model performance, whereas a value near to 0.5 indicates poor prediction accuracy.

3. Result

3.1. Input Parameters and Their Variability

TWI of the watershed ranged between 1.46 and 25.94, as illustrated in Figure 3a. Five classes—less than 6.13, 6.13–9.45, 9.45–12.77, 12.77–16.10, and more than 16.10—were generated after reclassification. Nearly 64% of the district area came under the first and second subclass of TWI and the remaining 36% came under other subclasses. TWI values show the parts in the study area that are more prone to water erosion.

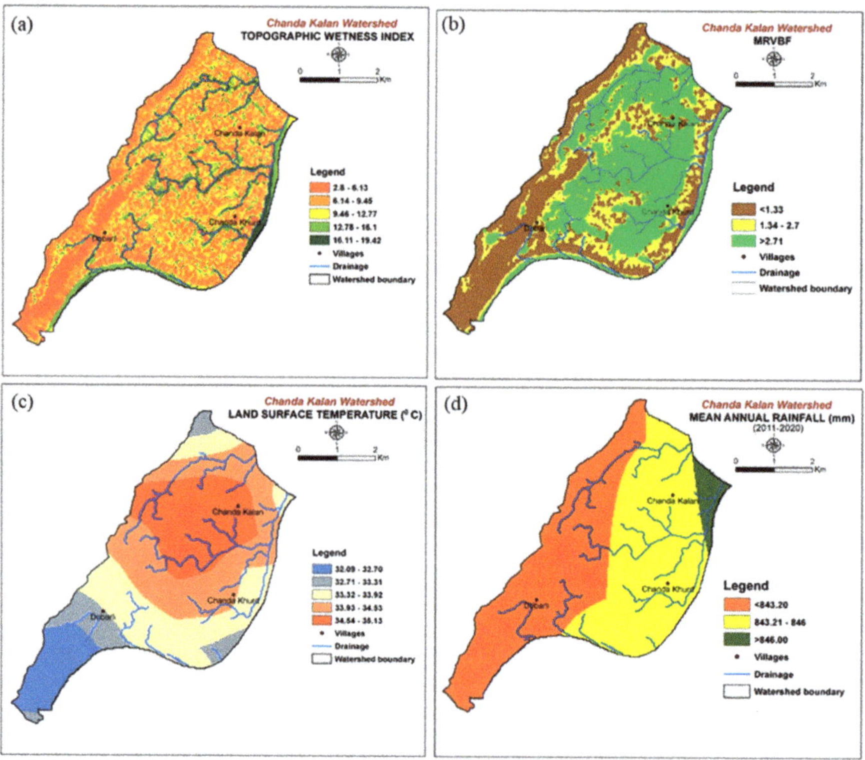

Figure 3. Thematic layers: (**a**) TWI, (**b**) MRVBF, (**c**) LST, (**d**) MAR.

In this study, MRVBF is classified into three classes, namely, first class (<1.33), second class (1.33–2.77), and third class (>2.77) (Figure 3b). The highest percentage of the study area comes under third class (43%), followed by first class (35%) and second class (22%). The land surface temperature of the study area divides the whole area into six subclasses: <32.7 °C, 32.70–33.30 °C, 33.30–33.91 °C, 33.91–34.52 °C, and >34.52 °C (Figure 3c). The highest area 801.54 ha (30.51%) comes under subclass LST 33.91–34.52 °C, followed by subclass LST >34.54 °C occupying 630.9 ha (24.02%), and the lowest area comes under subclass LST <32.7 °C of an area 271.89 (10.35%). Nearly 54.53% of the study area came under LST values >34.52 °C and 33.91–34.52 °C, which lies in the central part of the watershed, and the remaining 45.37% of the area came under other subclasses. The analysis of decadal (2011–2020) mean annual rainfall trends shows that the study area is divided into two subclasses (Figure 3d), where northern, northeastern, eastern, and the majority of the central area received relatively higher annual rainfall (>843.10 mm), covering an area of 1760.04 ha that accounts for 66.88% of the total area, while southwestern and southern parts received relatively less annual rainfall (<843.1 mm), covering an area of 871.65 ha that accounts for 33.12% of the total study area.

The spatial analysis of the mean NDVI values from 2011 to 2022 divides the whole area of the Chanda Kalan Watershed (2626.56 ha) into six subclasses: 0.15–0.20, 0.20–0.25, 0.25–0.30, 0.30–0.35, 0.35–0.40, and >0.40 (Figure 4a). The highest area, 926.19 ha (35.26%), comes under the low vegetal degradation category with NDVI values of 0.30 to 0.35, followed by an area of 851.13 ha (32.4%) covering maximum greenness with very low vegetal degradation with NDVI values of 0.35–0.40. The lowest area, 40.5 ha (1.54%), falls under the very severe vegetal degradation category with NDVI values of 0.15 to 0.2. Nearly 67.66% of the study area came under NDVI values 0.30–0.35 and 0.35–0.40, which lies in the southeast and southwest watershed, and the remaining 32.34% of the area came under other subclasses. The LULC map was prepared from Landsat 9 using supervised classification classes (Figure 4b). The study area of Chanda Kalan Watershed is divided into seven classes, namely, Agriculture, Bare Ground, Shrubs/Scrub, Open Forest, Ravines, Built Up, and Water Bodies, based on the LULC map, where the area under Agriculture covers the highest area of the watershed, which lies in the central and northeast part of the watershed. The soil textural map of the study area divides the whole area into three textural classes, namely, (i) clayey, (ii) fine loamy soil, and (iii) rock outcrops (Figure 4c). Most of the study area comes under clayey soil, particularly the central and the southeastern parts, while the northern and northeastern areas come under the fine loamy soil category and the southwestern area comes under the rock outcrop category. The study area is divided into five soil organic carbon class for soil depth of 0–15 cm based on soil organic carbon content (decigram/kg): <122.5, 122.5–176.5, 176.5–230.5, 230.5–284.5, and >284.5 (Figure 4d). The maximum area of 1707.84 ha covering 65.26% comes under the subclass with SOC content <122.5 decigram/kg and lies in the northwest and central part of the watershed, while the minimum area with <284.5 decigram/kg covers 57.24 ha (2.19%) and lies in the lower part of the watershed.

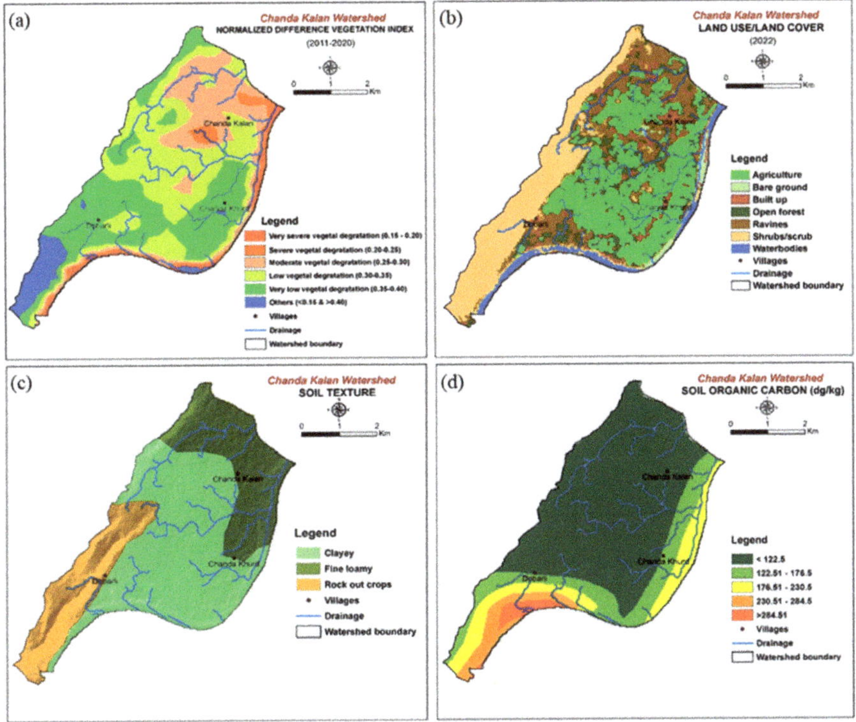

Figure 4. Thematic layers: (**a**) NDVI, (**b**) LULC, (**c**) soil texture, (**d**) soil organic carbon.

3.2. Land Degradation Vulnerability

Consistency ratios for each thematic layer (Table 4), normalized matrix (Table 5), and subcategories of each thematic layer (Table 6) were calculated before the integration of thematic layers. The results revealed that the judgement matrices utilized in the investigation were accurate (CR < 0.10) and had reasonable consistency. The reclassified thematic layers are combined using the weighted overlay approach based on their respective weight. In this study, five LDVZ categories, namely, very low, low, moderate, high, and very high, were identified through the AHP- and GIS-based modeling approach. The quantile breaks were used for the above classification of the integrated product. The results represent that about 1444.68 hectares of the total study area (55%) are under very low to low classes of land degradation vulnerability, and these lands covered almost half the area of the watershed (Figure 5). About 530.01 hectares (20.52%) of the watershed came under the moderate class of the LDVZ and covered mainly southern to southeastern parts of the watershed. High and very high classes of LDVZ zones covered about 607.95 hectares (23.54%) of the watershed. These classes covered the area mostly the northern and somewhat central and southern parts of the study area. These two classes showed high to very high severity of land degradation, such as ravines and gullies.

Table 4. AHP pairwise comparison matrix for thematic layers.

	LULC	NDVI	ST	LST	TWI	MAR	SOC	MRVBF	Weight	CR
LULC	1	3	5	5	7	7	8	9	0.38	0.098
NDVI	0.3	1	2	5	5	6	7	8	0.23	
ST	0.2	0.5	1	3	3	4	6	7	0.15	
LST	0.2	0.2	0.3	1	2	3	3	4	0.08	
TWI	0.1	0.2	0.3	0.5	1	2	2	3	0.06	
MAR	0.1	0.2	0.3	0.3	0.5	1	2	3	0.05	
SOC	0.1	0.1	0.2	0.3	0.5	0.5	1	2	0.03	
MRVBF	0.1	0.1	0.1	0.3	0.3	0.3	0.5	1	0.02	

LULC, land use/land cover; NDVI, Normalized Difference Vegetation Index; ST, soil texture; LST, land surface temperature; TWI, Topographic Wetness Index; MAR, mean annual rainfall; SOC, soil organic carbon; MRVBF, Multi-Resolution Index of Valley Bottom Flatness.

Table 5. Normalized matrix for thematic layers.

	LULC	NDVI	Texture	LST	TWI	MAR	SOC	MRVBF
LULC	0.44	0.56	0.54	0.32	0.36	0.29	0.27	0.24
NDVI	0.15	0.19	0.22	0.32	0.26	0.25	0.24	0.22
Texture	0.09	0.09	0.11	0.19	0.16	0.17	0.20	0.19
LST	0.09	0.04	0.04	0.06	0.10	0.13	0.10	0.11
TWI	0.06	0.04	0.04	0.03	0.05	0.08	0.07	0.08
MAR	0.06	0.03	0.03	0.02	0.03	0.04	0.07	0.08
SOC	0.06	0.03	0.02	0.02	0.03	0.02	0.03	0.05
MRVBF	0.05	0.02	0.02	0.02	0.02	0.01	0.02	0.03

Table 6. Weighting of sub-classes.

Thematic Layer	Subclass	Weight	CR
LULC	Ravines	0.511	0.085
	Bare ground	0.292	
	Shrub/scrub	0.097	
	open forest	0.062	
	Agriculture	0.039	
NDVI	0.15–0.20	0.449	0.095
	0.20–0.25	0.275	
	0.25–0.30	0.120	
	0.30–0.35	0.076	
	0.35–0.40	0.050	
	>0.40	0.032	
Soil texture	Rock outcrops	0.633	0.062
	Fine loamy	0.260	
	Clayey	0.106	
LST	<32.7	0.062	0.026
	32.70–33.30	0.099	
	33.30–33.91	0.161	
	33.91–34.52	0.262	
	>34.52	0.416	
TWI	<6.13	0.520	0.08
	6.13–9.45	0.220	
	9.45–12.77	0.149	
	12.77–16.10	0.073	
	>16.10	0.038	
MAR	<843.2	0.082	0.042
	843.2–846	0.343	
	>846	0.575	
SOC	<122.5	0.487	0.045
	122.5–176.5	0.256	
	176.5–230.5	0.133	
	230.5–284.5	0.081	
	>284.5	0.044	
MRVBF	<1.33	0.633	0.062
	1.33–2.70	0.260	
	>2.70	0.106	

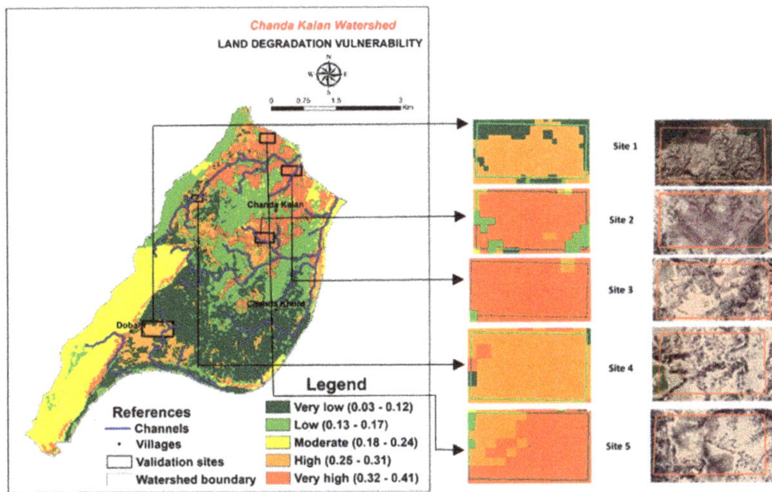

Figure 5. Land degradation vulnerable zones and validation with Google Earth images (Dark green color indicates low vulnerability and deep red indicates higher vulnerability).

3.3. *Validation of Land Degradation Vulnerability Zones*

Validation of LDVZ of the study area was conducted by the visual validation method. In this process, validation was performed with the help of high-resolution Google Earth images. Five sites from the degraded part of the study area were validated with the high-resolution Google Earth images. The visual assessment of high and very high classes using high-resolution Google Earth images of growing season of 2022 indicated that the degree of land degradation (ravines, gullies, and bare grounds) in the selected watershed is in accordance with the outcomes of the model used in this study (Figures 5 and 6).

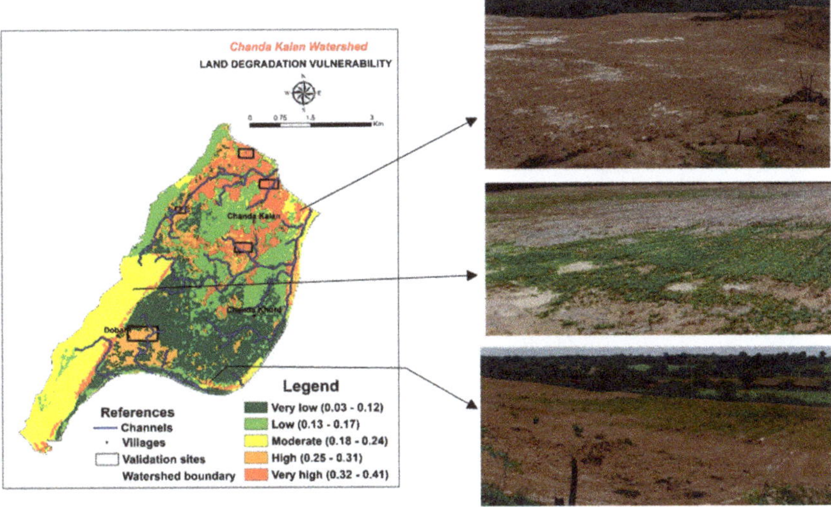

Figure 6. Validation of land degradation vulnerable zones with field photographs.

Figure 7 shows the ROC curve of the LDVZ map generated using the AHP method. The AUC value of the ROC curve was found to be 86%. Hence, it was determined that the AHP model derives reasonable results in predicting land degradation vulnerability zones in the study area.

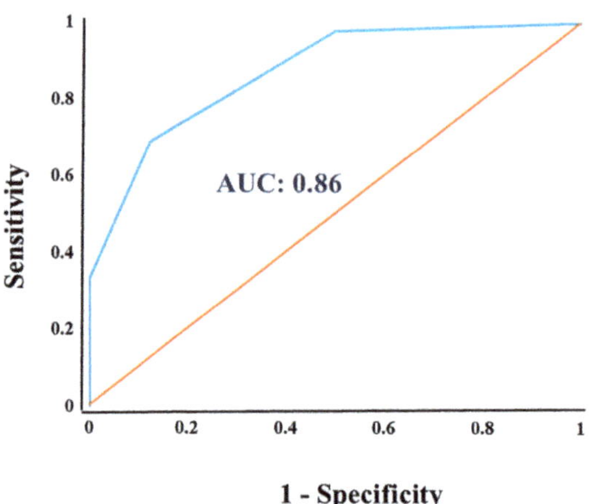

Figure 7. ROC curve of the LDVZ map using AHP model.

4. Discussion

Land degradation is generally considered one of world's most serious environmental issues. In India, the western state of Rajasthan is a part of the Thar Desert, where degradation is a severe issue. The Chambal River valley in the state of Rajasthan is one of the severely affected regions in the country, where gully erosion/ravines have major physical and economic implications [18,19,36]. For sustainable agricultural planning and development, the identification of vulnerable hotspots to soil/water erosion is the need of the hour. Therefore, the present research was carried out to identify hot spots of land degradation in a small watershed using an AHP- and GIS-based modeling approach. Previous research has found that only a few variables play an important role in the assessment of land degradation [13,37]. In the present investigation, LULC, NDVI, TWI, MRVBF, LST, MAR, soil texture, and SOC were considered for the mapping of land degradation vulnerable zones. LULC and NDVI were taken as the most influential layers for land degradation vulnerability. Land use/land cover implies man-made and natural modification of the land surface, and it is a major cause of land degradation [38,39]. The NDVI has long been recognized as a useful measure for determining the greenness of flora and it is well accepted in science that a decrease in NDVI is a sign of land degradation and is closely linked to climatic conditions [40,41]. The most basic soil physical property, on the other hand, is soil texture, which impacts hydraulic properties and surface soil loss [42].

LST is an important parameter in the semi-arid region as it is directly linked with soil moisture availability and indirectly linked with the flora conditions of the study area [43,44]. A rise in the LST might result in a reduction in vegetative greenness and an increase in land degradation. Increased rainfall during the monsoon season increases the risk of topsoil loss by higher water velocity, which causes more soil erosion [37]. SOC is a universal biomarker of soil degradation since its decrease may have severe consequences for soil-derived ecosystem services [45]. Similarly, TWI is one of the most important terrain parameters and plays a significant role in assessing land degradation vulnerability. A higher TWI value is associated with good vegetation cover, and vice versa [46]. As a result, vegetation cover promotes infiltration, reduces surface runoff, and thus greatly delays the incidence of soil erosion [47]. Therefore, thematic layers were given weights based on their importance. The AHP model assigned the weightage of each factor, i.e., LULC (0.38), NDVI (0.23), soil texture (0.15), LST (0.08), TWI (0.06), MAR (0.05), SOC (0.03), and MRVBF (0.02). The higher the index value, the more exposed the area is to land degradation, whereas the lower the value, the less vulnerable it is. Consistency ratios for each thematic layer and subclasses of each thematic layer were calculated before the integration of thematic layers. The computed CR value was less than 0.1, which shows that all the parameters' assumptions about their impact on soil erosion are valid.

Research findings showed that five land degradation vulnerability zones (LDVZ) namely, very low, low, moderate, high, and very high, were identified in the study area. Very low, low, moderate, high, and very high classes covered 27%, 29%, 20%, 11%, and 12% of the area of the watershed, respectively. Parmar et al. (2021) [48] also conducted a study to assess land degradation vulnerability using the geospatial technique in the Kutch district of Gujarat, India, and the results revealed that 67% of the land area has high vulnerability to land degradation, and 27% of the area falls under the moderate class. Similarly, an assessment of potential land degradation using the geospatial technique and multi-influencing factor technique was carried by Senapati et al. (2020) [49] in the Akarsa Watershed, West Bengal, and they also classified the study area into five land degraded zones. The analysis revealed that the very low to low categories of land degradation vulnerability covered almost half of the area of the watershed, and this portion of the study area is associated with good vegetative coverage with open forest, very low vegetative degradation, adequate rainfall (843.21–846 mm), and well-drained soils deep in nature with clayey texture.

All the environmental covariables are linked with each other, e.g., adequate rainfall positively correlates with NDVI and LULC and a good amount of LULC links with optimum SOC content and better soil health, which directly relate to a lower chance of land degradation [50]. The moderate class of LDVZ was related to very less vegetative coverage in the scrub/shrub class of LU/LC, with normal rainfall (<843.20 mm) and rock outcrops. This class also represented a low to medium MRVBF value with low TWI. High and very high classes of LDVZ covered about 607.95 hectares (23.54%) of the watershed. These two classes showed high to very high severity of land degradation, such as ravines and gullies. This section of the watershed had no or very little vegetation cover, higher rainfall (>846 mm), high valley bottom flatness, higher LST, and clay to fine loamy texture soils with low soil organic carbon. In this research, the AHP- and GIS-based modeling approach showed its potential for the assessment of vulnerability to land degradation by compiling different parameters. Validation of LDVZ was carried out with the help of Google Earth images of high resolution and the results were very well in agreement with the AHP–GIS model-based approach. Similar work has been conducted by several researchers [12,13,29,51–54] with respect to assessment and mapping of LDVZ based on an AHP and GIS modeling approach and their validation with Google Earth imageries.

This study has identified areas that are more prone to land degradation, which can help prioritize and implement soil water conservation practices to reduce the consequences of degradation. Farmers should be encouraged to grow cover crops and crop rotation practices to maintain soil quality over time. Farmers should maintain crop residue and biomass over soil surface after harvesting to avoid exposing the topsoil. Furthermore, the findings of this research may be useful in developing better soil and water management policies. Although this study was carried out at the watershed level, it should be replicated to the sub-district or district level. The pedological parameters used in this study are available at coarse resolution, which caused some challenge and gaps in the results. Future research should concentrate on high-resolution satellite and soil survey data to delineate degradation zones with higher accuracy.

5. Conclusions

In the study, LULC, NDVI, soil texture, LST, MAR, TWI, SOC, and MRVBF were considered major contributing factors in the identification of land degradation vulnerability zones through the GIS- and AHP-based model. The AHP- and GIS-based modeling shows that about 607.95 hectares of the total study area are in the high and very high categories of LDV, and 530.01 hectares are in the moderate LDV category. Validation of moderate, high, and very high LDV classes using high-resolution Google Earth imagery demonstrates that the degree of land degradation features of Google Earth imagery of the selected study area was in agreement with the AHP–GIS model-based approach. This study demonstrates the potential of high-resolution satellite data and the robustness of GIS-based spatial modeling in obtaining accurate, reliable, and cost-effective results for the assessment of land degradation in semi-arid ecosystems. The prevalence and severity of LDV were determined using AHP- and GIS-based modeling, which will be extremely useful in recommending soil conservation and management measures that are suited for each site, particularly in highly and extremely vulnerable regions, for long-term land resources management. These data were derived from satellite data that could cause some challenges and gaps in the results. Therefore, macro- and micro-scale observations are required to account for the high environmental variability and to distinguish between the influences of anthropogenic actions and climate variability on land degradation processes.

Author Contributions: Conceptualization, B.Y. and L.C.M.; methodology, G.P.O.R. and N.K.; software, B.L.T. and L.C.M.; validation, S.P. and B.L.M.; formal analysis, B.Y. and L.C.M.; investigation, B.Y.; resources, B.S.D.; data curation, N.K. and S.V.S.; writing—original draft preparation, B.Y., L.C.M., and S.P.; writing—review and editing, P.K.J., G.P.O.R., and S.V.S.; visualization, P.K.J. and B.L.T.; supervision, B.L.M. and B.S.D. All authors have read and agreed to the published version of the manuscript.

Funding: This research received no external funding.

Institutional Review Board Statement: Not applicable.

Informed Consent Statement: Not applicable.

Data Availability Statement: Not applicable.

Acknowledgments: The authors are grateful to the ICAR NBSS&LUP and DST for help and support.

Conflicts of Interest: The authors declare that there is no conflict of interest, either financially or otherwise.

References

1. Reddy, G.P.O.; Sarkar, D. Land degradation, environment, and food security. *Geogr. You* **2012**, *12*, 6–9.
2. Tagore, G.S.; Bairagi, G.D.; Sharma, N.K.; Sharma, R.; Bhelawe, S.; Verma, P.K. Mapping of Degraded Lands Using Remote Sensing and GIS Techniques. *J. Agric. Phys.* **2012**, *12*, 29–36.
3. Barbero-Sierra, C.; Marques, M.J.; Ruiz-Pérez, M.; Escadafal, R.; Exbrayat, W. How is desertification research addressed in Spain? Land versus soil approaches. *Land Degrad. Dev.* **2015**, *26*, 423–432. [CrossRef]
4. Fleskens, L.; Stringer, L.C. Land management and policy responses to mitigate desertification and land degradation. *Land Degrad. Dev.* **2014**, *25*, 1–4. [CrossRef]
5. Bai, Z.G.; Dent, D.L.; Olsson, L.; Schaepman, M.E. Proxy global assessment of land degradation. *Soil Use Manag.* **2008**, *24*, 223–234. [CrossRef]
6. Dubovyk, O. The Role of Remote Sensing in Land Degradation Assessments: Opportunities and Challenges. *Eur. J. Remote Sens.* **2017**, *50*, 601–613. [CrossRef]
7. Han, W.; Liu, G.; Su, X.; Wu, X.; Chen, L. Assessment of potential land degradation and recommendations for management in the south subtropical region, Southwest China. *Land Degrad. Dev.* **2019**, *30*, 979–990. [CrossRef]
8. SDSN. Indicators and a Monitoring Framework for the Sustainable Development Goals: Launching a Data Revolution for the SDGs. A Report to the Secretary-General of the United Nations by the Leadership Council of the Sustainable Development Solutions Network. 2015. Available online: https://sustainabledevelopment.un.org/content/documents/2013150612-FINAL-SDSN-Indicator-Report1.pdf (accessed on 20 February 2022).
9. Wunder, S.; Kaphengst, T.; Frelih-Larsen, A. Implementing land degradation neutrality (SDG 15.3) at national level: General approach, indicator selection and experiences from Germany. In *International Yearbook of Soil Law and Policy*; Springer: Cham, Switzerland, 2018; pp. 191–219. [CrossRef]
10. Masoudi, M.; Joka, P.; Pradhan, B. A new approach for land degradation and desertification assessment using geospatial techniques. *Nat. Hazards Earth Syst. Sci.* **2018**, *18*, 1133–1140. [CrossRef]
11. Pacheco, F.A.L.; Fernandes, L.F.S.; Valle, R.F.; Valera, C.A.; Pissarra, T.C.T. Land degradation: Multiple environmental consequences and routes to neutrality. *Curr. Opin. Environ. Sci.* **2018**, *5*, 79–86. [CrossRef]
12. Romshoo, S.A.; Amin, M.; Sastry, K.L.N.; Parmar, M. Integration of social, economic, and environmental factors in GIS for land degradation vulnerability assessment in the Pir Panjal Himalaya, Kashmir, India. *Appl. Geogr.* **2020**, *125*, 102307. [CrossRef]
13. Sandeep, P.; Reddy, G.P.O.; Jegankumar, R.; Arun Kumar, K.C. Modeling, and assessment of land degradation vulnerability in semi-arid ecosystem of Southern India using temporal satellite data, AHP and GIS. *Environ. Model. Assess.* **2021**, *26*, 143–154. [CrossRef]
14. Maji, A.K.; Reddy, G.P.O.; Sarkar, D. *Degraded and Wastelands of India, Status, and Spatial Distribution*; ICAR and NAAS Publication: Cochin, India, 2010; pp. 1–158.
15. NRSA. *Mapping Salt-Affected Soils of India on 1:250,000*; NRSA: Hyderabad, India, 1996.
16. Government of India. Desertification and Land Degradation Atlas of India. Indian Space ResearchOrganization, Bengaluru. 2016. Available online: https://vedas.sac.gov.in/vedas/downloads/atlas/DSM/Desertification_Atlas_2016_SAC_ISRO.pdf (accessed on 20 February 2022).
17. Singh, K.; Tewari, S.K. Does the road to land degradation neutrality in India is paved with restoration science? *Restor. Ecol.* **2021**, *30*, e13585. [CrossRef]
18. Pani, P. Controlling gully erosion: An analysis of land reclamation processes in Chambal Valley, India. *Dev. Pract.* **2016**, *26*, 1047–1059. [CrossRef]
19. Verma, S.K.; Singh, A.K.; Singh, A.; Prasad, J. Greening of Chambal ravine: Site-specific approach for sustainable development. *J. Soil Water Conserv.* **2021**, *20*, 375–385. [CrossRef]
20. AbdelRahman, M.A.E. Estimating soil fertility status in physically degraded land using GIS and remote sensing techniques in Chamarajanagar district, Karnataka, India. *Egypt. J. Remote Sens. Space Sci.* **2016**, *19*, 95–108. [CrossRef]
21. Sujatha, G.; Dwivedi, R.S.; Sreenivas, K.; Venkataratnam, L. Mapping and monitoring of degraded lands in part of Jaunpur district of Uttar Pradesh using temporal spaceborne multispectral data. *Int. J. Remote Sens.* **2000**, *21*, 519–531. [CrossRef]
22. AbdelRahman, M.A.E.; Tahoun, S.A.; Abdel Bary, E.A.; Arafat, S.M. Detecting Land Degradation Processes Using Geo Statistical Approach in Port Said, Egypt. *Zagazig J. Agric. Res.* **2008**, *35*, 1361–1379.

23. Reddy, G.P.O.; Kumar, N.; Singh, S.K. Remote Sensing and GIS in Mapping and Monitoring of Land Degradation. In *Geospatial Technologies in Land Resources Mapping, Monitoring and Management*; Reddy, G.P.O., Singh, S.K., Eds.; Geotechnologies and the Environment; Springer: Cham, Switzerland, 2018; Volume 21, pp. 401–424. [CrossRef]
24. Jong, R.; Bruin, S.; Schaepman, M.; Dent, D. Quantitative mapping of global land degradation using earth observations. *Int. J. Remote Sens.* **2011**, *32*, 6823–6847. [CrossRef]
25. Higginbottom, T.; Symeonakis, E. Assessing land degradation and desertification using vegetation index data: Current frameworks and future directions. *Remote Sens.* **2014**, *6*, 9552–9575. [CrossRef]
26. Chowdhury, A.; Jha, M.K.; Chowdary, V.M.; Mal, B.C. Integrated remote sensing and GIS-based approach for assessing groundwater potential in West Medinipur district, West Bengal, India. *Int. J. Remote Sens.* **2009**, *30*, 231–250. [CrossRef]
27. Nithya, C.N.; Srinivas, Y.; Magesh, N.S.; Kaliraj, S. Assessment of groundwater potential zones in Chittar basin, Southern India using GIS based AHP technique. *Remote Sens. Appl. Soc. Environ.* **2019**, *15*, 100248. [CrossRef]
28. Saaty, T.L. How to make a decision: The analytic hierarchy process. *Interfaces* **1994**, *26*, 19–43. [CrossRef]
29. Tolche, A.D.; Gurara, M.A.; Pham, Q.B.; Anh, D.T. Modelling and accessing land degradation vulnerability using remote sensing techniques and the analytical hierarchy process approach. *Geocarto Int.* **2021**, *36*, 1–21. [CrossRef]
30. CGWB. *Ground Water Brochure, Bundi District, Rajasthan. Ministry of Water Resources, River Development & Ganga Rejuvenation*; Government of India: New Delhi, India, 2018.
31. Mukherjee, I.; Singh, U.K. Delineation of groundwater potential zones in a drought-prone semi-arid region of east India using GIS and analytical hierarchical process techniques. *Catena* **2020**, *194*, 104681. [CrossRef]
32. Ghorbani Nejad, S.; Falah, F.; Daneshfar, M.; Haghizadeh, A.; Rahmati, O. Delineation of groundwater potential zones using remote sensing and GIS-based data-driven models. *Geocarto Int.* **2017**, *32*, 167–187. [CrossRef]
33. Meng, X.; Zhu, Y.; Yin, M.; Liu, D. The impact of land use and rainfall patterns on the soil loss of the hillslope. *Sci. Rep.* **2021**, *11*, 16341. [CrossRef]
34. Jönsson, P.; Eklundh, L. TIMESAT—A program for analyzing time-series of satellite sensor data. *Comput. Geosci.* **2004**, *30*, 833–845. [CrossRef]
35. Morales, F.; de Vries, W.T. Establishment of land use suitability mapping criteria using analytic hierarchy process (AHP) with practitioners and beneficiaries. *Land* **2021**, *10*, 235. [CrossRef]
36. Ranga, V.; Poesen, J.; Van Rompaey, A.; Mohapatra, S.N.; Pani, P. Detection and analysis of badlands dynamics in the Chambal River Valley (India), during the last 40 (1971–2010) years. *Environ. Earth Sci.* **2016**, *75*, 183. [CrossRef]
37. Aslam, B.; Maqsoom, A.; Alaloul, W.S.; Musarat, M.A.; Jabbar, T.; Zafar, A. Soil erosion susceptibility mapping using a GIS-based multi-criteria decision approach: Case of district Chitral, Pakistan. *Ain Shams Eng. J.* **2021**, *12*, 1637–1649. [CrossRef]
38. Kidane, M.; Bezie, A.; Kesete, N.; Tolessa, T. The impact of land use and land cover (LULC) dynamics on soil erosion and sediment yield in Ethiopia. *Heliyon* **2019**, *5*, e02981. [CrossRef]
39. Tolessa, T.; Dechassa, C.; Simane, B.; Alamerew, B.; Kidane, M. Land use/land cover dynamics in response to various driving forces in Didessa sub-basin, Ethiopia. *GeoJournal* **2020**, *85*, 747–760. [CrossRef]
40. Emmanuel, O. *Effects of Deforestation on Land Degradation*; LAP LAMBERT Academic Publishing: Ado Ekiti, Nigeria, 2017.
41. Lennart, O.; Humberto, B. *Climate Change and Land: An IPCC Special Report on Climate Change, Desertification, Land Degradation, Sustainable Land Management, Food Security, and Greenhouse Gas Fluxes in Terrestrial Ecosystems (Sweden: Land Degradation)*; Moreno, J.M., Ed.; IPCC: Geneva, Switzerland, 2019.
42. Tóth, G.; Hermann, T.; da Silva, M.R.; Montanarella, L. Monitoring soil for sustainable development and land degradation neutrality. *Environ. Monit. Assess.* **2018**, *190*, 57. [CrossRef] [PubMed]
43. Faramarzi, M.; Heidarizadi, Z.; Mohamadi, A.; Heydari, M. Detection of vegetation changes in relation to normalized difference vegetation index (NDVI) in semi-arid rangeland in western Iran. *J. Agric. Sci. Technol.* **2018**, *20*, 51–60.
44. Yang, L.; Sun, G.; Zhi, L.; Zhao, J. Negative soil moisture-precipitation feedback in dry and wet regions. *Sci. Rep.* **2018**, *8*, 4026. [CrossRef]
45. Lorenz, K.; Lal, R.; Ehlers, K. Soil organic carbon stock as an indicator for monitoring land and soil degradation in relation to United Nations' Sustainable Development Goals. *Land Degrad. Dev.* **2019**, *30*, 824–838. [CrossRef]
46. Seutloali, K.E.; Dube, T.; Mutanga, O. Assessing and mapping the severity of soil erosion using the 30-m Landsat multispectral satellite data in the former South African homelands of Transkei. *Phys. Chem. Earth Parts A/B/C* **2017**, *100*, 296–304. [CrossRef]
47. Zhou, P.; Luukkanen, O.; Tokola, T.; Nieminen, J. Effect of vegetation cover on soil erosion in a mountainous watershed. *Catena* **2008**, *75*, 319–325. [CrossRef]
48. Parmar, M.; Bhawsar, Z.; Kotecha, M.; Shukla, A.; Rajawat, A.S. Assessment of land degradation vulnerability using geospatial technique: A case study of Kachchh District of Gujarat, India. *J. Indian Soc. Remote Sens.* **2021**, *49*, 1661–1675. [CrossRef]
49. Senapati, U.; Das, T.K. Assessment of Potential Land Degradation in Akarsa Watershed, West Bengal, Using GIS, and Multi-influencing Factor Technique. In *Gully Erosion Studies from India and Surrounding Regions*; Springer: Cham, Switzerland, 2020; pp. 187–205. [CrossRef]
50. AbdelRahman, M.A.; Natarajan, A.; Hegde, R.; Prakash, S.S. Assessment of land degradation using comprehensive geostatistical approach and remote sensing data in GIS-model builder. *Egypt. J. Remote Sens. Space Sci.* **2019**, *22*, 323–334. [CrossRef]
51. Mihi, A.; Benarfa, N.; Arar, A. Assessing and mapping water erosion-prone areas in northeastern Algeria using analytic hierarchy process, USLE/RUSLE equation, GIS, and remote sensing. *Appl. Geomat.* **2020**, *12*, 179–191. [CrossRef]

52. Abuzaid, A.S.; AbdelRahman, M.A.; Fadl, M.E.; Scopa, A. Land degradation vulnerability mapping in a newly reclaimed desert oasis in a hyper-arid agro-ecosystem using AHP and geospatial techniques. *Agronomy* **2021**, *11*, 1426. [CrossRef]
53. Sar, N.; Chatterjee, S.; Das Adhikari, M. Integrated remote sensing and GIS based spatial modelling through analytical hierarchy process (AHP) for water logging hazard, vulnerability, and risk assessment in Keleghai river basin, India. *Model. Earth Syst. Environ.* **2015**, *1*, 31. [CrossRef]
54. Kartic Kumar, M.; Annadurai, R.; Ravichandran, P.T.; Arumugam, K. Mapping of landslide susceptibility using analytical hierarchy process at Kothagiri Taluk, Tamil Nadu, India. *Int. J. Appl. Eng. Res.* **2015**, *10*, 5503–5523.

Article

Spatial Mapping of Soil Salinity Using Machine Learning and Remote Sensing in Kot Addu, Pakistan

Yasin ul Haq [1,*,†], Muhammad Shahbaz [1,†], H. M. Shahzad Asif [1], Ali Al-Laith [2] and Wesam H. Alsabban [3]

[1] Department of Computer Science, University of Engineering and Technology, Lahore 54000, Pakistan; m.shahbaz@uet.edu.pk (M.S.); shehzad@uet.edu.pk (H.M.S.A.)

[2] Computer Science Department, Copenhagen University, 2100 Copenhagen, Denmark; alal@di.ku.dk

[3] Information Systems Department, Faculty of Computer and Information Systems, Umm Al-Qura University, Makkah 24231, Saudi Arabia; whsabban@uqu.edu.sa

* Correspondence: yaseenulhaq@uet.edu.pk; Tel.: +92-334-9951063

† These authors contributed equally to this work.

Abstract: The accumulation of salt through natural causes and human artifice, such as saline inundation or mineral weathering, is marked as salinization, but the hindrance toward spatial mapping of soil salinity has somewhat remained a consistent riddle despite decades of efforts. The purpose of the current study is the spatial mapping of soil salinity in Kot Addu (situated in the south of the Punjab province, Pakistan) using Landsat 8 data in five advanced machine learning regression models, i.e., Random Forest Regressor, AdaBoost Regressor, Decision Tree Regressor, Partial Least Squares Regression and Ridge Regressor. For this purpose, spectral data were obtained between 20 and 27 of January 2017 and a field survey was carried out to gather a total of fifty-five soil samples. To evaluate and compare the model's performances, the coefficient of determination (R^2), Mean Squared Error (MSE), Mean Absolute Error (MAE) and the Root-Mean-Squared Error (RMSE) were used. Spectral data of band values, salinity indices and vegetation indices were employed to study the salinity of soil. The results revealed that the Random Forest Regressor outperformed the other models in terms of prediction, achieving an R^2 of 0.94, MAE of 1.42 dS/m, MSE of 3.58 dS/m and RMSE of 1.89 dS/m when using the Differential Vegetation Index (DVI). Alternatively, when using the Soil Adjusted Vegetation Index (SAVI), the Random Forest Regressor achieved an R^2 of 0.93, MAE of 1.46 dS/m, MSE of 3.90 dS/m and RMSE of 1.97 dS/m. Hence, remote sensing technology with machine learning models is an efficient method for the assessment of soil salinity at local scales. This study will contribute to mitigating osmotic stress and minimizing the risk of soil erosion by providing early warnings regarding soil salinity. Additionally, it will assist agriculture officers in estimating soil salinity levels within a shorter time frame and at a reduced cost, enabling effective resource allocation.

Keywords: DVI; machine learning; remote sensing; random forest; spatial mapping; soil salinity; salinity indices; vegetation indices

Citation: Haq, Y.u.; Shahbaz, M.; Asif, H.M.S.; Al-Laith, A.; Alsabban, W.H. Spatial Mapping of Soil Salinity Using Machine Learning and Remote Sensing in Kot Addu, Pakistan. *Sustainability* **2023**, *15*, 12943. https://doi.org/10.3390/su151712943

Academic Editors: Antonio Ganga, Blaž Repe and Mario Elia

Received: 18 July 2023
Revised: 24 August 2023
Accepted: 25 August 2023
Published: 28 August 2023

1. Introduction

Soil salinization refers to the accumulation of salts in the soil, leading to adverse effects on water quality, agricultural yield, soil composition and economic growth [1]. This type of land degradation, known as salinization, is particularly prevalent in arid and semi-arid regions where evaporation rates exceed precipitation rates [2]. According to the findings of the Food and Agriculture Organization (FAO), a total of 831 million hectares (mha) of land area are affected by salt, with 397 mha classified as saline soils and 434 mha as sodic soils [3]. Soil salinity negatively impacts both water and soil quality [4], leading to adverse consequences for agricultural production. Salinity obstructs plant development and poses challenges for the sustainable use of land resources [5–7], resulting in an annual decline of 1–2% in Pakistan. Based on FAO estimates from 2010, it is approximated that around 60% of the world's farmlands are significantly afflicted by salinization. Additionally, approximately 3% of the world's resources are impacted by salt [5,7]. The detrimental impacts of salinity can be further intensified by climate change, drought, water resource scarcity and changes in land use [8]. According to the FAO report from 2011, salinization has had adverse effects on 25% of Pakistan's irrigated lands. Consequently, a substantial portion of agricultural land, specifically 1.40 mha, has been abandoned due to the impacts of salinization [7]. In areas where salinization has affected soils, crop losses ranging from 30% to 60% have been observed, leading to the abandonment of approximately 20% to 30% of these affected regions, as reported by the World Bank in 2006 [6].

With the world's population growth and the anticipated demand for large agricultural lands, it has become critically important to monitor soil salinization in real-time and detect its early warning signs to improve the land utilization [5]. Making informed decisions regarding the proper reclamation and management of such lands necessitates ongoing salinity monitoring [9]. The traditional techniques for measuring salinity are laboratory analysis and field survey. Salinity monitoring and mapping typically involve conducting a comprehensive soil survey and extrapolating data obtained from analytical samples. The spatial mapping of the salinity of soil in an area requires dense sampling, which is a laborious, expensive and hectic job [10–14]. Indeed, remote sensing offers a faster, more cost-effective and accurate approach to examining and plotting soil salinity [5,15,16].

Almost 65 years ago, both color and black-and-white images were employed to identify soil salinity and gather information about various elements present on the Earth's surface [17]. In the present day, satellite remote sensing offers a cost-effective means to explore salinization across several geographical and time spans. Remote sensing employs the reflected electromagnetic energy from the land surface to acquire data pertaining to diverse objects at varying levels of intricacy. The salinity of soil can be adequately assessed using this approach. The emergence of GIS and remote sensing techniques has provided an opportunity for technology to potentially replace or supplement traditional approaches in soil salinity assessment. In terms of predicting accuracy, employing spectral reflectance derived directly from sensors or applying spectral transformations like PCA [18,19], tasseled cap transformation [20] and spectral indices [21,22], has yielded promising outcomes. Many researchers highlight the worth of spectral reflectance in remote sensing investigations, considering it a fundamental perception in the field [23–28]. Numerous research initiatives have been dedicated to digital soil mapping, employing diverse forms of satellite data, and employing geostatistical or statistical methodologies.

In the Tafilalet plain of Morocco, a study demonstrated the Successfulness of Landsat 8 OLI imagery for modeling the salinity of soil [29]. The findings revealed that the RMSE varied from 0.62 to 0.80 dS/m, whereas the R^2 varied between 0.53 and 0.75. Another research study conducted by Hihi et al. [30] utilized a simple linear regression model and found a moderate correlation ($R^2 = 0.48$) between spectral indices and electrical conductivity (EC) extracted from a Sentinel 2 MSI imagery. Similarly, an association was found between the canopy temperature received from MODIS data and soil salinity in a study conducted in Uzbekistan [31]. According to Hoa et al. [32], employing the Gaussian processes technique on SAR Sentinel-1 imagery, along with modern ML models, resulted in an exceptionally accurate model with an R^2 of 0.808. This model effectively captures the correlation with satellite data and EC. Taghadosi et al. [33] indicated the effectiveness of radiance images obtained from VH and VV polarizations of SAR Sentinel-1 data in differentiating the soil's salinity. They achieved the most accurate technique by applying the SVR approach with an RBF kernel, which resulted in an R^2 of 0.9783 and an RMSE of 0.3561. In Algeria [11] demonstrated the value of EO-1 ALI and Landsat ETM$^+$ images in recognizing and defining sodic and saline soils. In their study on the temporal–spatial variation in the salinity of soil in Libya, Zurqani et al. [34] utilized the temporal data of Landsat images covering a period of 29 years (1972–2001), in conjunction with ground truth data. To enhance precision, [35] advocated the adoption of hyperspectral photography. They introduced an innovative salinity index gained from EO-1 data and achieved $R^2 = 0.873$ in univariate regression analysis. Similarly, Sahbeni [36] utilized Sentinel 2 MSI data and multiple linear regression to simulate the dispersion of salinity of the soil in the Great Hungarian Plain. In their findings, Sahbeni [36] investigated the effectiveness of geographically weighted regression (GWR), MLR, and RFR models for predicting salinity of the soil. They found that topographic factors and geographical position have an essential task in modeling methods. Additionally, they observed that satellite imagery captured in the arid season is particularly useful for predicting EC. The results indicated that the final model achieved a significant level of intermediate accuracy, having an RMSE of 0.1942 g/kg and an R^2 of 0.51. It was found that the MLR model exhibited the least level of accuracy compared to the GWR and RFR models. However, the RFR model demonstrated superior estimated accuracy compared to the GWR technique. In a different investigation by [37], eight distinct prototypes for the mapping of salinity of the soil in a desert region were explored by spectral signatures measurements and Landsat 8 OLI data. Significant analysis conducted on methods built on VNIR bands yielded poor results, with an exceptionally high RMSE of 0.65 or higher and an R^2 of 0.41. Conversely, the models based on SWIR bands have given valuable findings, with an R^2 of 0.97 and an RMSE of 0.13.

Similarly, it was observed by Zhang X. and Huang B. [38] that spectral transformations and smoothing techniques had an impact on the accuracy of models used for predicting soil salinity based on soil-reflected spectra. Among the different methods evaluated, the Principal Component Regression model with the median filtering data smoothing method demonstrated the precise outcomes, with an R^2 of 0.7206 and an RMSE of 0.3929. In addition to the salinity index, such as the NDSI, vegetation indices like the NDVI and SAVI have also demonstrated their effectiveness in indirectly detecting soil salinity. These indices utilize markers such as vegetation health and halophytic plant identification, as noted by [39–42]. These indices provide a strong relationship with the EC of the soil, making them valuable in identifying areas damaged by salt.

Salinization impacts the district of kot Addu heavily and it seemed crucial to use satellite data for monitoring soil salinity in the district. Consequent to that, the district has been marked as a model with an objective of mapping soil salinity in the region. The approach may not only protect the land of agriculture rather to save the additional areas at risk through the proposed approach. The monitoring and mapping process aims to reduce the risk of soil erosion by promptly alerting stakeholders about soil salinity levels. By utilizing various spectral indices, the study seeks to identify salt-affected areas and evaluate the effectiveness of these indices in this specific context.

The objectives of this research are:

(i)　To determine how accurate the soil salinity can be predicted from the Landsat 8 OLI sensor and field measurements by using ML regression models (Random Forest Regressor (RFR), AdaBoost Regressor (ABR), Decision Tree Regressor(DTR), Ridge Regressor (RR) and Partial Least Squares Regression (PLSR)).

(ii)　To determine the optimal spectral indices for the purpose of soil salinity mapping.

(iii)　To identify the most efficient machine learning model to determine the salinity of soil by Landsat 8 OLI imagery.

2. Material and Method

2.1. Study Area

Kot Addu city is situated in the center of Pakistan, in the southern region of the Punjab Province, in the District of Muzaffargarh Figure 1. Sugarcane, wheat and cotton are the principal crops farmed in the alluvial plain that surrounds the city; rice, maize, mash, ground nuts, bajra and oil seeds (rapeseed) are cultivated in a very small ratio. Frequently, certain areas remain flooded. Mangoes, citrus, dates and pomegranates are the principal fruit trees grown; however, many citrus and mango farms also have a small amount of space for pears, dates and bananas [43]. Most of the agricultural lands in this region are mildly to moderately salinized. Kot Addu experiences mild winter and scorching summer seasons due to its desert climate. The city has encountered some of the most severe climate conditions in Pakistan. The peak temperature ever noted was roughly 51 °C (324.15 °K), and the lowest temperature ever noted was roughly −1 °C (272.15 °K). About 127 mm of rain falls every year on average (5.0 in.). Such a climate exacerbates the salinity issue; therefore, research is being conducted to track the salinity issue in this region.

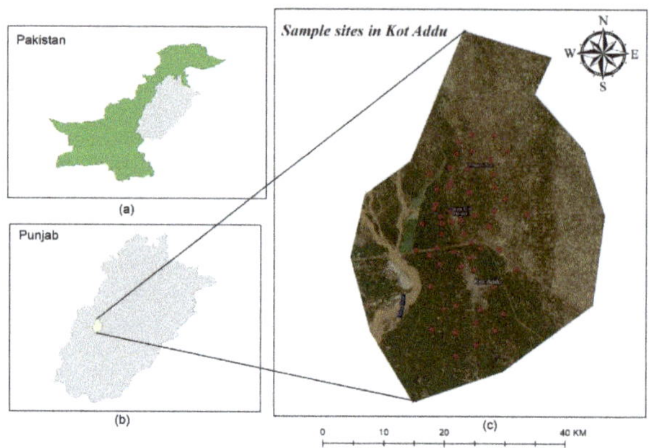

Figure 1. Site map of Kot Addu.

2.2. Satellite Data

The Landsat 8 data used in this study was obtained from January 10 to 30, 2017. Images with a cloud cover of less than 5 percent were exclusively chosen. Spectral indices were calculated utilizing spectral band values, as shown in Table 1. Band 8 panchromatic was excluded due to its proximity and susceptibility to cloud interference.

Table 1. Spectral indices for current study.

Spectral Indices	Expression	Reference
Normalized Difference Vegetation Index (NDVI)	$\frac{(NIR-R)}{(NIR+R)}$	[44]
Normalized Difference Salinity Index (NDSI)	$\frac{(R-NIR)}{(R+NIR)}$	[45]
Soil Adjusted Vegetation Index (SAVI)	$\frac{(NIR-R)}{(NIR+R)+L*(1-L)}$	[46]
Simple Ratio (SR)	$\frac{(R-NIR)}{(G+NIR)}$	[47]
Differential Vegetation Index (DVI)	$(NIR-R)$	[48]
Ratio Spectral Index (RSI)	$\frac{R}{NIR}$	[49]
Mosaic Simple Ratio (MSR)	$\frac{NIR}{R}$	[50]
Vegetation Soil Salinity Index (VSSI)	$2*G-5*(R+NIR)G-5*(R+NIR)$	[22]
Salinity Index (SI1)	$\frac{(B)}{(R)}$	[51]
Salinity Index (SI2)	$\frac{(B-R)}{B+R}$	[52]
Salinity Index (SI3)	$\frac{(G*R)}{B}$	[52]
Salinity Index (SI4)	$\frac{NIR*R}{G}$	[52]

2.3. Soil Sample

To gather soil samples, a survey study was carried out by the Soil and Environmental Sciences Institute at Agriculture University Faisalabad in the Kot Addu region of the

Muzaffargarh District in January 2017 [43]. The main objective was to monitor the salt status across the expansive 32,457-acre study area. A total of fifty-five soil samples were randomly collected from the top 15 cm of the soil surface using an auger. Geographical coordinates for each sample location were recorded using a GPS device. The collected soil samples were properly labeled, placed in polybags, and sent to the Soil and Water Testing Laboratory at the University of Agriculture, Faisalabad, for further analysis [43]. The soil specimens underwent grinding, air drying and filtration through a 2 mm sieve. The sieved soil was then transformed into a soil-saturated paste using purified water and left to rest overnight. The concentrated solution extracted from this paste was measured for electrical conductivity (EC) using a conductivity meter, to determine the soil EC [43].

2.4. Spectral Indices

Using Arc Map 10.3, twelve spectral indices were calculated from the preprocessed data based on the literature review [22,44–52]. The selection of these indices for the study was facilitated by their potential link to the detection of salinity.

The formulas for spectral indices are listed in Table 1.

After computing the spectral indices and forming the dataset, the subsequent step involved applying the standard scaler technique. This was carried out to ensure uniform contributions from the features before proceeding with the training of the machine learning models. The objective is to comprehend the association among satellite data (spectral indices) and the salt content of the soil specimens.

The methodology of the investigation represent in Figure 2.

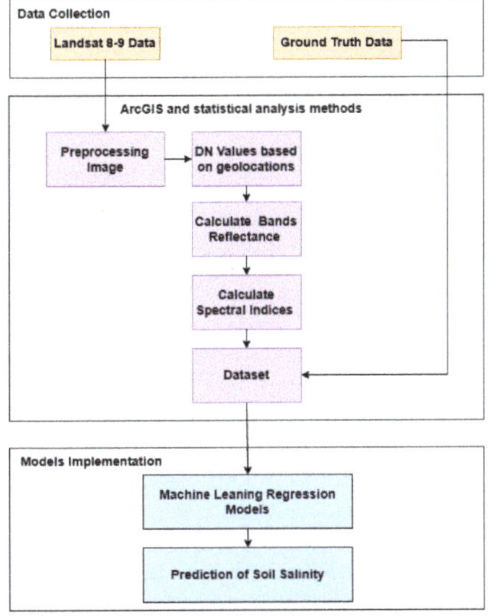

Figure 2. Methodology of current study.

2.5. Random Forest Regressor

Breiman [53] and Cutler and Stevens [54] proposed an RF algorithm that works on the basis of several decision trees which are not correlated with each other [55]. It works as a classifier if labels are given, and like a regressor for continuous or numeric values. Random samples are selected from the calibration set to build the decision tree. For the complete division of variable space, random selections are made from n inputs to split each node in the decision tree. The average result of all decision trees is the final value of the RF model. Significant attention needs to be given to the tuning of the RF model for prediction, specifically in regards to the count of decision trees (ntree), the count of randomly sampled variables as candidates for each split (mtree) and the least count of specimens essential for a node to be considered a leaf (nodesize). The RF model is more stable, having a higher value of ntree. The nodesize value is determined by repeated testing (nodesize = 1). The The Residual Sum of Square (RSS) formula is used in the RFR model for regression. In this study, the RFR model is implemented by using the "RandomForestRegressor" from sklearn in jupyter notebook.

$$RSS = \sum_{left} (y_i - y_L^*)^2 + \sum_{right} (y_i - y_R^*)^2 \tag{1}$$

where y_L^* is the mean y value for the left node and y_R^* is the mean y value for the right node.

2.6. AdaBoost Regressor

AdaBoost, known as Adaptive Boosting, is an ensemble learning technique utilized for classification and regression tasks. For regression problems, it is referred to as AdaBoost Regressor. This widely used machine learning algorithm blends the predictions of several weak learners, typically shallow decision trees, to construct a robust ensemble model. The fundamental concept behind the ABR involves iteratively training a sequence of weak learners, with each subsequent learner emphasizing examples that previous ones struggled to predict accurately. This adaptive nature enables the model to enhance its performance through successive iterations.

2.7. Decision Tree Regressor

The Decision Tree Regressor functions as a supervised machine learning technique utilized for regression purposes. Unlike classification challenges that seek to predict categorical labels, regression tasks focus on forecasting continuous numerical values. A decision tree takes the form of a tree-like model in which internal nodes indicate decisions based on specific features, while leaf nodes correspond to the predicted output values. In the case of the DTR, the value at each leaf node is derived from the average (or another measure) of the target values from the training samples that lead to that particular leaf.

2.8. Ridge Regressor

The Ridge Regressor is a supervised linear regression algorithm, commonly employed in machine learning tasks, particularly when dealing with multicollinearity among predictor variables. It is a modified version of standard linear regression that incorporates a penalty term in the cost function to prevent overfitting and enhance generalization. In ordinary linear regression, the objective is to determine coefficients for predictor variables that minimize the sum of squared differences between predicted and actual target values. However, when predictor variables are highly correlated, the coefficients might grow large, leading to heightened sensitivity to noise and potential overfitting. To tackle this issue, the RR introduces an L2 regularization term to the linear regression cost function. This regularization term imposes a penalty based on the squared magnitudes of the coefficients. By applying this penalty, the RR encourages the model not only to fit the data but also to maintain small coefficients, effectively reducing the influence of individual predictors.

2.9. Partial Least Squares Regression

Partial Least Squares Regression is a statistical technique utilized to model the association between a group of independent variables (X) and a dependent variable (Y). It proves particularly valuable when working with datasets that exhibit high dimensionality or when there is a possibility of multicollinearity among the predictor variables. The primary objective of PLSR is to discover a reduced representation of both the independent and dependent variables by forming new latent variables (also called components) as linear combinations of the original variables. These components are crafted in a manner that maximizes the covariance between the independent and dependent variables within each subsequent component.

2.10. Evaluation

The regression algorithms received spectral indices as input, with the observed soil salinity serving as the target value. Overall, 55 soil specimens were gathered between January 20 and 27 January 2017. The dataset was then distributed into training and testing sets, with a proportion of 80% for calibration and 20% for validation. Out of the 55 samples, 45 were utilized for training the model, while the remaining 10 were used for testing. Various statistical parameters, including the R^2, MAE, MSE, and RMSE, were employed to assess the proficiency of the ML techniques.

$$R^2 = 1 - \frac{\sum_{k=1}^{n}(y_k - \hat{y}_k)^2}{\sum_{k=1}^{n}(y_k - \bar{y}_k)^2} \tag{2}$$

$$MAE = \frac{1}{n}\sum_{k=1}^{n}|y_k - \hat{y}_k| \tag{3}$$

$$MSE = \frac{1}{n}\sum_{k=1}^{n}(y_k - \hat{y}_k)^2 \tag{4}$$

$$RMSE = \sqrt{\frac{1}{n}\sum_{k=1}^{n}(y_k - \hat{y}_k)^2} \tag{5}$$

Here, y_k is the kth observed salinity, $\hat{y}_k$ is the kth predicted salinity, $\bar{y}_k$ is the mean salinity of all the soil samples, and n is the total count of specimens.

3. Results

3.1. Models' Performance

This study predicted the salinity of soil using spectral indices, and various statistical parameters, including the R^2, MAE, MSE and RMSE, were employed to assess the performance of ML models. Among these parameters, R^2 played a significant role in evaluating the model's performance. It is a measure of how well the model has learned the data, and a higher R^2 value reveals a better fit of the model to the data. Twelve different spectral indices were used in the study, and their R^2 values were compared to determine which index is more efficient in mapping the salinity of soil. In Figure 3, a comprehensive analysis of R^2 values across a variety of spectral indices, including the NDVI, SAVI, MSR, NDSI, SR, DVI and RSI, is presented. The evaluation encompassed the utilization of multiple models, specifically the RFR, ADR, DTR, RR and PLSR. After a thorough analysis of the results, it becomes clear that the RFR model exhibited exceptional performance, particularly in relation to the DVI and SAVI spectral indices. Impressively, it achieved noteworthy R^2 values of 0.94 and 0.93 for these indices, respectively. In contrast, the implementation of the ADR model led to a distinct decline in R^2 values for these same indices, producing R^2 values of 0.59 and 0.56 for DVI and SAVI, respectively. Furthermore, the RFR model displayed superior R^2 values across several other spectral indices, including NDVI, MSR, NDSI, SR, and RSI, when compared to the ADR model. Conversely, the ADR model presented lower R^2 values for certain spectral indices like DTR, RR, and PLSR, when compared to the RFR model. As a result, the RFR model stands as a more suitable choice for accurately mapping soil salinity in this study. For a detailed account of the R^2 values for different spectral indices, refer to Figure 3.

The appropriate selection of spectral indices is crucial when mapping soil salinity using satellite data. Spectral indices play a fundamental role as input features for ML models in remote sensing applications. In the context of this study, the spectral indices VSSI, SI1, SI2, SI3 and SI4 displayed lower R^2 values, as depicted in Figure 4. These lower R^2 values indicate that, when utilizing these spectral indices as input features for the ML models, the models performed poorly, and their predictions did not closely match the actual soil salinity levels.

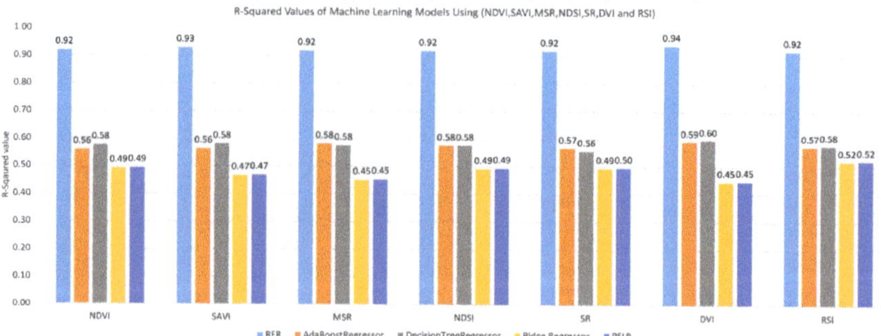

Figure 3. Comparison of R^2 of machine learning models using spectral indices NDVI, SAVI, MSR, NDSI, SR, DVI and RSI.

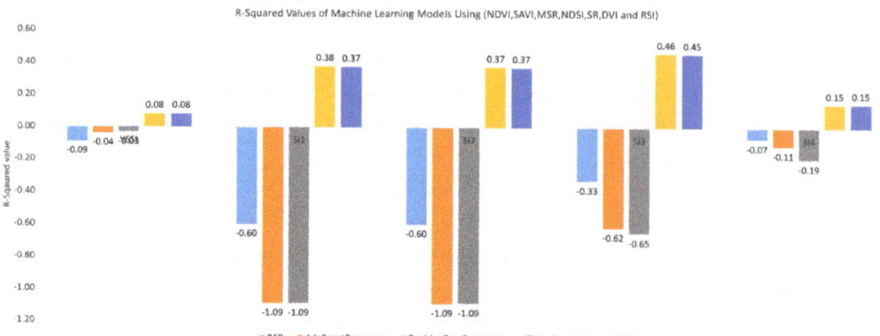

Figure 4. Comparison of R^2 of machine learning models using spectral indices VSSI, SI1, SI2, SI3 and SI4.

3.2. Model Calibration and Validation

For calibrating ML models using Landsat 8 OLI data, a total of twelve spectral indices were employed. The models were trained using 80% of the available samples, while the remaining 20% was used for testing. The implementation was carried out in Python, and four statistical methods (R^2, MAE, MSE and RMSE) were utilized to assess the models' performance. The results presented in Figure 3 indicate that certain spectral indices, namely DVI, SAVI, NDVI, MSR, NDSI, SR and RSI, achieved high R^2 values, reflecting a strong correlation between the predictions made by the ML models and the actual soil salinity levels. On the contrary, Figure 5 reveals that specific spectral indices, such as VSSI, S1, S2, S3 and S4, produced very low R^2 values. These findings indicate that the ML models utilizing these indices were not effective in accurately predicting soil salinity levels, as their predictions did not closely match the actual data. Conversely, the exceptional performance of the RFR technique is particularly noteworthy, especially when integrating the DVI spectral index. The RFR model accomplished a significant R^2 value of 0.94, signifying a robust correlation between its predictions and the observed soil salinity levels. Furthermore, the model exhibited minimal errors with an MAE of 1.42, MSE of 3.58, and RMSE of 1.89. Collectively, these metrics emphasize the RFR model's exceptional capability to precisely forecast soil salinity levels. Contrasting the error rates across various models using the DVI spectral index further highlights the superiority of the RFR model. The RFR model outperforms other models, illustrating notably reduced error rates in its soil salinity predictions, particularly when making use of the DVI spectral index. Remarkably, the RFR model attains an MAE of 1.42, an MSE of 3.58, and an RMSE of 1.89. In contrast, the ADR model shows higher error rates, presenting an MAE of 2.89, MSE of 22.70, and RMSE of 4.76. Similarly, the DTR and RR models demonstrate elevated errors, with MAE values of 2.75 and 4.32, along with corresponding MSE values of 22.40 and 30.78, respectively. Furthermore, the PLSR model exhibits comparable error rates to the RR model, featuring an MAE of 4.32, MSE of 30.67, and RMSE of 5.54. The significant difference in error rates between the RFR model and the alternative models highlights the RFR model's superiority in precisely forecasting soil salinity, particularly when utilizing the DVI spectral index as an incorporated feature.

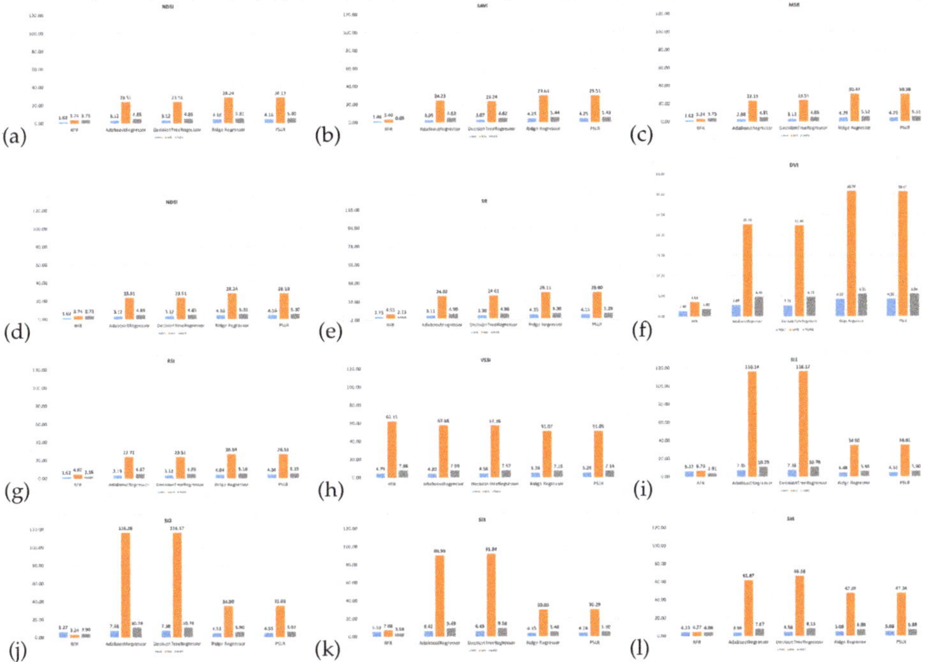

Figure 5. Errors comparison of ML Models (**a**) NDVI, (**b**) SAVI, (**c**) MSR, (**d**) NDSI, (**e**) SR, (**f**) DVI, (**g**) RSI, (**h**) VSSI, (**i**) SI1, (**j**) SI2, (**k**) SI3, (**l**) SI4.

The line graphs depicting the RFR (Figure 6), ABR (Figure 7), DTR (Figure 8), RR (Figure 9), and PLSR (Figure 10) illustrate the visualization of the observed and estimated soil salinity values. On the x-axis, sample numbers are represented, while the y-axis displays both actual and predicted EC values, distinguished by different colors. Within the used spectral indices, encompassing DVI, SAVI, NDVI, MSR, NDSI, SR, and RSI, a close alignment is observed between the lines representing actual and predicted values in the line graphs. This close alignment indicates that these spectral indices yield better performance when used with the RFR model, resulting in more accurate predictions of soil salinity. On the other hand, spectral indices VSSI, SI1, SI2, SI3 and SI4 exhibit larger discrepancies between the actual and predicted lines in the line graphs. This observation suggests that these spectral indices are associated with higher error rates when analyzed with the RFR model, indicating that they may not be as effective in accurately predicting soil salinity. Overall, the visual analysis of the line graphs supports the finding that the RFR model, especially when utilizing distinct spectral indices, outperforms other ML models and offers more reliable predictions of soil salinity levels.

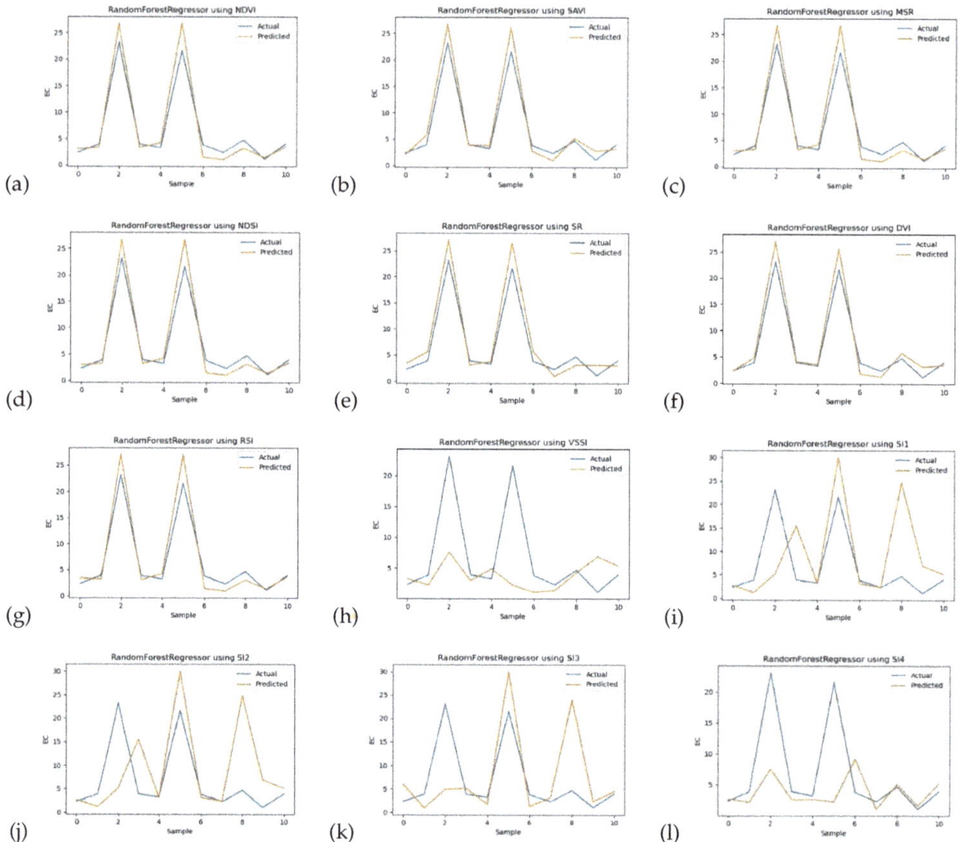

Figure 6. Line graphs of EC values using Random Forest Regressor (**a**) NDVI, (**b**) SAVI, (**c**) MSR, (**d**) NDSI, (**e**) SR, (**f**) DVI, (**g**) RSI, (**h**) VSSI, (**i**) SI1, (**j**) SI2, (**k**) SI3, (**l**) SI4.

In Figure 6a–g, the predicted and actual lines using the RFR model are positioned closely together, indicating a strong alignment and accurate predictions. This finding suggests that certain spectral indices utilized in this section had lower error rates and resulted in the best model fit for predicting the salinity of soil. The close proximity of the lines in these graphs signifies that the model's predictions closely match the actual soil salinity values, leading to more reliable results. However, a different scenario is observed in Figure 6i–l where spectral indices VSSI, SI1, SI2, SI3 and SI4 are used. In these graphs, the predicted and actual lines are more dispersed, indicating higher error rates and less accurate predictions by the RFR model. This outcome implies that, when mapping soil salinity, these specific spectral indices did not perform well in capturing the true salinity variations in the soil, resulting in less reliable predictions.

In Figure 7a–g, the predicted and observed lines obtained through the ABR model align closely when utilizing spectral indices such as the NDVI, SAVI, MSR, NDSI, SR, DVI and RSI. This alignment implies that these spectral indices demonstrate reduced error rates and yield more precise predictions for soil salinity mapping. The proximity of the lines in these graphs signifies the strong performance of the ABR model with these specific spectral indices, effectively capturing authentic salinity fluctuations within the soil and yielding dependable forecasts. Conversely, in Figure 7i–l, where the spectral indices VSSI, SI1, SI2, SI3 and SI4 are employed, the predicted and observed lines exhibit greater dispersion.

This observation suggests elevated error rates associated with these spectral indices when employed for soil salinity mapping. The wider distribution of lines indicates that the ABR model's efficacy with these indices is diminished, resulting in less precise predictions and increased uncertainties.

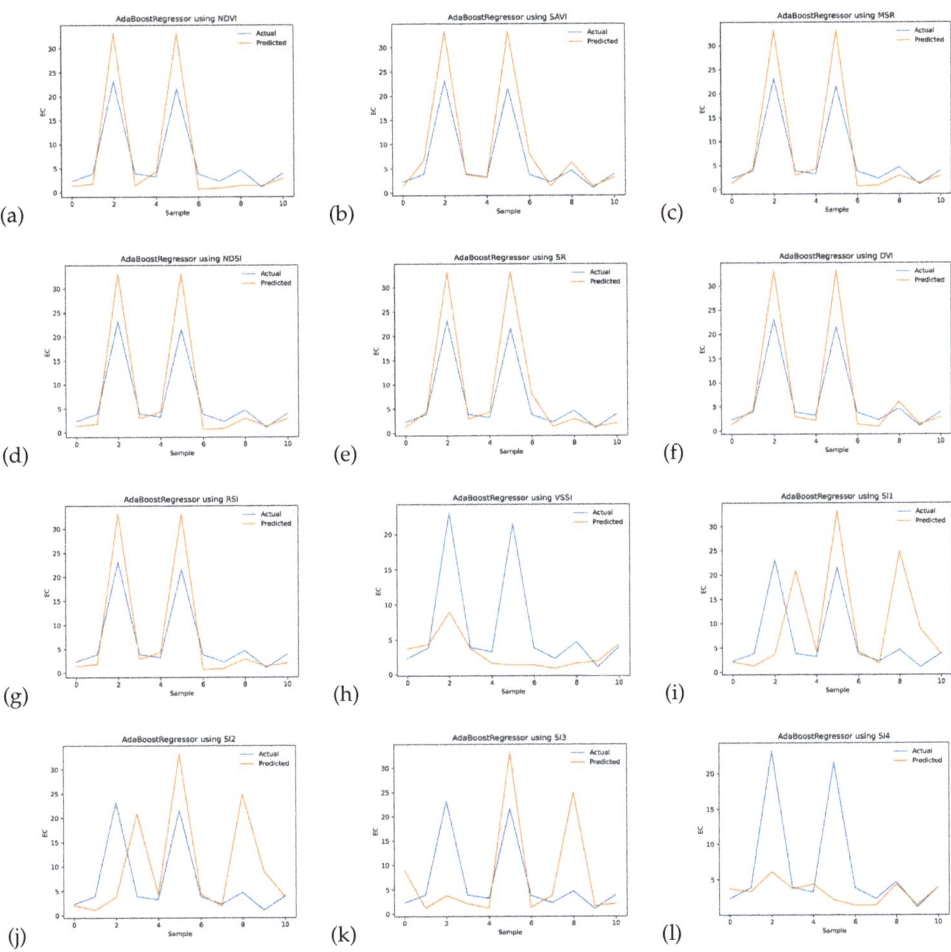

Figure 7. Line graphs of EC values using AdaBoost Regressor (**a**) NDVI, (**b**) SAVI, (**c**) MSR, (**d**) NDSI, (**e**) SR, (**f**) DVI, (**g**) RSI, (**h**) VSSI, (**i**) SI1, (**j**) SI2, (**k**) SI3, (**l**) SI4.

In Figure 8a–g, the predicted and actual lines using the DTR (Decision Tree Regression) model are not as closely aligned as those in Figure 6a–g, which indicates that the DTR model has higher error rates compared to the RFR (Random Forest Regression) model when predicting soil salinity. The spread between the predicted and actual lines in Figure 8a–g suggests that the DTR model did not perform as well as the RFR model, leading to less accurate predictions and higher uncertainties. However, when comparing the DTR model (Figure 8) to the RR model (Figure 9) and the PLSR model (Figure 10), the predicted and actual lines are closer. This observation indicates that the DTR model has lower error rates compared to the RR and PLSR models when using the same spectral indices for predicting soil salinity.

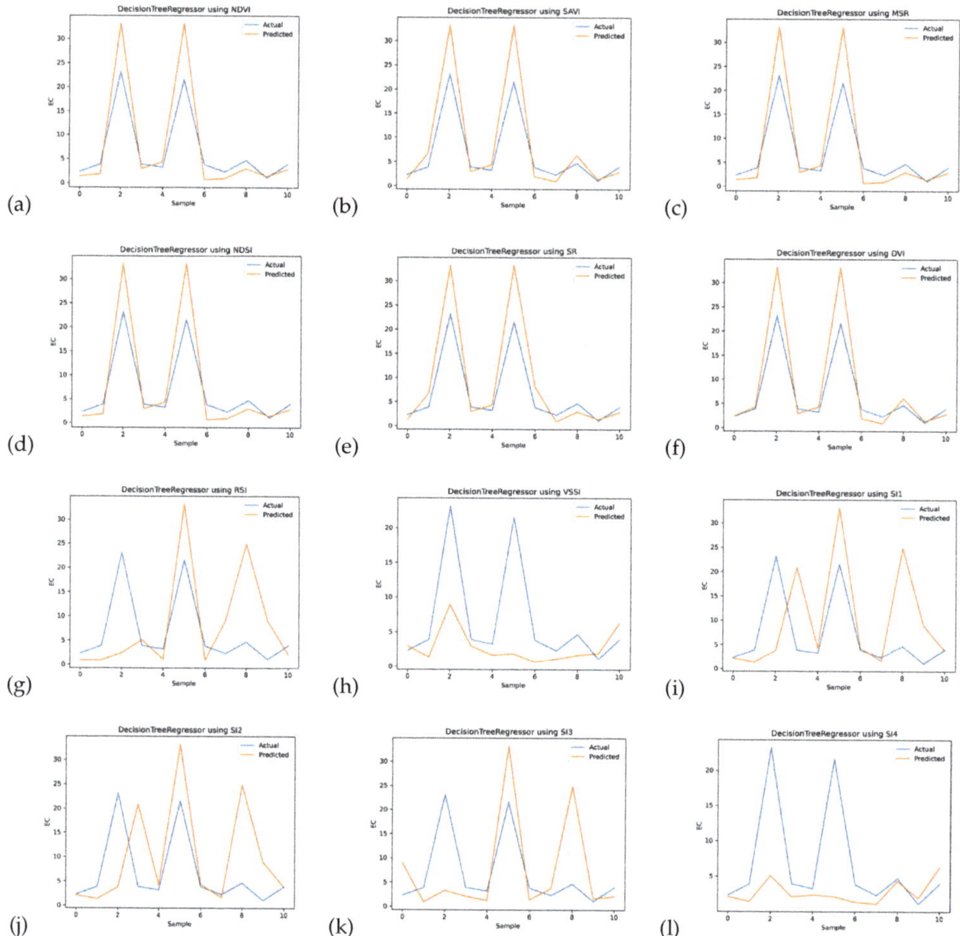

Figure 8. Line graphs of EC values using Decision Tree Regressor (**a**) NDVI, (**b**) SAVI, (**c**) MSR, (**d**) NDSI, (**e**) SR, (**f**) DVI, (**g**) RSI, (**h**) VSSI, (**i**) SI1, (**j**) SI2, (**k**) SI3, (**l**) SI4.

When comparing the predicted and actual lines of the PLSR model (Figure 10) to those of the RFR (Figure 6), ABR (Figure 7) and DTR (Figure 8) models, the gap is more pronounced. This visual analysis underscores that the PLSR model exhibits the highest error rate among the models under consideration. The comparative evaluation of line graphs underscores that the PLSR model is characterized by elevated error rates and less accurate predictions compared to the other models.

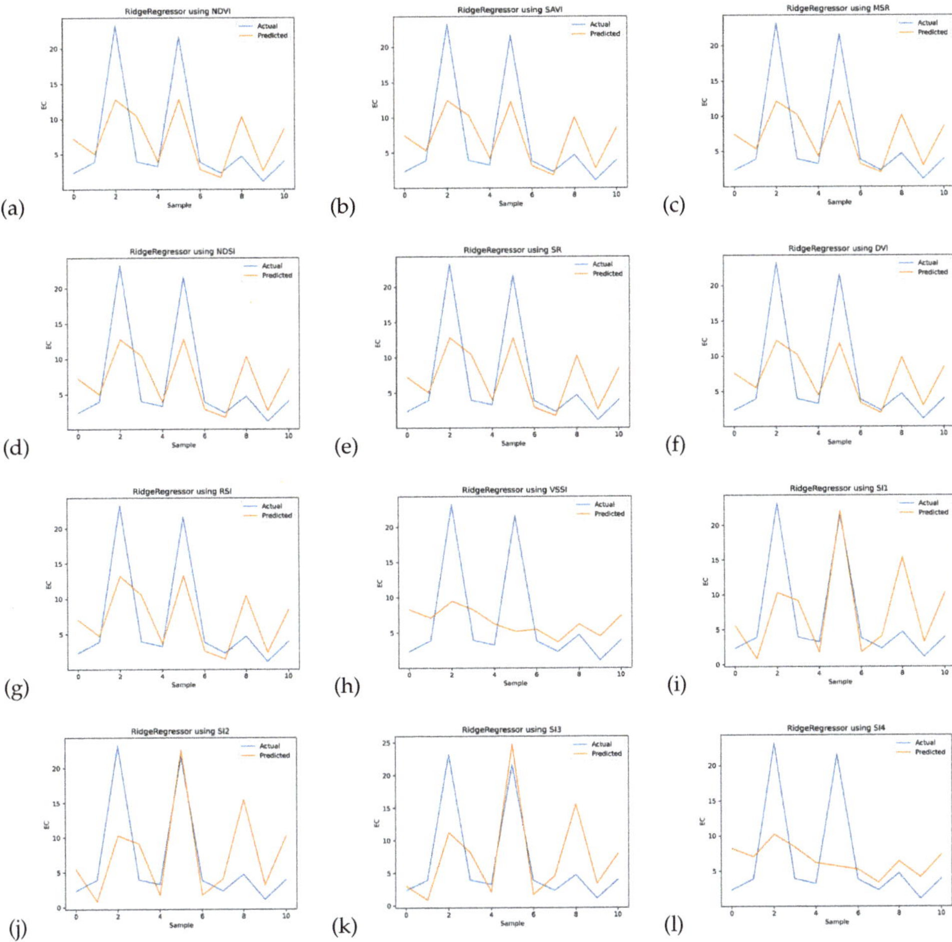

Figure 9. Line graphs of EC values using Ridge Regressor (**a**) NDVI, (**b**) SAVI, (**c**) MSR, (**d**) NDSI, (**e**) SR, (**f**) DVI, (**g**) RSI, (**h**) VSSI, (**i**) SI1, (**j**) SI2, (**k**) SI3, (**l**) SI4.

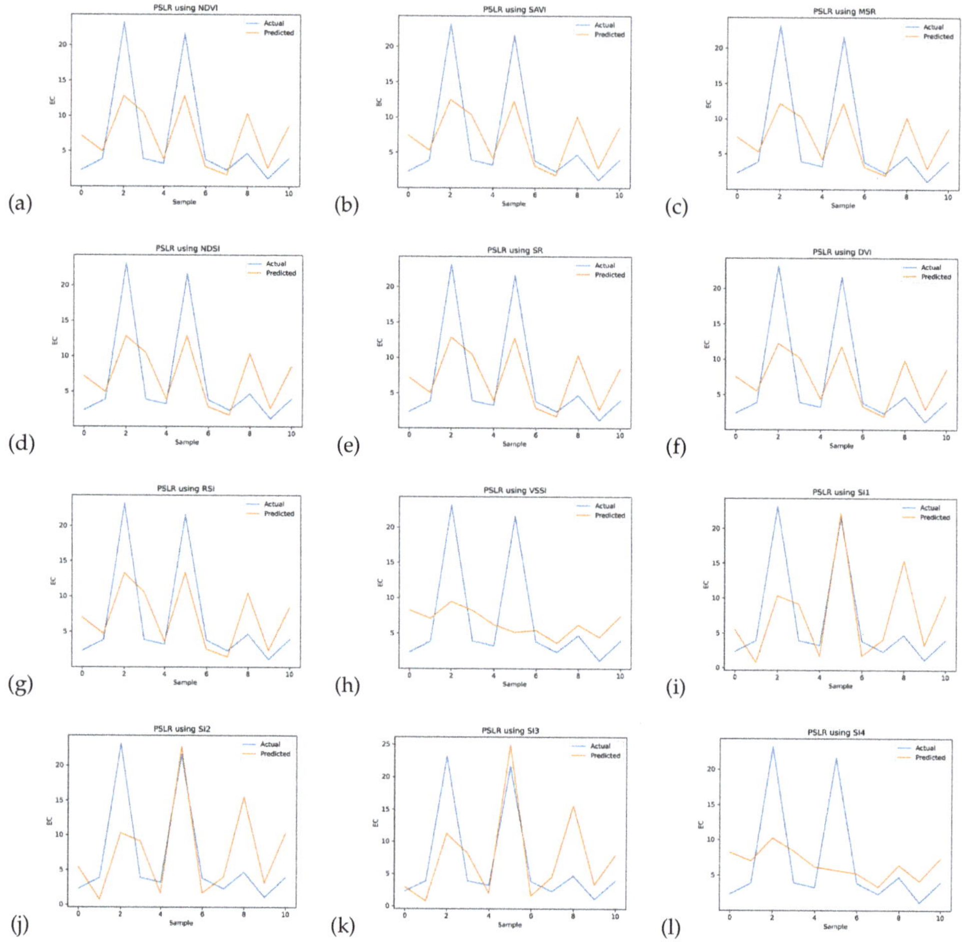

Figure 10. Line graphs of EC values using Partial Least Square Regressor (**a**) NDVI, (**b**) SAVI, (**c**) MSR, (**d**) NDSI, (**e**) SR, (**f**) DVI, (**g**) RSI, (**h**) VSSI, (**i**) SI1, (**j**) SI2, (**k**) SI3, (**l**) SI4.

4. Discussion

The first research objective was to assess the accuracy of predicting soil salinity using Landsat 8 OLI data with the aid of ML models. We successfully achieved soil salinity predictions by employing these ML models in conjunction with various spectral indices. The RFR model performed exceptionally well with an R^2 of 0.93 (Figure 3), MAE of 1.46, MSE of 3.90 and RMSE of 1.97 when using the SAVI spectral index as shown in (Figure 5). However, upon further analysis, we found that the RFR model provided even more precise predictions with an R^2 of 0.94 (Figure 3), MAE of 1.42, MSE of 3.58 and RMSE of 1.89 when utilizing the DVI spectral index (Figure 5). The predicted and actual lines using the RFR model were closely aligned in Figure 6f, indicating the model's strong performance in accurately predicting soil salinity. Conversely, when using the SAVI spectral index, the ADT model exhibited higher error rates with an R^2 of 0.56, MAE of 3.03, MSE of 24.23 and RMSE of 4.92. However, the ADT model showed improved predictions with an R^2 of 0.59, MAE of 2.89, MSE of 22.70 and RMSE of 4.76 when utilizing the DVI spectral index. Similarly, the DTR model revealed increased error rates, presenting an R^2 value of 0.58, a MAE of 3.07, an MSE of 23.24 and an RMSE of 4.82 when making use of the SAVI spectral index.

On the other hand, the DTR model illustrated enhanced performance, achieving an R^2 of 0.59, a MAE of 2.75, an MSE of 22.40 and an RMSE of 4.73 through the utilization of the DVI spectral index. In contrast, the RR model demonstrated diminished R^2 values and elevated error rates, indicating a suboptimal fit when employing the SAVI spectral index, resulting in an R^2 of 0.47, a MAE of 4.25, an MSE of 29.63 and an RMSE of 5.44. A slight improvement in results was noted as the RR model integrated the DVI spectral index, resulting in an R^2 value of 0.47, a MAE of 4.25, an MSE of 29.63, and an RMSE of 5.44. Similarly, the utilization of the SAVI spectral index by the PLSR model led to higher error rates, yielding an R^2 of 0.47, a MAE of 4.25, an MSE of 29.51, and an RMSE of 5.43. A slight increase in error rates was observed when the PLSR model applied the DVI spectral index, resulting in an R^2 of 0.47, an MAE of 4.32, an MSE of 30.67 and an RMSE of 5.54.

Conversely, when utilizing the VSSI, SI1, SI2, SI3, and SI4 spectral indices for soil salinity prediction, the ML models exhibited high error rates and notably low R^2 values. These outcomes clearly indicated that these specific spectral indices were not well-suited for the models and they had significant errors in predicting the actual salinity values, as depicted in Figure 4. The large discrepancies between the forecast and real salinity values for these spectral indices suggest that they may not capture the essential information needed for accurate soil salinity predictions, highlighting the importance of careful selection of appropriate spectral indices for improving model performance. The RFR model outperformed the ADR, DTR, RR and PLSR models due to its implementation of random sampling [55], superior fitting on small datasets [56] and consequent enhancement in decision-making accuracy [54]. Overall, the RFR model outperformed the other models in predicting soil salinity, especially when using the DVI spectral index. The selection of appropriate spectral indices is crucial in optimizing the performance of the ML models for accurate soil salinity predictions.

The second research objective aimed to identify the most efficient spectral indices for mapping the salinity of soil using Landsat 8 OLI data. For this purpose, twelve spectral indices were employed for mapping the salinity of soil. Higher R^2 values and minimum error rates were found during the prediction of salinity of soil by seven specific spectral indices (DVI, SAVI, NDVI, NDSI, MSR, SR, and RSI). Likewise, notable correlation between NDSI, NDVI, and SAVI indices to the level of salinity of soil was revealed in preliminary analysis [38]. More importantly, a unique demonstration of performance was made by DVI index with an impressive R^2 value of 0.94. Moreover, it demonstrated outstanding accuracy in predicting the salinity of soil with the minimal values of MAE of 1.42, an MSE of 3.58, and an RMSE of 1.89 (Figure 6f). This underscores its strong alignment with the RFR model and its significant contribution to precise soil salinity predictions. The findings indicated that spectral indices utilizing the NIR band [37] displayed greater accuracy in predicting soil salinity compared to those relying on visible bands. In contrast, the VSSI, SI1, SI2, SI3 and SI4 indices exhibited lower R^2 values (Figure 4) and higher error rates (Figure 4), indicating their reduced effectiveness in accurately estimating soil salinity.

The third research objective was to identify the most efficient ML model for determining the salinity of soil using Landsat 8 OLI data. The performance of five ML techniques were compared based on R^2, MAE, MSE and RMSE metrics. The results clearly demonstrated that the RFR model, specifically using the DVI index, outperformed the other ML models, including the ABR, DTR, RR and PLSR, when using the selected spectral indices (excluding VSSI, SI1, SI2, SI3 and SI4) from Landsat 8 OLI data to forecast the salinity of soil. The RFR model achieved an impressive R^2 of 0.94, indicating a strong correlation between predicted and observed soil salinity values, and exhibited minimum error rates as depicted in Figure 3. When comparing the predicted and actual salinity lines, the RFR model (Figure 6) demonstrated a closer alignment, highlighting its superior performance compared to the ABR (Figure 7), DTR (Figure 8), RR (Figure 9) and PLSR (Figure 10) models. This indicates that the RFR model provided more accurate predictions and better captured the variations in soil salinity, making it the most efficient ML model for this specific task [54,55]. These findings emphasize the significance of selecting an appropriate

ML algorithm and spectral indices when mapping soil salinity using satellite data, as it directly impacts the accuracy and reliability of the predictions.

This study makes a valuable addition to the application of ML models for the mapping of salinity of soil based on spectral indices, because we have successfully predicted the salinity of soil by Landsat 8 imagery with the highest 0.94 of R^2. We can determine the soil salinity of barren land by just entering the geo-coordinates. It can be useful to predict the soil salinity of barren land in Pakistan and the global region. If the soil salinity of the land is determined in time, then we can take necessary steps to decrease the soil erosion. The proper soil salinity value (EC < 4) has a significant impact on the growth of the plants. The soil salinity level is an important factor in deciding the suitable crop for cultivation on a specific plot of land. So, this study helps to improve agriculture management practices.

5. Conclusions

The salinity of soil is a significant global issue, notably in arid and semiarid regions, and it poses detrimental effects on food security worldwide. It deteriorates the soil environment and has adverse impacts on climate, hydrology, agriculture, geochemistry and the economy. Regular monitoring of the scale and intensity of soil salinity is crucial to mitigate these environmental risks. Hence, remote sensing presents a suitable option for remotely determining the salinity of soil in specific regions. The current study demonstrates the effectiveness of Landsat 8 OLI data, specifically spectral indices, in characterizing and evaluating soil salinity. Satellite data reveal that saline soils exhibit higher reflectance in the visible, NIR and SWIR spectra compared to regular soils. While salinity indices SI1, SI2, SI3 and SI4 are not useful in vegetated regions, they can be employed for comprehensive observations and the assessment of saline soils. Among the five ML models utilized for estimating the salinity of soil derived from spectral indices, the RFR model demonstrated the best performance in evaluating salinity in study area with an R^2 0.94, MAE 1.42, MSE 3.58 and RMSE 1.89 as compared to ADB, DTR, RR and PLSR. Among the twelve spectral indices utilized, DVI, SAVI, NDVI, NDSI, MSR, SR and RSI exhibited significant correlations with soil salinity. These indices are recommended to spatially map the salinity of soil. On the other hand, VSSI, SI1, SI2, SI3 and SI4 had unclear correlations with the salinity of soil. Based on the results, it is possible to develop a web-based application that allows users to map soil salinity by entering the geocoordinates. Such an application would be beneficial for farmers and agricultural management, enabling them to make informed decisions regarding crop selection to mitigate monetary losses caused by climate change. This method offers a cost-effective and efficient approach for detecting soil salinity in specific regions. Furthermore, since Landsat 8 OLI data are freely available, the results of this study can be optimized by increasing the dataset and applying deep learning models.

Author Contributions: Conceptualization, Y.u.H. and M.S.; methodology, Y.u.H.; software, Y.u.H.; validation, Y.u.H., M.S. and H.M.S.A.; formal analysis, Y.u.H., M.S., H.M.S.A. and A.A.-L.; investigation, Y.u.H., M.S.; resources, Y.u.H., M.S., H.M.S.A., A.A.-L. and W.H.A.; data curation, Y.u.H.; writing—original draft preparation, Y.u.H.; writing—review and editing, Y.u.H., M.S., H.M.S.A., A.A.-L. and W.H.A.; visualization, Y.u.H., M.S. and H.M.S.A.; supervision, M.S. and H.M.S.A., A.A.-L. and W.H.A.; project administration, M.S. and H.M.S.A.; funding acquisition, W.H.A. All authors have read and agreed to the published version of the manuscript.

Funding: The research was sponsored by NRPU Project No. 17006, granted by the Higher Education Commission of Pakistan.

Institutional Review Board Statement: Not applicable.

Informed Consent Statement: Not applicable.

Data Availability Statement: Upon reasonable request, the underlying data for this article will be made available by the corresponding author.

Conflicts of Interest: The authors declare no conflict of interest.

References

1. Lhissoui, R.; El Harti, A.; Chokmani, K. Mapping soil salinity in irrigated land using optical remote sensing data. *Eurasian J. Soil Sci.* **2014**, *3*, 82–88.
2. Asfaw, E.; Suryabhagavan, K.; Argaw, M. Soil salinity modeling and mapping using remote sensing and GIS: The case of Wonji sugar cane irrigation farm, Ethiopia. *J. Saudi Soc. Agric. Sci.* **2018**, *17*, 250–258. [CrossRef]
3. FAO. Extent and Causes of Salt Affected Soils in Participating Countries. Global Network on Integrated Soil Management for Sustainable Use of Saltaffected Soils. 2000. Available online: http://www.fao.org/ag/agl/agll/spush/topic2.htm (accessed on 22 May 2023).
4. Zhu, J.K. Plant salt tolerance. *Trends Plant Sci.* **2001**, *6*, 66–71.
5. Metternicht, G.I.; Zinck, J. Remote sensing of soil salinity: Potentials and constraints. *Remote Sens. Environ.* **2003**, *85*, 1–20.
6. Zheng, Z.; Zhang, F.; Ma, F.; Chai, X.; Zhu, Z.; Shi, J.; Zhang, S. Spatiotemporal changes in soil salinity in a drip-irrigated field. *Geoderma* **2009**, *149*, 243–248. [CrossRef]
7. Yao, R.; Yang, J. Quantitative evaluation of soil salinity and its spatial distribution using electromagnetic induction method. *Agric. Water Manag.* **2010**, *97*, 1961–1970. [CrossRef]
8. Scudiero, E.; Corwin, D.; Anderson, R.; Yemoto, K.; Clary, W.; Wang, Z.; Skaggs, T. Remote sensing is a viable tool for mapping soil salinity in agricultural lands. *Calif. Agric.* **2017**, *71*, 231–238. [CrossRef]
9. Bilgili, A.V.; Cullu, M.A.; van Es, H.; Aydemir, A.; Aydemir, S. The use of hyperspectral visible and near infrared reflectance spectroscopy for the characterization of salt-affected soils in the Harran Plain, Turkey. *Arid Land Res. Manag.* **2011**, *25*, 19–37.
10. Brunner, P.; Li, H.; Kinzelbach, W.; Li, W. Generating soil electrical conductivity maps at regional level by integrating measurements on the ground and remote sensing data. *Int. J. Remote Sens.* **2007**, *28*, 3341–3361. [CrossRef]
11. Dehaan, R.; Taylor, G. Image-derived spectral endmembers as indicators of salinisation. *Int. J. Remote Sens.* **2003**, *24*, 775–794. [CrossRef]
12. Nanni, M.R.; Demattê, J.A.M. Spectral reflectance methodology in comparison to traditional soil analysis. *Soil Sci. Soc. Am. J.* **2006**, *70*, 393–407. [CrossRef]
13. Haq, Y.U.; Shahbaz, M.; Asif, H.S.; Al-Laith, A.; Alsabban, W.; Aziz, M.H. Identification of soil type in Pakistan using remote sensing and machine learning. *PeerJ Comput. Sci.* **2022**, *8*, e1109.
14. Ramzan, Z.; Asif, H.S.; Yousuf, I.; Shahbaz, M. A Multimodal Data Fusion and Deep Neural Networks Based Technique for Tea Yield Estimation in Pakistan Using Satellite Imagery. *IEEE Access* **2023**, *11*, 42578–42594. [CrossRef]
15. Farifteh, J.; Farshad, A.; George, R. Assessing salt-affected soils using remote sensing, solute modelling, and geophysics. *Geoderma* **2006**, *130*, 191–206.
16. Liu, Y.; Zhang, F.; Wang, C.; Wu, S.; Liu, J.; Xu, A.; Pan, K.; Pan, X. Estimating the soil salinity over partially vegetated surfaces from multispectral remote sensing image using non-negative matrix factorization. *Geoderma* **2019**, *354*, 113887. [CrossRef]
17. Dale, P.; Hulsman, K.; Chandica, A. Classification of reflectance on colour infrared aerial photographs and sub-tropical salt-marsh vegetation types. *Int. J. Remote Sens.* **1986**, *7*, 1783–1788. [CrossRef]
18. ten Caten, A.; Dalmolin, R.S.D.; de Araujo Pedron, F.; Santos, M.d.L.M. Principal components as predictor variables in digital mapping of soil classes/Componentes principais como preditores no mapeamento digital de classes de solos. *Ciência Rural* **2011**, *41*, 1170–1177. [CrossRef]
19. Sahu, S.; Prasad, M.; Tripathy, B. PCA Classification technique of remote sensing analysis of colour composite image of chillika lagoon, Odisha. *Int. J. Adv. Res. Comput. Sci. Softw. Eng.* **2015**, *5*, 513–518.
20. Li, B.; Ti, C.; Zhao, Y.; Yan, X. Estimating soil moisture with Landsat data and its application in extracting the spatial distribution of winter flooded paddies. *Remote Sens.* **2016**, *8*, 38. [CrossRef]
21. Wu, W. The generalized difference vegetation index (GDVI) for dryland characterization. *Remote Sens.* **2014**, *6*, 1211–1233. [CrossRef]
22. Dehni, A.; Lounis, M. Remote sensing techniques for salt affected soil mapping: Application to the Oran region of Algeria. *Procedia Eng.* **2012**, *33*, 188–198. [CrossRef]
23. van Leeuwen, W.J. Visible, near-IR, and shortwave IR spectral characteristics of terrestrial surfaces. In *The SAGE Handbook of Remote Sensing*; Sage Publications Ltd.: New York, NY, USA, 2009.
24. Aceves, E.Á.; Guevara, H.J.P.; Enríquez, A.C.; Gaxiola, J.; Cervantes, M.; Barrientos, J.H.; Herrera, L.E.; Guevara, V.M.P.; Samuel, C.L. Determining salinity and ion soil using satellite image processing. *Pol. J. Environ. Stud.* **2019**, *28*, 1549–1560. [CrossRef]
25. Ling, C.; Liu, H.; Ju, H.; Zhang, H.; You, J.; Li, W. A study on spectral signature analysis of wetland vegetation based on ground imaging spectrum data. *J. Physics Conf. Ser.* **2017**, *910*, 12045 [CrossRef]
26. Ji, W.; Adamchuk, V.; Chen, S.; Biswas, A.; Leclerc, M.; Rossel, R.V. The use of proximal soil sensor data fusion and digital soil mapping for precision agriculture. *Pedometrics* **2017**, *2017*, 298.
27. Li, Z.; Li, Y.; Xing, A.; Zhuo, Z.; Zhang, S.; Zhang, Y.; Huang, Y. Spatial prediction of soil salinity in a semiarid oasis: Environmental sensitive variable selection and model comparison. *Chin. Geogr. Sci.* **2019**, *29*, 784–797. [CrossRef]

28. Pandey, P.; Singh, S.; Khan, M.S.; Semwal, M. Non-invasive Estimation of Foliar Nitrogen Concentration Using Spectral Characteristics of Menthol Mint (*Mentha arvensis* L.). *Front. Plant Sci.* **2022**, *13*, 680282. [CrossRef]
29. El Hafyani, M.; Essahlaoui, A.; El Baghdadi, M.; Teodoro, A.C.; Mohajane, M.; El Hmaidi, A.; El Ouali, A. Modeling and mapping of soil salinity in Tafilalet plain (Morocco). *Arab. J. Geosci.* **2019**, *12*, 1–7. [CrossRef]
30. Hihi, S.; Rabah, Z.B.; Bouaziz, M.; Chtourou, M.Y.; Bouaziz, S. Prediction of soil salinity using remote sensing tools and linear regression model. *Adv. Remote Sens.* **2019**, *8*, 77–88. [CrossRef]
31. Ivushkin, K.; Bartholomeus, H.; Bregt, A.K.; Pulatov, A. Satellite thermography for soil salinity assessment of cropped areas in Uzbekistan. *Land Degrad. Dev.* **2017**, *28*, 870–877. [CrossRef]
32. Hoa, P.V.; Giang, N.V.; Binh, N.A.; Hai, L.V.H.; Pham, T.D.; Hasanlou, M.; Tien Bui, D. Soil salinity mapping using SAR Sentinel-1 data and advanced machine learning algorithms: A case study at Ben Tre Province of the Mekong River Delta (Vietnam). *Remote Sens.* **2019**, *11*, 128. [CrossRef]
33. Taghadosi, M.M.; Hasanlou, M.; Eftekhari, K. Soil salinity mapping using dual-polarized SAR Sentinel-1 imagery. *Int. J. Remote Sens.* **2019**, *40*, 237–252. [CrossRef]
34. Zurqani, H.; Nwer, B.; Rhoma, A. Assessment of spatial and temporal variations of soil salinity using remote sensing and geographic information system in Libya. In Proceedings of the 1st Annual International Conference on Geological and Earth Sciences, Singapore, 3–4 December 2012; pp. 3–4.
35. Yong-Ling, W.; Peng, G.; Zhi-Liang, Z. A spectral index for estimating soil salinity in the Yellow River Delta Region of China using EO-1 Hyperion data. *Pedosphere* **2010**, *20*, 378–388.
36. Sahbeni, G. Soil salinity mapping using Landsat 8 OLI data and regression modeling in the Great Hungarian Plain. *SN Appl. Sci.* **2021**, *3*, 587. [CrossRef]
37. Al-Ali, Z.; Bannari, A.; Rhinane, H.; El-Battay, A.; Shahid, S.A.; Hameid, N. Validation and comparison of physical models for soil salinity mapping over an arid landscape using spectral reflectance measurements and Landsat-OLI data. *Remote Sens.* **2021**, *13*, 494. [CrossRef]
38. Zhang, X.; Huang, B. Prediction of soil salinity with soil-reflected spectra: A comparison of two regression methods. *Sci. Rep.* **2019**, *9*, 5067. [CrossRef]
39. Alhammadi, M.; Glenn, E. Detecting date palm trees health and vegetation greenness change on the eastern coast of the United Arab Emirates using SAVI. *Int. J. Remote Sens.* **2008**, *29*, 1745–1765. [CrossRef]
40. Iqbal, F. Detection of salt affected soil in rice-wheat area using satellite image. *Afr. J. Agric. Res.* **2011**, *6*, 4973–4982.
41. Zhang, T.T.; Zeng, S.L.; Gao, Y.; Ouyang, Z.T.; Li, B.; Fang, C.M.; Zhao, B. Using hyperspectral vegetation indices as a proxy to monitor soil salinity. *Ecol. Indic.* **2011**, *11*, 1552–1562. [CrossRef]
42. Aldakheel, Y.Y. Assessing NDVI spatial pattern as related to irrigation and soil salinity management in Al-Hassa Oasis, Saudi Arabia. *J. Indian Soc. Remote Sens.* **2011**, *39*, 171–180. [CrossRef]
43. Ijaz, M.; Ahmad, H.R.; Bibi, S.; Ayub, M.A.; Khalid, S. Soil salinity detection and monitoring using Landsat data: A case study from Kot Addu, Pakistan. *Arab. J. Geosci.* **2020**, *13*, 1–9. [CrossRef]
44. Rouse, J.W.; Haas, R.H.; Schell, J.A.; Deering, D.W. Monitoring vegetation systems in the Great Plains with ERTS. *NASA Spec. Publ.* **1974**, *351*, 309.
45. Khan, N.M.; Rastoskuev, V.V.; Sato, Y.; Shiozawa, S. Assessment of hydrosaline land degradation by using a simple approach of remote sensing indicators. *Agric. Water Manag.* **2005**, *77*, 96–109. [CrossRef]
46. Huete, A.R. A soil-adjusted vegetation index (SAVI). *Remote Sens. Environ.* **1988**, *25*, 295–309. [CrossRef]
47. Abbas, A.; Khan, S. Using remote sensing techniques for appraisal of irrigated soil salinity. In Proceedings of the International Congress on Modelling and Simulation (MODSIM), Christenchurch, New Zealand, 10–13 December 2007; Modelling and Simulation Society of Australia and New Zealand: Christenchurch, New Zealand, 2007; pp. 2632–2638.
48. Basso, F.; Bove, E.; Dumontet, S.; Ferrara, A.; Pisante, M.; Quaranta, G.; Taberner, M. Evaluating environmental sensitivity at the basin scale through the use of geographic information systems and remotely sensed data: An example covering the Agri basin (Southern Italy). *Catena* **2000**, *40*, 19–35. [CrossRef]
49. Kahaer, Y.; Tashpolat, N. Estimating salt concentrations based on optimized spectral indices in soils with regional heterogeneity. *J. Spectrosc.* **2019**, *2019*, 2402749. [CrossRef]
50. Vogelmann, J.; Rock, B. Spectral characterization of suspected acid deposition damage in red spruce (Picea Rubens) stands from Vermont. In Proceedings of the Airborne Imaging Spectrometer Data Analysis Workshop, Jet Propulsion Laboratory, Pasadena, CA, USA, 8–10 April 1985.
51. Bannari, A.; Guedon, A.; El-Harti, A.; Cherkaoui, F.; El-Ghmari, A. Characterization of slightly and moderately saline and sodic soils in irrigated agricultural land using simulated data of advanced land imaging (EO-1) sensor. *Commun. Soil Sci. Plant Anal.* **2008**, *39*, 2795–2811. [CrossRef]
52. Abbas, A.; Khan, S.; Hussain, N.; Hanjra, M.A.; Akbar, S. Characterizing soil salinity in irrigated agriculture using a remote sensing approach. *Phys. Chem. Earth Parts A/B/C* **2013**, *55*, 43–52. [CrossRef]
53. Breiman, L. Random forests. *Mach. Learn.* **2001**, *45*, 5–32. [CrossRef]
54. Cutler, A.; Stevens, J.R. [23] random forests for microarrays. *Methods Enzymol.* **2006**, *411*, 422–432.

55. Meng, X.; Bao, Y.; Ye, Q.; Liu, H.; Zhang, X.; Tang, H.; Zhang, X. Soil organic matter prediction model with satellite hyperspectral image based on optimized denoising method. *Remote Sens.* **2021**, *13*, 2273. [CrossRef]
56. Liaw, A.; Wiener, M. Classification and regression by randomForest. *R News* **2002**, *2*, 18–22.

Article

GIS-Based Soil Erosion Risk Assessment in the Watersheds of Bukidnon, Philippines Using the RUSLE Model

Indie G. Dapin [1,2,*] **and Victor B. Ella** [2]

1   Department of Agricultural and Biosystems Engineering, College of Engineering, Central Mindanao University, Musuan, Maramag 8710, Philippines
2   Land and Water Resources Engineering Division, Institute of Agricultural and Biosystems Engineering, College of Engineering and Agro-Industrial Technology, University of the Philippines Los Baños, Los Baños 4031, Philippines
*   Correspondence: indiedapin@cmu.edu.ph; Tel.: +63-936-7411-138

Abstract: The sustainability of watersheds for supplying water and for carbon sequestration and other environmental services depends to a large extent on their susceptibility to soil erosion, particularly under changing climate. This study aimed to assess the risk of soil erosion in the watersheds in Bukidnon, Philippines, determine the spatial distribution of soil loss based on recent land cover maps, and predict soil loss under various rainfall scenarios based on recently reported climate change projections. The soil erosion risk assessment and soil loss prediction made use of GIS and the RUSLE model, while the rainfall scenarios were formulated based on PAGASA's prediction of drier years for Bukidnon in the early-future to late-future. Results showed that a general increase in soil loss was observed in 2015, over the period from 2010 to 2020, although some watershed clusters also showed a declining trend of soil erosion, particularly the Agusan-Cugman and Maridugao watershed clusters. Nearly 60% of Bukidnon has high to very severe soil loss rates. Under extreme rainfall change scenario with 12.61% less annual rainfall, the soil loss changes were only +1.37% and −2.87% in the category of none-to-slight and very severe, respectively. Results showed that a decrease in rainfall would have little effect on resolving the excessive soil erosion problem in Bukidnon. Results of this study suggest that having more vegetative land cover and employing soil conservation measures may prove to be effective in minimizing the risk of soil erosion in the watersheds. This study provides valuable information to enhance the sustainability of the watersheds. The erosion-prone areas identified will help decision-makers identify priority areas for soil conservation and environmental protection.

Keywords: RUSLE; GIS; climate change; soil erosion; erosion risk assessment

Citation: Dapin, I.G.; Ella, V.B. GIS-Based Soil Erosion Risk Assessment in the Watersheds of Bukidnon, Philippines Using the RUSLE Model. *Sustainability* **2023**, *15*, 3325. https://doi.org/10.3390/su15043325

Academic Editors: Mario Elia, Antonio Ganga and Blaž Repe

Received: 21 December 2022
Revised: 30 January 2023
Accepted: 2 February 2023
Published: 11 February 2023

1. Introduction

Soil erosion by water is a naturally occurring process associated with the hydrologic cycle. At small spatial scales, soil particle detachment by rainfall impact may predominate, but at larger spatial scales other water erosion processes are likely to dominate. Erosion of soil from catchment areas and sediment deposition in waterways can reduce storage capacity in the reservoir and degrade downstream water quality. Increasing soil erosion can also decrease soil fertility and crop yield [1], severely threatening local, national, and global food production systems and environmental sustainability [2,3]. Soil erosion severely restricts agricultural land use by reducing the productive potential of soils. It also contributes to water pollution by introducing suspended matter and nutrients into bodies of water. In addition, land that has been eroded becomes susceptible to other environmental impacts. Soil erosion is primarily caused by unsustainable agricultural practices, forest clearance, overgrazing, mining, and construction activities [4].

Soil erosion is considered one of the worst environmental issues in the Philippines [2,5]. Many parts of the country are highly susceptible to soil erosion because of their steep topographic conditions, which are compounded by the occurrence of severe rainfall events, land cover degradation, poor farming practices, and other soil-related factors [2]. Reduction in reservoir storage capacity is significantly related to increased soil erosion in the catchment areas. This is true in the case of the Pulangi IV reservoir in Bukidnon, Philippines. It was reported by Deutsch et al. [6] that the Pulangi IV reservoir was silting up at a rate of about one meter per year at the dam and that sedimentation had reduced the reservoir capacity by approximately 50%. Moreover, this contributed to the premature deterioration of hydropower turbines and frequent power outages [6]. Bukidnon is a watershed–landlocked province and the catchment area of various rivers in Mindanao.

To some extent, soil erosion can be mitigated. Better land use planning can help reduce the long-term threat of soil erosion [7]. There is a need for a straightforward and practical approach to estimating and mapping soil erosion risk that uses readily available data to improve water and soil conservation initiatives [3]. Estimating the risk of soil loss and its spatial distribution is critical for a successful assessment of soil erosion. Erosion-induced soil loss can be calculated using a prediction model such as the Revised Universal Soil Loss Equation (RUSLE) [8]. RUSLE is an empirical model for estimating soil erosion and is practical for predicting soil loss on hillslopes [2,9]. The model has been widely adopted due to its simple and straightforward computational input requirements compared to other conceptual and process-based models [10]. Numerous studies on soil erosion risk assessment and soil loss prediction around the world have been conducted using RUSLE in conjunction with Geographic Information Systems (GIS) and remote sensing techniques [1–4,8,10–14]. GIS has great potential in soil erosion inventory for soil erosion modeling and erosion hazard assessment [1] because it facilitates the manipulation, integration, analysis, and display of large amounts of spatial data, and can provide spatial distribution information on erosion [15,16].

According to the climate projection of the Philippine Atmospheric, Geophysical, and Astronomical Services Administration (PAGASA) [17], Mindanao will generally have a decreasing rainfall trend in 2050 (2036–2065). In 2050, seasonal rainfall in Bukidnon is projected to decrease significantly under high- and medium-range emission scenarios, most notably during the December–January–February (DJF) and March–April–May (MAM) seasons. Overall, the projected annual rainfall will reduce as well. The equivalent annual rainfall change based on the projected seasonal rainfall change under high-range and medium-range emission scenarios were estimated at −1.74% and −8.32%, respectively. A more recent report about the Philippine's climate extremes revealed that the projected climate will be drier across the country, with more severe conditions expected in Visayas and Mindanao. Bukidnon, in the early-future (2020–2039), mid-future (2046–2065), and late-future (2080–2099), is still projected to have drier years [18]. Therefore, it is critical to incorporate climate projection scenarios when assessing the future risk of soil erosion.

While previous studies have attempted to characterize the soil erosion characteristics in some watersheds in Bukidnon [5,19,20], no study on GIS-based soil erosion assessments in all Bukidnon watersheds exists in the published literature, particularly in a predictive mode based on recent land use data and on recently reported climate change scenarios by PAGASA. Adornado and Yoshida [21] have previously used the RUSLE model to assess soil erosion in Bukidnon. However, the land cover in Bukidnon has changed significantly over time, and the climatic conditions are also expected to change. Additionally, technological advancements have occurred in recent years, land cover maps are updated regularly, the DEM data has much higher resolution, and satellite precipitation data are already available. Thus, this study aims to generate a more updated GIS-based soil erosion risk assessment in the watersheds in Bukidnon, assess the spatial distribution of soil loss based on the land cover maps in 2010, 2015, and 2020, and predict soil loss under various rainfall scenarios based on recently reported climate change projections.

2. Materials and Methods

2.1. Study Area

The province of Bukidnon is a landlocked province in the middle of Mindanao Island, southern Philippines, and is geographically located between $7°18.42'–8°38.22'$ N latitude and $124°15.6'–125°31.74'$ E longitude, as shown in Figure 1. The topography of the province is predominantly hilly and mountainous, especially in the eastern portion, and the other two mountain ranges in the west, have an average elevation of 915 m, and a range of 22 to 2867 m above sea level (see Figure 2a). It has rolling uplands, deep canyons, valleys alternating with the low plains, and terrain characterized by deep ravines and dense forest mountains in several mountain ranges. Due to its relatively high elevation, Bukidnon remains relatively cool and moist throughout the year. Bukidnon has a developing agricultural-based economy and is primarily a producer of rice, corn, sugar, coffee, rubber, flower, fruits, vegetables, poultry, and livestock.

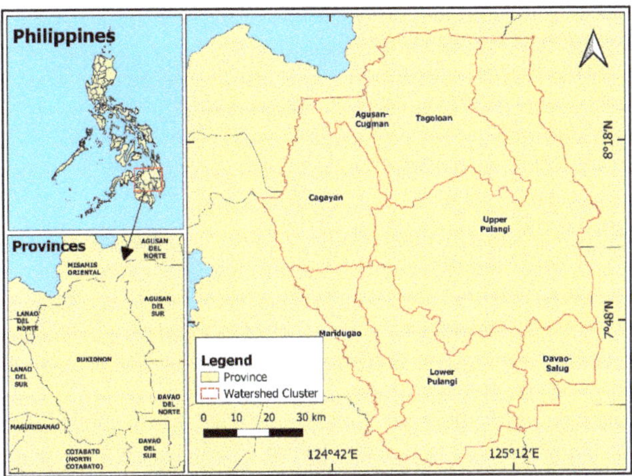

Figure 1. Location map of the province of Bukidnon with watershed cluster map.

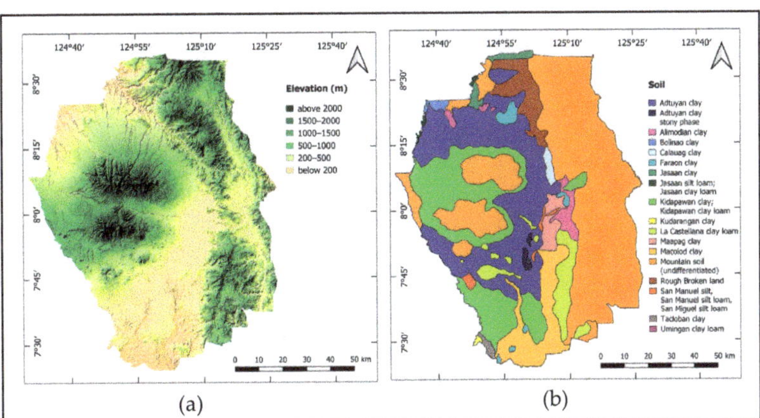

Figure 2. The (**a**) elevation map and (**b**) soil map of Bukidnon.

2.2. Datasets

For rainfall, the TRMM_3B42_Daily v7 was used in this study. These data are a daily accumulated precipitation product generated from the research-quality three-hourly Tropical Rainfall Measuring Mission (TRMM) Multi-Satellite Precipitation Analysis (TMPA) [22].

Accordingly, it is produced at the National Aeronautics and Space Administration (NASA) Goddard Earth Sciences Data and Information Services Center (GES DISC), as a value-added product. A simple summation of valid retrievals in a grid cell was applied for the day data. These rainfall data in millimeters are available in raster format with a resolution of $0.25° \times 0.25°$ (https://disc.gsfc.nasa.gov/datasets/TRMM_3B42_Daily_7/summary (accessed on 10 November 2022)). For this study, the annual accumulated TRMM_3B42_Daily v7 was taken from the Geospatial Interactive Online Visualization ANd aNalysis Infrastructure (GIOVANNI) website (https://giovanni.gsfc.nasa.gov/giovanni (accessed on 10 November 2022)). GIOVANNI is a web-based application developed by the GES DISC that provides access to Earth science remote sensing data (https://gpm.nasa.gov/data/sources/giovanni (accessed on 10 November 2022)). The TRMM 3B42_Daily v7 was used in this study because it has the least overall monthly bias and most closely matches the rainfall distribution observed at weather stations, particularly the dry days and torrential rain days, across the entire Philippines, compared to other gridded rainfall products [23]. During this study, the available TRMM_3B42_Daily v7 data were from 1998 to 2019.

The Interferometric Synthetic Aperture Radar (IfSAR) Digital Elevation Model (DEM) with 5×5 m resolution was used to delineate the province's major river watersheds. This was obtained from the National Mapping and Resource Information Authority (NAMRIA) of the Philippines. The DEM was projected into WGS 84/UTM zone 51 N. The elevation map of Bukidnon is shown in Figure 2a.

The soil map of Bukidnon was extracted from the soil map of the Philippines and was downloaded from the Geoportal Philippines website (https://geoportal.gov.ph (accessed on 1 December 2022)). For some areas in the soil map with soil texture classified as undifferentiated, the Food and Agriculture Organization (FAO) Soil Map of the Philippines [24] and the Harmonized World Soil Database (HWSD) [25] were used as a reference to determine the soil textural class in those areas. The soil map of Bukidnon is shown in Figure 2b.

Three land cover maps for 2010, 2015, and 2020 were used in this study. The land cover maps of Bukidnon were extracted from the 2010, 2015, and 2020 land cover maps of the Philippines. The land cover maps and the administrative boundary map were downloaded via the Geoportal Philippines website (https://geoportal.gov.ph (accessed on 1 December 2022)), managed by NAMRIA. The land cover maps are shown in Figure 3.

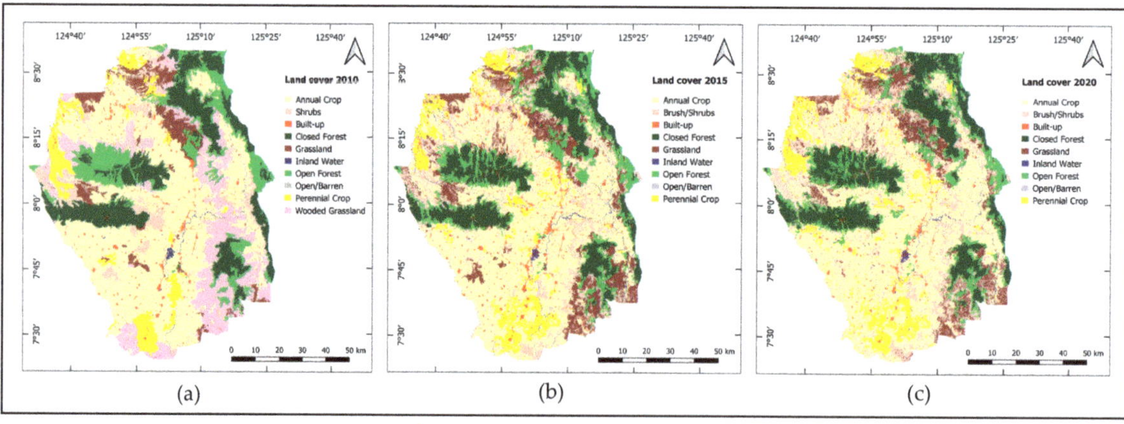

Figure 3. The land cover maps of Bukidnon in (**a**) 2010, (**b**) 2015, and (**c**) 2020.

2.3. RUSLE

The Revised Universal Soil Loss Equation, or the RUSLE model, is a well-known and widely used empirical model for estimating soil erosion [9]. It was developed using the five following factors to estimate the average annual soil loss (A): rainfall erosivity factor

(R), soil erodibility factor (K), slope length and steepness factor (LS), cover management factor (C), and conservation practice factor (P). RUSLE is expressed as:

$$A = R \times K \times LS \times C \times P; \text{ in t/ha/y} \tag{1}$$

The rainfall erosivity factor (R) quantifies the kinetic energy impact of rainfall and predicts the rate and amount of runoff associated with that precipitation [9]. Originally, to determine the R-factor, rainfall intensity data is required [10,26]. Due to the unavailability of rainfall intensity records for Bukidnon, the equation for R-factor (Equation (2)) used by Salvacion [2] in Marinduque, Blanco and Nadaoka [27] in Laguna Lake Watershed, and Adornado et al. [28] in Quezon, Philippines was adopted. Accordingly, this simplified equation produced a result that was acceptable for tropical and humid subtropical climatic zones [9,28]. The equation is as follows:

$$R = 38.5 + 0.35P \tag{2}$$

where R is the rainfall erosivity factor (MJ mm/ha/h/y), and P is the average annual precipitation (mm). Three sets of R-factor maps were generated with three different sets of average annual precipitation maps. The three sets of the average annual precipitation maps represented the average annual precipitation of 1998–2009, 2003–2014, and 2008–2019. Each period consists of an equal number of years, with a seven-year overlap between subsequent periods. This was done because this study used three land cover maps at different periods, as previously mentioned. Furthermore, the calculation of the average annual precipitation using no less than ten years of precipitation data was based on the methodology of the previous studies [28–31].

The soil erodibility factor (K) indicates the soil's resistance to erosion caused by raindrop impact, as well as by the runoff generated from the rainfall. Soil erodibility is determined by geological and soil characteristics such as structure, texture, parent material, porosity, and organic matter content. Regardless of the soil concentration of sand and clay, the silt content is directly related to soil erodibility [9,26]. This study adapted the K-factor values from David [32] for each soil textural class, which was also adapted by Salvacion [2].

The LS-factor in RUSLE is the slope length (L) and steepness (S) factors combined to reflect the effect of regional topography on the rate of soil erosion. Cumulative runoff increases in both amount and rate as the slope lengthens. As the land slope increases, the runoff velocity increases proportionately, resulting in massive erosion [9,28]. The LS-factor was generated using Equation (4). Equation (4) is a method developed by Moore and Burch [33] and Moore and Wilson [34] and was used by Hrabalíková and Janeček [35] in which it was proven as one of the better options to calculate the LS-factor. According to Andreoli [11], this expression is appropriate for areas with complex topography, including plateaus, terraced ledges, and mountains, because it considers the convergence and divergence of the flows. Using the DEM, flow accumulation and slope were generated in Quantum GIS (QGIS) through the r.flow Geographic Resources Analysis Support System (GRASS) algorithm, and slope algorithm of the Geospatial Data Abstraction Library (GDAL), respectively, in the QGIS toolbox. The equations are as follows:

$$A_s = FA * cell\ size \tag{3}$$

$$LS = (A_s/22.1)^m * \left(\frac{\sin \beta}{0.0896}\right)^n \tag{4}$$

where A_s is the specific catchment area, FA is the flow accumulation in each grid cell and its value corresponds to the number of flowlines that traverse that grid cell, cell size is the resolution of the grid (for this study 5 m), the value of m is 0.4, n is equal to 1.3, and β is the slope angle [9,11,35].

The C-factor values represent how cropping and management practices affect the erosion rate. It is inextricably linked to land use types and is a factor in reducing soil erosion vulnerability. It is defined as the ratio of soil loss from land under specific conditions to the equivalent loss from clean-tilled, continuous fallow. Essentially, vegetative cover prevents raindrops from colliding with the soil surface and dissipates the kinetic energy of rainfall before it reaches the soil surface, slowing down runoff, thus facilitating infiltration; hence, the amount and type of vegetation cover has a significant effect on soil loss. *C*-factor is directly related to the vegetation type, stage of growth, and percentage of cover [1,9]. In this study, the *C*-factor values from David [32] and Delgado and Canters [36] were used as a reference in assigning the *C*-factor for each land cover class of the three land cover maps.

The conservation or support practice factor (*P*) indicates the effects of implementations that decrease the rate and amount of runoff, thereby reducing the amount and rate of soil erosion. It indicates the proportion of soil loss caused by a particular support practice compared to soil loss caused by upward and downward slope, contour farming, and tillage. Primary support practices include strip cropping, contour farming, terracing, cross-slope cultivation, and grassed waterways. *P*-factor values are calculated as the ratio of soil loss caused by a specific support practice to soil loss caused by row farming in both upward and downward slope conditions [9]. The value of the *P*-factor ranges from 0 to 1, where a value close to 0 indicates good conservation practice while values approaching 1 indicate poor or no erosion control practice [37]. Since no records document the extent and adoption of conservation practices in the province, though some may have adopted them, a value of 1 was assigned for the *P*-factor for the entire province.

2.4. Annual Rainfall Change Scenario

Initially, a baseline average annual rainfall scenario was established before formulating annual rainfall change scenarios. Since the accumulated annual TRMM_3B42_Daily data used in this study was from 1998 to 2019, the same period was used in generating the baseline scenario condition. Thus, the average annual rainfall for the period 1998–2019 was used to generate the *R*-factor of the baseline scenario.

As previously mentioned, Bukidnon is expected to experience drier years in the future. In Bukidnon, the projected rainfall on the early-future (2020–2039), mid-future (2046–2065), and late-future (2080–2099), and under both the moderate emission (RCP4.5) and high emission (RCP8.5) scenarios, range from −3.70 to −12.61% from the baseline value [18]. On this basis, three annual rainfall scenarios were generated (Table 1) to assess the risk of soil erosion under the projected rainfall conditions. Rather than using the six scenarios proposed by DOST-PAGASA et al. [18] to represent each future category and emission scenario, only three scenarios were considered because all projected annual rainfall values are consistently lower than the baseline value. To generate the rainfall amount of each scenario, the amount of rainfall equivalent to the percent change in rainfall, as shown in Table 1, was subtracted from the baseline average annual rainfall scenario. The results in each scenario were used to obtain the corresponding *R*-factor.

Table 1. Rainfall change scenarios.

Scenario	Rainfall Change (%)	Description
R1	−3.70	Dry: Early future (2020–2039) under High emissions (RCP8.5)
R2	−8.30	Very dry: Late future (2080–2099) under Moderate emission (RCP4.5)
R3	−12.61	Extremely dry: Late future (2080–2099) under High emission (RCP8.5)

3. Results

3.1. RUSLE Factor Distribution

GIS was used in this study to generate the RUSLE factors, calculate the average annual soil erosion rates, and produce maps showing the distribution of these factors, the soil erosion rates, and the soil erosion risk map in Bukidnon. Cell-by-cell calculations of the mean annual soil erosion rates [1] were conducted; thus, the RUSLE factors were prepared in raster format using QGIS. Figures 4–6 show the map of the RUSLE factors used in calculating the annual soil erosion rates.

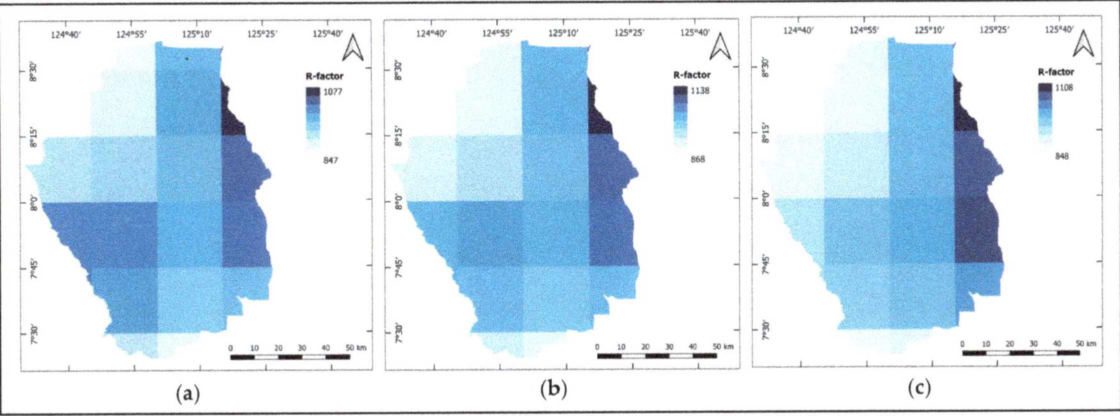

Figure 4. The *R*-factor map of Bukidnon derived from the average annual rainfall for the period (**a**) 1998–2009, (**b**) 2003–2014, and (**c**) 2008–2019.

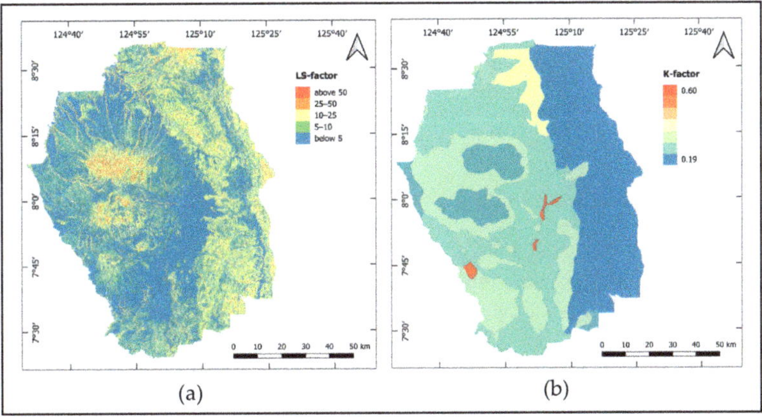

Figure 5. The (**a**) *LS*-factor and (**b**) K-factor map of Bukidnon.

As previously mentioned, the TRMM_3B42_Daily accumulated annual precipitation data in raster format downloaded from GIOVANNI were used as the dataset to generate the average annual rainfall in the province. Since three land cover maps were used in the study, three maps for the average annual precipitation were also created. Each represents the average for 1998–2009, 2003–2014, and 2008–2019, respectively. Using the average annual rainfall from 1998–2019 as a reference, it was observed that most of the northern and western parts of the province in 1998–2009 were wetter while the eastern areas were drier. In addition, the average annual rainfall in 2003–2014 was generally higher in all areas of the province compared to other time periods. In 2008–2019, however, the province was predominantly wetter, with a few drier areas on the western side. The average annual precipitation maps were used to calculate the *R*-factor maps using Equation (2), thus resulting in three *R*-factor maps, as shown in Figure 4. Generally, values of the *R*-factor are much higher in the northeastern and eastern parts of the province, but lower values of the *R*-factor can be found in the northwestern part of the province, as shown in Figure 4.

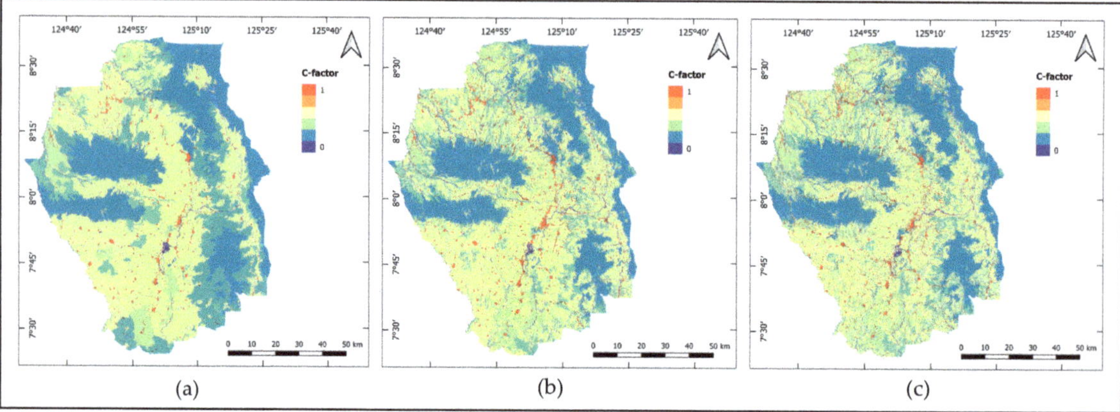

Figure 6. The *C*-factor maps of Bukidnon for (**a**) 2010, (**b**) 2015, and (**c**) 2020.

Using Equations (3) and (4), the *LS*-factor of the province was obtained. The spatial distribution of the *LS*-factor is shown in Figure 5a. Approximately, 48.3% of the province's *LS*-factors are below 5, followed by 22.98% within 5–10, 24.16% within 10–25, 4.12% within 25–50, and only approximately 0.54% above 50. There are certain areas in the province with *LS*-factors over 500, but they only represent approximately 0.00008% of the province's total area. Higher *LS*-factor values can be found in the mountain ranges, in deep canyons, and in nearby river networks, whereas lower values can be found in the plains and perhaps the cropland areas in the province. Extremely high values of the *LS*-factor are expected, given that the range elevation in the province is relatively high, having a standard deviation of 445 m. Additionally, the slope in the province ranges from 0 to 77° with 16.08° as the average and 12.27° as standard deviation, and approximately 20.33% of the province is mountainous and extremely steep, with slopes exceeding 26.6° or 50%, resulting in some extremely high *LS*-factor. Most of the *LS*-factors over 50 are in areas with a slope above 50%. *LS*-factors over 50 were also observed by Salvacion [2] in Marinduque, Philippines. Using the same expression of the *LS*-factor used in this study, Andreoli [11] was able to obtain higher *LS*-factors over 350 and others over 2000 [38,39].

Figure 5b shows the soil erodibility factor (*K*) of the province, which ranges between 0.19 and 0.6. Lower K-factor values are more prevalent in the eastern side of the province, while higher values are generally located in the northern side and a few other areas. Most of the province has a lower K-factor, ranging between 0.19 and 0.3.

Three *C*-factor maps were generated based on the land cover maps of 2010, 2015, and 2020. As shown in Figure 6, the *C*-factor values range from 0 to 1. During *C*-factor classification based on the land cover classes, the 0 value was assigned for water bodies while 1 was set for barren and built-up areas. The rest of the land cover classes were reclassified based on data available from the existing literature [2,32,36,40]. As land cover changes over time, so does the *C*-factor distribution in the province.

3.2. Soil Loss

Figure 7 shows the spatial distribution of soil erosion rate in Bukidnon during 2010, 2015, and 2020. Because the province is a watershed area of the neighboring provinces, the Bukidnon watershed areas were clustered into seven watershed clusters based on the classification considered by Rola et al. [41], as shown previously in Figure 1. These watershed clusters include Tagoloan in the north; Agusan-Cugman, and Cagayan in the northwest; Upper Pulangi in the central and eastern side; Maridugao in the southwest; Lower Pulangi in the south; and Davao-Salug in the southeastern side of Bukidnon.

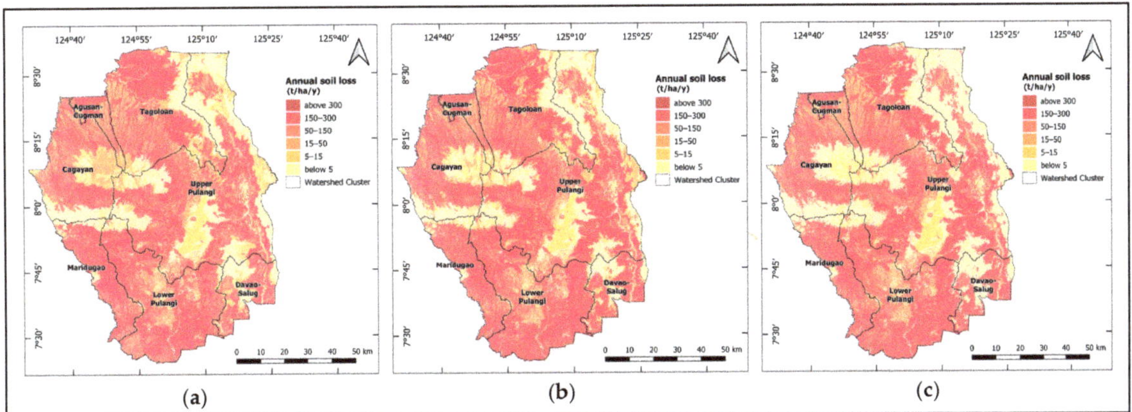

Figure 7. The annual soil loss map of Bukidnon in (**a**) 2010, (**b**) 2015, and (**c**) 2020.

On average, the RUSLE model-predicted soil erosion rates in Bukidnon range between 312 to 363 t/ha/y based on the periods under consideration. The predicted values range from 0 to 114,275 t/ha/y. The predicted soil erosion rates may appear to be exceedingly high. However, a study conducted in Marinduque, Philippines, showed that the RUSLE predicted soil erosion rates is around 120 t/ha/y on average, and can vary from 0 to as high as 20,767 t/ha/y [2]. These high erosion rates were observed in an island province that has a drier climate and less complex topography than Bukidnon. Additionally, the plot experiment on soil erosion in a few selected areas in Bukidnon revealed that a month of 5 mm rainfall can cause up to 229 t/ha of soil loss on certain plots with a slope of 15%. Although soil deposition can also occur, reaching 400 t/ha in some plots [42], this could also mean that somewhere near the plots an accumulated soil loss equal to that amount is also possible. Thus, the RUSLE-predicted soil erosion rates obtained in this study are comparable with the empirical evidence obtained from previous studies in the Philippines. As illustrated in Figure 7, areas prone to soil erosion are primarily found along the buffer zones of the major mountain ranges in Bukidnon. These areas are frequently found in deep river canyons and rolling areas that were previously used for crop production. The predicted soil erosion rates in these areas are extremely high, exceeding 300 t/ha/y on average. It can be observed in Figure 7 that the areas with lower soil erosion rates are the plains and mountain ranges that have good vegetative cover, the protected areas of the province.

Figure 8 illustrates the mean value of the average annual soil erosion rate for each watershed cluster. The thin lines in the graph represent the magnitude of standard deviations on top of the mean soil erosion rates. The predicted soil loss in 2020 is lower than in 2015 in most areas of the watershed clusters, except for Tagoloan. This decline may be attributable to a wetter climate in 2003–2014, compared to 2008–2019, that was used in deriving the *R*-factor. The exception in Tagoloan may be attributed to the decrease in vegetative cover in 2020, especially in the forest areas, despite the climate in 2008–2019 being drier than in 2003–2014. The soil erosion rates in the Maridugao watershed cluster are slightly decreasing. In contrast, the Upper and Lower Pulangi, and the Davao-Salug watershed clusters have an increasing soil erosion rate from 2010 to 2020, of which the highest can be observed during 2015. The increase in soil erosion in the Upper Pulangi watershed cluster may have contributed more to the accumulation of sediments in the Pulangi reservoir.

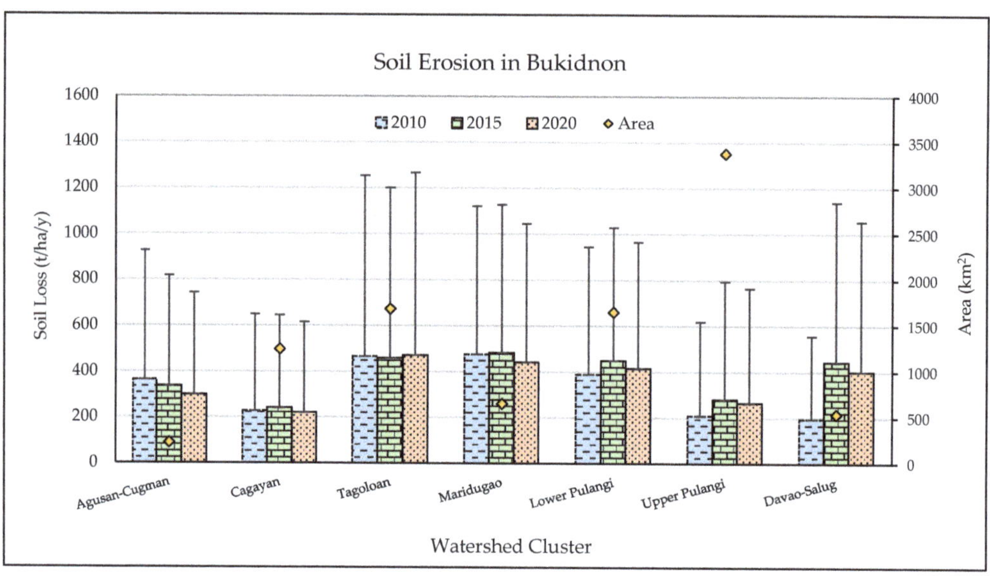

Figure 8. Mean value of the average annual soil erosion rate in Bukidnon.

As shown in Table 2 and Figure 8, Maridugao and Tagoloan have the highest mean soil erosion rates among the watershed clusters. From 2010 to 2020, the Tagoloan watershed had the highest maximum soil erosion rates. This can be attributed to a combination of higher *LS*-factor and *R*-factor values in the region. In fact, the cluster with the highest *LS*-factor can be found in Tagoloan watershed cluster. In contrast, the Davao-Saug cluster in 2010, the Agusan-Cugman cluster in 2015, and the Cagayan watershed cluster in 2020 all have lower maximum values of soil loss rates. Table 2 contains statistical values for soil erosion rates predicted by RUSLE by watershed cluster.

Table 2. Statistics of the predicted RUSLE erosion rates per watershed cluster.

Watershed Cluster	Soil Erosion Rates (t/ha/y)									
	Mean			Minimum	Maximum			Standard Deviation		
	2010	2015	2020		2010	2015	2020	2010	2015	2020
Agusan-Cugman	363	338	300	0	42,192	29,127	51,003	564	479	443
Cagayan	227	241	220	0	46,693	47,834	35,126	421	403	394
Tagoloan	466	458	472	0	84,508	91,421	114,275	789	742	795
Maridugao	475	482	441	0	42,155	37,769	41,749	645	646	605
Lower Pulangi	389	449	414	0	58,496	61,037	49,218	556	580	551
Upper Pulangi	209	281	265	0	46,396	54,213	47,484	409	514	498
Davao-Salug	196	442	401	0	25,556	55,941	44,924	361	697	654

Several areas in Bukidnon had extremely high values of the predicted soil erosion rates, as shown in Figure 7. This could be the result of setting the *P*-factor value equal to one for the entire province and could also be attributed to the resolution of the DEM that was used to generate the *LS*-factor. The resolution may affect the computation of the *LS*-factor. Hrabalíková and Janeček [35] predicted soil loss rates closer to those observed while using a similar *LS*-factor expression to that used in this study, and a DEM with 1 m resolution. Hence, to avoid overestimating the predicted soil loss, a much higher DEM resolution would be preferable, as would documentation of conservation practices in the area.

Based on the predicted values of soil erosion rates shown in Table 3, the areas with very severe soil erosion in Bukidnon decreased slightly in 2020 compared to 2015, by about

1.77%, equivalent to 16,564 ha. Though it can be observed that there was a slight increase in the severe areas from 2015 to 2020, the moderate and high soil erosion areas had decreased by 2.03, and 3.66%, respectively. This decrease has led to an increase in the extent of areas under none to slight soil loss categories by about 7.1%, equivalent to 66,445 ha. When the results in Table 3 were compared with those reported by Adornado and Yoshida [21], the extent of severe and very severe areas had increased by about twofold. The extent of severe areas went from 6.66% to 13.45% on average, while very severe areas went from 15.19% to as high as 34.16% on average. The twofold increase on these areas has been prevalent since 2010. This may be due to the decrease in the areas classified as none to slight, high, and very high soil loss categories. The differences of the results may also be attributed to the nature of the model inputs used in the earlier study. In that study, a DEM was derived from 100 m interval contour lines to obtain an *LS*-factor. The generated DEM they used had less details about the topography of the province and this could have affected the calculation of the *LS*-factor. Additionally, their *C*-factor was generated from an analog land cover map and the ASTER image available during that time [21]. Nevertheless, both results indicate the potential severity of soil erosion in Bukidnon as the rates of the very high to very severe soil erosion areas have not dropped significantly from 2015 to 2020, based on the predicted soil erosion rates, and this poses a serious problem on the sustainability of land and water resources in this province.

Table 3. The extent of erosion in Bukidnon [1].

Category	Soil Loss Rates (t/ha/y)	Area Affected by Erosion (%)		
		2010	2015	2020
None to slight	0–5	18.79	18.40	25.50
Moderate	5–15	9.55	9.68	7.65
High	15–50	11.14	10.12	6.46
Very high	50–150	15.34	12.20	12.32
Severe	150–300	14.87	12.62	12.87
Very severe	>300	30.31	36.97	35.20

Note: [1] Bukidnon total area = 935,846 ha, based on the calculation using GIS.

3.3. Soil Loss under Annual Rainfall Change Scenarios

As expected, the predicted rate of soil loss (Table 4) decreased under the very severe category while it increased under none to slight, high, very high, and severe soil loss categories, in all the rainfall scenarios, as all generated rainfall scenarios had negative percent changes. The areas categorized as none to slight increased with a range of 0.4 to 1.37% against the baseline scenario, equivalent to 3743 to 12,821 ha. The extent of the very severe areas fell within the range of 0.79 to 2.87%, which means around 7393 to 26,858 ha of land will no longer experience very severe soil loss if annual rainfall decreases by 3.17 to 12.61% in future decades. However, this reduction in very severe areas will also result in an expansion of areas classified as having high, very high, and severe soil loss. In addition, approximately no more than 0.9% (8423 ha) of the province of Bukidnon may experience a decrease in the moderate soil loss rates areas. The extent of very severe soil loss areas may reduce significantly but remains small compared to the total size of the province.

Tables 5 and 6 depict the extent of areas under the baseline scenario and the third rainfall change scenario (R3) at various soil loss categories for each watershed cluster. The extent of areas for each soil loss category are expressed in percent relative to the total area of each watershed cluster. The results indicate that the extent of the none to slight soil loss category for most of the watershed clusters will increase relative to baseline values under the R3 scenario. By contrast, the extent of the very severe areas will decrease relative to baseline values under the R3 scenario. The extent of the none to slight soil loss categories will increase between 0.48 and 1.98%, whereas the extent of the very severe areas will decrease between 2.49 and 3.89% of the area of the watershed clusters.

Table 4. The extent of soil erosion under rainfall change scenarios in Bukidnon [1].

Category	Area Affected by Erosion (%)			
	Baseline (B)	R1	R2	R3
None to slight	25.64	26.04	26.54	27.01
Moderate	7.55	7.29	6.96	6.65
High	6.50	6.65	6.86	7.07
Very high	12.39	12.67	13.04	13.41
Severe	12.97	13.20	13.49	13.77
Very severe	34.96	34.16	33.12	32.09

Note: [1] Total area considered = 935,846 ha, based on the calculation using GIS.

Table 5. Percent (%) of the extent of soil erosion under baseline rainfall scenario in Bukidnon [1].

Watershed Cluster	Area (ha)	Extent of Soil Erosion (%)					
		None to Slight	Moderate	High	Very High	Severe	Very Severe
Agusan-Cugman	22,202	9.43	7.54	17.40	19.06	13.98	32.61
Cagayan	124,202	31.38	9.13	6.80	14.56	14.06	24.08
Tagoloan	167,959	21.95	6.10	7.36	11.94	10.83	41.86
Maridugao	65,233	14.54	3.55	5.87	14.46	16.43	45.14
Lower Pulangi	164,739	10.10	4.09	7.93	15.88	17.34	44.69
Upper Pulangi	337,527	34.39	10.65	5.34	10.36	11.29	28.03
Davao-Salug	53,986	36.87	4.53	2.32	5.65	9.71	40.97

Note: [1] Total area considered = 935,846 ha, based on the calculation using GIS.

Table 6. Percent (%) of the extent of soil erosion under R3 rainfall change scenario in Bukidnon.

Watershed Cluster	Area (ha)	Extent of Soil Erosion (%)					
		None to Slight	Moderate	High	Very High	Severe	Very Severe
Agusan-Cugman	22,202	10.34	7.80	18.40	19.31	14.40	29.77
Cagayan	124,202	32.99	7.94	7.55	15.78	14.34	21.39
Tagoloan	167,959	23.19	5.32	8.06	12.49	11.65	39.33
Maridugao	65,233	15.04	3.42	6.59	15.98	17.40	41.55
Lower Pulangi	164,739	10.58	4.15	8.70	17.27	18.52	40.80
Upper Pulangi	337,527	36.38	9.14	5.69	11.33	11.99	25.54
Davao-Salug	53,986	38.29	3.31	2.49	6.56	11.15	38.26

As shown in Table 6, the Upper Pulangi, Davao-Salug, and Cagayan watershed clusters will consistently and significantly have larger areas with none to slight soil loss. On the other hand, Lower Pulangi, Maridugao, and Agusan-Cugman watershed clusters will have a smaller proportion of areas with none to slight soil loss but will have a greater proportion of areas with very severe soil loss in future decades. Relative to the watershed size, an extremely large extent of area with very severe soil loss will be in the Lower Pulangi watershed cluster. The distribution of soil loss, particularly in Maridugao and Lower Pulangi watershed clusters, appears to be negatively skewed. This means that there is a greater proportion of areas that have experienced very high to very severe soil loss, as shown in Table 5, and this will still happen, as shown in Table 6, despite an overall decrease in the very severe areas and an increase in the none to slight soil loss areas, as shown in Table 4. Significant variation can be found when comparing the soil erosion rates of the baseline scenario (Table 5) and the R3 scenario (Table 6) in all soil loss categories, except for the none to slight and very severe categories. The extent of areas under the moderate soil loss category in Agusan-Cugman and Lower Pulangi will potentially increase while the rest of the watershed clusters will decrease. All the watershed clusters will experience increases in the high, very high and severe soil loss rate categories.

The predicted soil loss using the RUSLE model at various time periods and under the rainfall change scenarios provided insights on the soil loss behavior in Bukidnon watershed areas. GIS was crucial in identifying the distribution of soil loss both in provincial and watershed cluster scales. The predicted soil loss under various rainfall change scenario may serve as baseline information for determining the potential soil loss in future decades under future rainfall conditions. Results showed that a reduction of approximately 12.61% of the annual rainfall could reduce the extent of very severe areas up to 2.87% and could increase the none to slight soil loss areas at 1.37% of the total area of Bukidnon. This implies that the reduction in rainfall alone will have little effect on the severity of erosion in the province. This suggests that increasing land cover and adapting soil conservation measures may be a more effective way to mitigate the severity of soil erosion than just anticipating for the province to experience less rainfall. Moreover, it is necessary to map and monitor areas adopting and implementing soil conservation measures and practices in order to compare the soil erosion in the areas with and without conservation measures. Not only will this help reduce soil loss, but it will also allow for a more accurate prediction of soil erosion. Nevertheless, the maps presented in this study may help planners in identifying priority areas for soil conservation measures, particularly those with excessive soil loss. Neighboring provinces downstream of the major rivers in Bukidnon should consider coordinating their efforts to implement conservation measures with those in the province of Bukidnon, as they are not exempted from the effects of excessive soil erosion in Bukidnon.

4. Discussion

The RUSLE model has proven to be a practical method for assessing soil erosion risk at the watershed scale and the impact of the rainfall change to soil erosion-prone areas in the province of Bukidnon. Excessive soil loss occurs on steep hillslopes with less vegetation that are devoid of support and conservation practices. As observed during the generation of the LS-factor, the LS-factor is sensitive to the resolution of DEM. The higher the resolution of DEM, the wider the range of values of the LS-factor. This is because the expression of the LS-factor resulted in much higher maximum values at a higher resolution [39], which may add to the uncertainty of the predicted soil erosion rates. Michalopoulou et al. [39] suggested a method to avoid overestimation of the LS-factor; however, it has not been evaluated and validated whether this strategy prevents overestimation or can lead to underestimation of the LS-factor. In addition, different land cover classification of land cover maps at different time periods made it difficult to compare the predicted soil loss rates between each period and from earlier studies, as changes in land cover classification entail a different C-factor value. Nevertheless, the results indicated that predicted soil loss under drier rainfall change scenarios was less significant compared to the size of the province, implying that the severity of soil erosion in the province may not necessarily be reduced just by experiencing less rainfall alone. Increased land cover and the adoption of conservation measures such as conservation agriculture may reduce the risk of extreme soil erosion more than anticipating future rainfall reductions.

The new updated information generated in this study could serve as the basis for the formulation of policies geared towards soil conservation and environmental protection in the province. The approach developed and employed may also be extended to other erosion-prone provinces and regions in the country. The application and integration of RUSLE and GIS to identify the areas most vulnerable to soil erosion will allow policymakers and decision-makers to identify priority areas to focus in implementing soil erosion control measures in the future. It is highly recommended to use DEM with higher resolution, whenever available, and promote the implementation and documentation of soil conservation and support practices in the province. It is also recommended that future studies consider alternative expressions for generating the LS-factor that would limit its overestimation and underestimation, and to use field measurements for evaluation and validation purposes.

Author Contributions: Conceptualization, I.G.D. and V.B.E.; methodology, I.G.D.; formal analysis, I.G.D. and V.B.E.; writing—original draft preparation, I.G.D. and V.B.E.; writing—review and editing, I.G.D. and V.B.E.; visualization, I.G.D.; supervision, V.B.E.; funding acquisition, I.G.D. and V.B.E. All authors have read and agreed to the published version of the manuscript.

Funding: This study was funded by the Department of Science and Technology—Science Education Institute (DOST-SEI) and Engineering Research and Development for Technology (DOST-ERDT) of the Philippines.

Institutional Review Board Statement: The study was approved by the University of the Philippines Los Baños, Graduate School.

Informed Consent Statement: Not applicable.

Data Availability Statement: Not applicable.

Acknowledgments: This study was conducted under the auspices of DOST-SEI and ERDT program. Acknowledgement is due to all agencies that provided the needed data used in this study.

Conflicts of Interest: The authors declare no conflict of interest. The funders had no role in the design of the study; in the collection, analyses, or interpretation of data; in the writing of the manuscript; or in the decision to publish the results.

References

1. El Jazouli, A.; Barakat, A.; Ghafiri, A.; El Moutaki, S.; Ettaqy, A.; Khellouk, R. Soil erosion modeled with USLE, GIS, and remote sensing: A case study of Ikkour watershed in Middle Atlas (Morocco). *Geosci. Lett.* **2017**, *4*, 25. [CrossRef]
2. Salvacion, A.R. Delineating soil erosion risk in Marinduque, Philippines using RUSLE. *GeoJournal* **2022**, *87*, 423–435. [CrossRef]
3. Hu, Y.; Tian, G.; Mayer, A.L.; He, R. Risk assessment of soil erosion by application of remote sensing and GIS in Yanshan Reservoir catchment, China. *Nat. Hazards* **2015**, *79*, 277–289. [CrossRef]
4. Tania, M. Assessment of soil erosion risk in Fizes River Catchment using USLE model and GIS. *ProEnvironment* **2013**, *6*, 595–599.
5. Ella, V.B. Simulating soil erosion and sediment yield in small upland watersheds using the WEPP model. In *Land Use Change in Tropical Watersheds: Evidence, Causes and Remedies*; Coxhead, I., Shively, G., Eds.; CAB International: Wallingford, UK, 2005; pp. 9–125.
6. Deutsch, W.G.; Busby, A.L.; Orprecio, J.L.; Bago-Labis, J.P.; Cequiña, E.Y. Community-based hydrological and water quality assessments in Mindanao, Philippines. In *Forests, Water and People in the Humid Tropics*; Bonell, M., Bruijnzeel, L.A., Eds.; Cambridge University Press: Cambridge, UK, 2005; pp. 134–150.
7. Tehrany, M.S.; Shabani, F.; Javier, D.N.; Kumar, L. Soil erosion susceptibility mapping for current and 2100 climate conditions using evidential belief function and frequency ratio. *Geomat. Nat. Hazards Risk* **2017**, *8*, 1695–1714. [CrossRef]
8. Fathizad, H.; Karimi, H.; Alibakhshi, S.M. The estimation of erosion and sediment by using the RUSLE model and RS and GIS techniques (Case study: Arid and semi-arid regions of Doviraj, Ilam province, Iran). *Int. J. Agric. Crop Sci.* **2014**, *7*, 304–314.
9. Ghosal, K.; Das Bhattacharya, S. A review of RUSLE Model. *J. Indian Soc. Remote Sens.* **2020**, *48*, 689–707. [CrossRef]
10. Gashaw, T.; Tulu, T.; Argaw, M. Erosion risk assessment for prioritization of conservation measures in *Geleda watershed*, Blue Nile basin, Ethiopia. *Environ. Sys. Res.* **2018**, *6*, 1. [CrossRef]
11. Andreoli, R. Modeling erosion risk using the RUSLE equation. In *QGIS and Applications in Water and Risks*; Baghdadi, N., Mallet, C., Zribi, M., Eds.; Wiley-Blackwell: Hoboken, NJ, USA, 2018; pp. 245–282.
12. Bekele, B.; Gemi, Y. Soil erosion risk and sediment yield assessment with universal soil loss equation and GIS: In Dijo watershed, Rift valley Basin of Ethiopia. *Model. Earth Syst. Environ.* **2021**, *7*, 273–281. [CrossRef]
13. Belasri, A.; Lakhouili, A. Estimation of soil erosion risk using the universal soil loss equation (USLE) and geo-information technology in Oued El Makhazine Watershed, Morocco. *J. Geogr. Inf. Syst.* **2016**, *8*, 98–107. [CrossRef]
14. Mhangara, P.; Kakembo, V.; Lim, K.J. Soil erosion risk assessment of the Keiskamma catchment, South Africa using GIS and remote sensing. *Environ. Earth Sci.* **2012**, *65*, 2087–2102. [CrossRef]
15. Duarte, L.; Teodoro, A.C.; Gonçalves, J.A.; Soares, D.; Cunha, M. Assessing soil erosion risk using RUSLE through a GIS open source desktop and web application. *Environ. Monit. Assess.* **2016**, *188*, 351. [CrossRef] [PubMed]
16. Tayebi, M.; Tayebi, M.H.; Sameni, A. Soil erosion risk assessment using GIS and CORINE model: A case study from western Shiraz, Iran. *Arch. Agron. Soil Sci.* **2017**, *63*, 1163–1175. [CrossRef]
17. PAGASA. *Climate Change in the Philippines*; PAGASA: Quezon City, Philippines, 2011.
18. DOST-PAGASA; Manila Observatory; Ateneo de Manila University. *Philippine Climate Extremes Report 2020: Observed and Projected Climate Extremes in the Philippines to Support Informed Decisions on Climate Change Adaptation and Risk Management*; Philippine Atmospheric, Geophysical and Astronomical Services Administration: Quezon City, Philippines, 2021; 145p.
19. Paningbatan, E.P. Identifying soil erosion hotspots in the Manupali river watershed. In *Land Use Change in Tropical Watersheds: Evidence, Causes and Remedies*; Coxhead, I., Shively, G., Eds.; CAB International: Wallingford, UK, 2005; pp. 126–132.

20. Alibuyog, N.R.; Ella, V.B.; Reyes, M.R.; Srinivasan, R.; Heatwole, C.; Dillaha, T. Predicting the effects of land use change on runoff and sediment yield in Manupali river subwatersheds using the SWAT model. *Int. Agric. Eng. J.* **2009**, *18*, 15–25.
21. Adornado, H.; Yoshida, M. Assessing the adverse impacts of climate change: A case study in the Philippines. *J. Dev. Sus. Agric.* **2010**, *5*, 141–146.
22. Goddard Earth Sciences Data and Information Services Center. TRMM (TMPA-RT) Near Real-Time Precipitation L3 1 day 0.25 degree × 0.25 degree V7, Edited by Andrey Savtchenko, Greenbelt, MD, Goddard Earth Sciences Data and Information Services Center (GES DISC) 2016. Available online: https://disc.gsfc.nasa.gov/datasets/TRMM_3B42RT_Daily_7/summary (accessed on 20 April 2021).
23. Peralta, J.C.A.C.; Narisma, G.T.T.; Cruz, F.A.T. Validation of high-resolution gridded rainfall datasets for climate applications in the Philippines. *J. Hydrometeorol.* **2020**, *21*, 1571–1587. [CrossRef]
24. BSWM. *Updating the Harmonized World Soil Database (HWSD): Correlation of Philippine Soils into FAO's World Reference Base for Soil Resources (WRB)*; BSWM: Quezon City, Philippines, 2013.
25. FAO; IIASA; ISRIC; ISSCAS; JRC. *Harmonized World Soil Database (Version 1.2)*; FAO: Rome, Italy; IIASA: Laxenburg, Austria, 2012.
26. Wischmeier, W.H.; Smith, D.D. *Predicting Rainfall Erosion Losses: A Guide to Conservation Planning*; Agriculture Handbook No. 537; U.S. Department of Agriculture: Washington, DC, USA, 1978.
27. Blanco, A.C.; Nadaoka, K. A comparative assessment and estimation of relative soil erosion rates and patterns in Laguna Lake watershed using three models: Towards development of an erosion index system for integrated watershed-lake management. In Proceedings of the Symposium on Infrastructure Development and the Environment, Quezon City, Philippines, 7–8 December 2006.
28. Adornado, H.A.; Yoshida, M.; Apolinares, H.A. Erosion vulnerability assessment in REINA, Quezon Province, Philippines with raster-based tool built within GIS environment. *Agric. Inf. Res.* **2009**, *18*, 24–31. [CrossRef]
29. Chatterjee, S.; Krishna, A.P.; Sharma, A.P. Geospatial assessment of soil erosion vulnerability at watershed level in some sections of the Upper Subarnarekha River basin, Jharkhand, India. *Environ. Earth Sci.* **2014**, *71*, 357–374. [CrossRef]
30. Kim, S.M.; Choi, Y.; Suh, J.; Oh, S.; Park, H.D.; Yoon, S.H. Estimation of soil erosion and sediment yield from mine tailing dumps using gis: A case study at the Samgwang mine, Korea. *Geosystem Eng.* **2012**, *15*, 2–9. [CrossRef]
31. Lee, G.S.; Lee, K.H. Scaling effect for estimating soil loss in the RUSLE model using remotely sensed geospatial data in Korea. *Hydrol. Earth Syst. Sci. Discuss* **2006**, *3*, 135–157.
32. David, W.P. Soil and Water Conservation Planning: Policy Issues and Recommendations. *J. Philipp. Dev.* **1988**, *15*, 47–84.
33. Moore, I.D.; Burch, G.J. Physical basis of the length-slope factor in the universal soil loss equation. *Soil Sci. Soc. Am. J.* **1986**, *50*, 1284–1288. [CrossRef]
34. Moore, I.D.; Wilson, J.P. Length-slope factors for the revised universal soil loss equation: Simplified method of estimation. *J. Soil Water Conserv.* **1992**, *47*, 423–428.
35. Hrabalíková, M.; Janeček, M. Comparison of different approaches to LS factor calculations based on a measured soil loss under simulated rainfall. *Soil Water Res.* **2017**, *12*, 69–77. [CrossRef]
36. Delgado, M.E.M.; Canters, F. Modeling the impacts of agroforestry systems on the spatial patterns of soil erosion risk in three catchments of Claveria, the Philippines. *Agroforest. Syst.* **2012**, *85*, 411–423. [CrossRef]
37. Rellini, I.; Scopesi, C.; Olivari, S.; Firpo, M.; Maerker, M. Assessment of soil erosion risk in a typical Mediterranean environment using a high resolution RUSLE approach (Portofino promontory, NW-Italy). *J. Maps* **2019**, *15*, 356–362. [CrossRef]
38. Gong, W.; Liu, T.; Duan, X.; Sun, Y.; Zhang, Y.; Tong, X.; Qiu, Z. Estimating the soil erosion response to land-use land-cover. *Water* **2022**, *14*, 742. [CrossRef]
39. Michalopoulou, M.; Depountis, N.; Nikolakopoulos, K.; Boumpoulis, V. The significance of digital elevation models in the calculation of LS factor and soil erosion. *Land* **2022**, *11*, 1592. [CrossRef]
40. FAO; IFAD. *Strategic Environmental Assessment, an Assessment of the Impact of Cassava Production and Processing on the Environment and Biodiversity*; FAO: Rome, Italy, 2001.
41. Rola, A.C.; Suminguit, V.J.; Sumbalan, A.T. *Realities of the Watershed Management Approach: The Manupali Watershed Experience*; PIDS Discussion Paper Series, No. 2004-23; Philippine Institute for Development Studies (PIDS): Makati City, Philippines, 2005.
42. Marin, R.A.; Jamis, C.V. Soil erosion status of the three sub-watersheds in Bukidnon Province, Philippines. *AES Bioflux* **2016**, *8*, 194–204.

Article

Assessing Landslide Susceptibility by Coupling Spatial Data Analysis and Logistic Model

Antonio Ganga [1], Mario Elia [2,*], Ersilia D'Ambrosio [2], Simona Tripaldi [3], Gian Franco Capra [1], Francesco Gentile [2] and Giovanni Sanesi [2]

[1] Dipartimento di Architettura, Design e Urbanistica, Università degli Studi di Sassari, Viale Piandanna n 4, 07100 Sassari, Italy; anto.ganga@gmail.com (A.G.); pedolnu@uniss.it (G.F.C.)

[2] Department of Agricultural and Environmental Sciences, University of Bari Aldo Moro, 70126 Bari, Italy; dambrosioersilia@libero.it (E.D.); francesco.gentile@uniba.it (F.G.); giovanni.sanesi@uniba.it (G.S.)

[3] Department of Earth and Geo-Environmental Sciences, University of Bari Aldo Moro, 70121 Bari, Italy; simona.tripaldi@uniba.it

* Correspondence: mario.elia@uniba.it

Abstract: Landslides represent one of the most critical issues for landscape managers. They can cause injuries and loss of human life and damage properties and infrastructure. The spatial and temporal distribution of these detrimental events makes them almost unpredictable. Studies on landslide susceptibility assessment can significantly contribute to prioritizing critical risk zones. Further, landslide prevention and mitigation and the relative importance of the affecting drivers acquire even more significance in areas characterized by seismicity. This study aimed to investigate the relationship between a set of environmental variables and the occurrence of landslide events in an area of the Apulia Region (Italy). Logistic regression was applied to a landslide-prone area in the Apulia Region (Italy) to identify the main causative factors using a large dataset of environmental predictors (47). The results of this case study show that the logistic regression achieved a good performance, with an AUC (Area Under Curve) >70%. Therefore, the model developed would be a useful tool to define and assess areas for landslide occurrence and contribute to implementing risk mitigation strategy and land use policy.

Keywords: landslide; logistic regression; environmental hazard; risk assessment

Citation: Ganga, A.; Elia, M.; D'Ambrosio, E.; Tripaldi, S.; Capra, G.F.; Gentile, F.; Sanesi, G. Assessing Landslide Susceptibility by Coupling Spatial Data Analysis and Logistic Model. *Sustainability* **2022**, *14*, 8426. https://doi.org/10.3390/su14148426

Academic Editor: Salvador García-Ayllón Veintimilla

Received: 11 May 2022
Accepted: 4 July 2022
Published: 9 July 2022

Publisher's Note: MDPI stays neutral with regard to jurisdictional claims in published maps and institutional affiliations.

1. Introduction

Landslides, one of the most devastating natural/human-induced disasters worldwide, cause injuries and loss of human life as well as damage to properties and infrastructure [1,2]. Haque et al. [3] estimated that from 1995 to 2014 in Europe, there were a total of 476 landslide events, with 1370 deaths and 740 injuries recorded. These alarming numbers illustrate the need to enhance national and regional efforts to prevent or minimize such impacts on human and natural assets.

With 620,808 recorded events, Italy is one of the most affected European countries [4]. Every year landslide events cause the displacement of hundreds of thousands of people, building and infrastructure damages, and conspicuous loss of cultural heritage. To show the magnitude of the recent status, 172 main events were recorded in 2017, 146 in 2016, 311 in 2015, 211 in 2014, and 112 in 2013. Related economic damages have been estimated at c.a. EUR 2.5 billion per year [4] in the period between 2014 and 2020.

Identifying critical zones becomes essential to managing landslide susceptibility, especially in areas where landslides severely threaten human and natural resources and the country's massive cultural heritage. In this regard, the scientific community can help decision-makers by developing ad hoc risk analysis models based on empirical evidence and drivers of landslide occurrence. Geomorphological and topographic conditions, seismicity, intense rainfall, and anthropogenic factors, such as road and railway infrastructures, urban sprawl, land cover changes, and agricultural practices, represent the main predisposing and triggering factors of landslide events [5–7].

Studies focusing on landslide susceptibility assessment can provide an outstanding contribution to these purposes since they estimate the spatial distribution of landslide occurrence probability in a given area based on predisposing factors [1]. The literature is characterized by inventory-based studies, data-driven (including bivariate and multivariate statistics) analysis, knowledge-driven works, and probabilistic, physically-based, and deterministic approaches [2]. Recently, several machine learning-based techniques have been tested to develop landslide monitoring tools. However, these methods still feature significant drawbacks for modeling complex algorithms, including uncertain model performances, scarce aptitude to detect local variability, and the use of "black boxes", which is not appreciated yet by the operative world. Consequently, most landscape managers still prefer more traditional and sound models, such as logistic regression [8].

In fact, the most widely used method at local and regional scales is logistic regression [2,9,10], which has an advantage over other multivariate statistical methods in that it is independent of data distribution and able to handle continuous, categorical, and binary data [11]. In a logistic regression approach, the binary dependent variables are related to a vast dataset of independent drivers, such as topographic parameters, land uses, and others [12]. This study aims to test logistic regression in a survey area (Apulia, Italy), characterized by frequent seismic events and historically affected by widespread landslides. Specifically, we aimed to (*i*) understand the relationship between selected environmental variables and the occurrence probability of landslide events and (*ii*) produce landslide susceptibility maps for management, mitigation, and prevention purposes.

The proposed analytical framework can be applied to other areas worldwide with similar environmental features. Consequently, results may be helpful for landscape/environmental planners involved in landslide risk mitigation and land-use planning activities.

2. Materials and Methods

2.1. Study Area

The research was carried out across the Dauni Mountains (902.9 km^2; Min. Lat. 41°4′32.01″, Min. Long.14°55′59″, Max. Lat. 41°55′35.27″, Max. Long. 15°31′51.6″) which are located in the western part of the Apulia Region (Italy) (Figure 1). They belong to the Dauno Sub-Apennine dominion, representing the marginal external front of the Southern Apennine chain [13,14]. The chain units, affected by intense tectonic deformations, are thrusted over the Bradanic Trough unit, which in turn overlays the Apulian Foreland units. Folds and thrust structures and eastward verging, along with clay-rich flysch formations, characterize the structural and geological framework of the area [15,16].

As reported by the parametric catalog of Italian earthquakes CPTI15 [17,18], the study area has low seismic activity. However, more intense seismic activity characterizes the nearby Southern Apennines and Gargano Promontory areas, located less than 50 km from the study area [19,20]. Due to the more frequent and more energetic seismic activity of the Southern Apennine chain, the study area may be affected by seismic shaking able to induce landslides, as reported by Del Gaudio et al. (2012), who created a seismic hazard map of an area of the Daunia.

The elevation ranges between 73 and 1135 m a.s.l. (mean 566 m a.s.l.). The morphology is typically hilly-mountainous, strongly shaped by landslides favored by lithological and soil features, seismicity, and the area steepness (*vide infra*). The most diffuse lithotype are clay-limestone (CL), sands and conglomerates (SC), and clayey and clayey-limestone (CCL) units. Soils are strongly influenced by both parent material and the whole environmental features, being more developed in CL and CCL units (Alfisols and Ultisols) [21,22] while less (Inceptisols and Entisols) in SC units and where landslide movements modify their original nature. Land uses are featured by common arable crops (70%, with olives as main tree crops), broad-leaved, coniferous, and mixed forests (21%). Urban areas are very limited (1%). Mean annual rainfall (1918–2007) ranges from 533 to 931 mm. Mean monthly temperatures range from 5 °C in January to 25 °C in August, negatively correlated with the altitude.

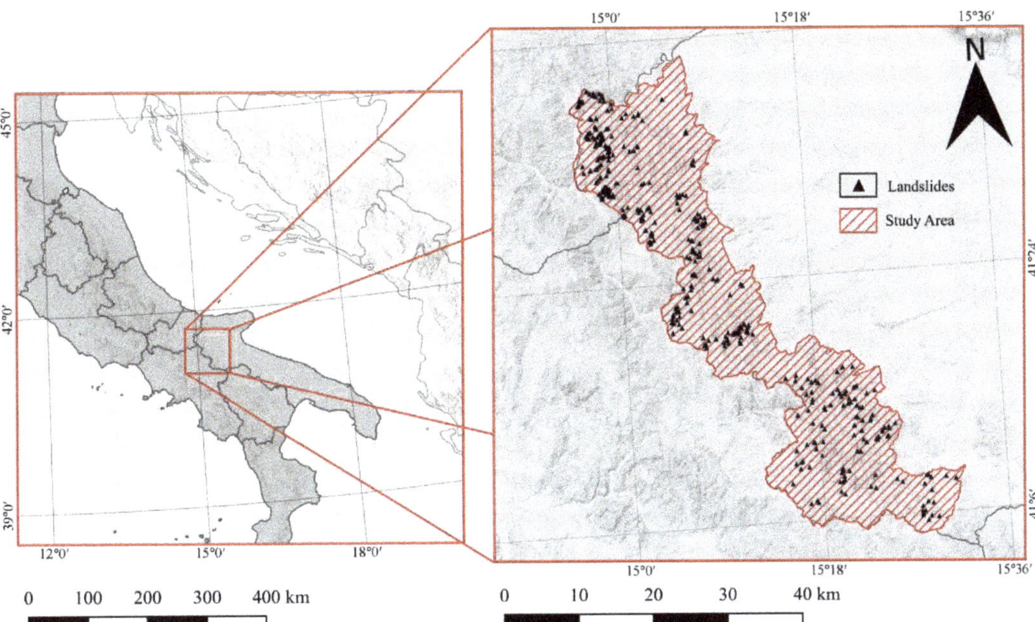

Figure 1. Dauni Mountains study area location with landslide detachment niches identification.

The study area is characterized by a high seismic hazard that, together with the previously reported environmental features, makes it extremely prone to landslide events; between 1918 and 2007, 515 mass movements were registered (Table 1).

Table 1. Number of landslides registered within the study area (1918–2007) with their main features.

Landslide Type	N.
Slow earthflow	228
Rotational/traslational landslide	143
Complex landslide	96
Rapid debris flow	18
Fall/topple	11
Area affected by numerous shallow landslide	11
Unclassified landslide	7
Sinkhole	1
Total number	515

2.2. Data Collection

We extracted data from multisource and multiscale geographic databases (Table S2). A total of 47 predictors were selected according to 7 main macro-class classification/factors: topography, seismic, geolithologic, pedologic, land use, morphologic, and climatic. Some of them are represented in the Figure S1.

The whole investigated area was divided into several grids of 250 square meters each. For each grid, the value of all predictors (vide supra) was calculated using GIS software. The seismic factor was evaluated using data from the national database and additionally evaluating the fault distance by applying the Euclidean distance algorithm.

Concerning the dependent variable, the location of landslide detachment niches was obtained from the landslides regional inventory. Since this database provides information concerning landslides verified between 1918 and 2007 without providing a specific occurrence date, the causative factors' assessment was limited to these years. Thus, climate causative factors were quantified as mean values for the same period. Meanwhile, the study area was reduced to subareas in which land use did not vary throughout this period in order not to consider the effect of land cover variation on slope stability. Available Corine Land Cover Maps (1990, 2000, 2006) were crossed, and only land cover persistence areas (582.2 km^2) were retained for further analysis. Overall, the total number of landslide detachment niches considered within the study area was 328.

Finally, the modified study area was divided into 9316 cells of 250 × 250 m, and, for each of them, the mean values of the causative factors were assessed. In addition, a dichotomic value (i.e., 0 or 1) linked to the absence or presence of landslide detachment niches was assigned to each cell as the dependent variable's value.

2.3. Causative Factor Selection

The first fundamental step was factor selection. Several factors occur in the process of landslide hazard assessment [1,2,23]. In fact, they are the result of interaction between intrinsic and external factors [24]. As a matter of fact, there are no definitive factor selection lists, and the tendency is often to rely on accessible databases [9]. However, the selected model allows for the use of a large number of independent variables, and therefore, a large number of databases have been considered, which are collected in the categories below (Figure 2).

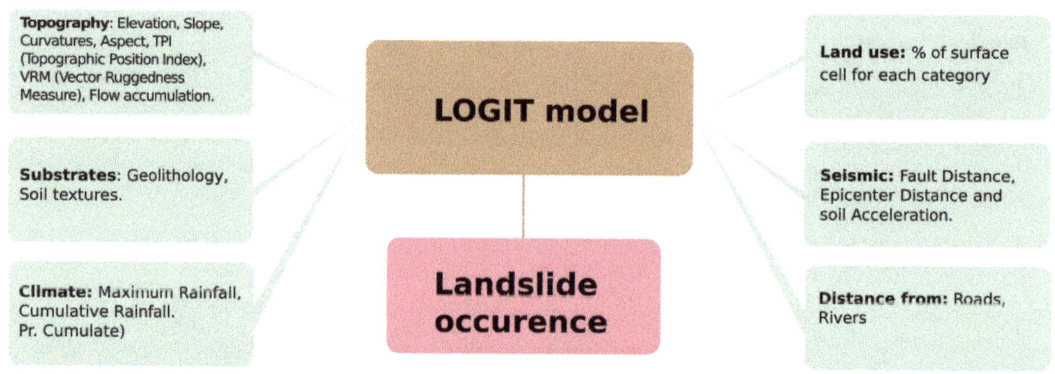

Figure 2. Flow chart of the methodological approach used.

The attached table (Table S1) shows all the calculated factors, the original source, and the spatial/temporal resolution. These data were acquired from public data infrastructures managed by individual regions, Italian Environmental Agencies (such as ISPRA), and the European Environmental Agency (EEA). According to [23], we classified the factors into (*i*) predisposing factors (e.g., topography and geological substrate), (*ii*) triggering factors (e.g., precipitations and earthquakes), and (*iii*) accelerating factors (e.g., human-altered systems).

Regarding predisposing factors, the following groups of predictors were considered:

- *Morphology.* The landform is one of the most important factors influencing landslide occurrence [25]. The slope angle is considered one of the main triggering factors for landslides [26,27]. The aspect is decisive for triggering landslide processes. It is strictly connected to climatic conditions [28] and the factor impacting slope stability [29]. The topographical indices are useful for describing and quantifying the intensity and characteristics of hydrological processes [25]. The main topographical indices are reported in Table S1.

- The topographic factors were calculated starting with a high-resolution digital terrain model (DTM) with 8-m resolution.
- *Geolitology.* The lithological features also define the resistance capacity of the substrate to external forces. Indeed, this is related to material strength because they have varied compositions and structures for different rock types [9,30,31].
- *Pedology* (texture). Soil texture is one of the soil's main physical features. Previous work has shown how soils of different textures can have different levels of susceptibility to the triggering of landslides [24]. Hamza et al. (2017) indicated that limestone has more probability of landslide occurrence, thus providing a hazard index value of 1.10 (Table 2). In contrast, the probability of occurrence of landslide for gypsum and sandstone is comparatively lower, with a hazard index ranging from 0.80 to 0.77.
- *Distance from river.* This parameter has a strong link to the erosion process in hilly regions [9]. Rivers and streams play an important role in landscape modification by modeling the landforms and shaping the lithological substrate [32]. The distance from the river is calculated as the Euclidean distance by the river stream extracted from a high resolution (1:10000) land use map.

Table 2. List of causative factors.

Class of Factors	Factors Selected
Topography	Elevation; Flow accumulation; Planform curvature; Profile curvature; Standard curvature; Slope; Aspect; Topographic indexes (Slope roughness, Terrain roughness, Topographic wetness index, Vector Ruggedness Measure, Topographic Position Index, Topographic factor).
Seismic Factors	Peak ground acceleration; Distance to active fault; Distance to earthquake epicenter.
Geolithology	Percentage of cover area of Terrace alluvium; Sandstones and clays; Clays and marls; Clays; Debris; Alluvium and river-lake deposits; Beaches; Lakes and glaciers; Sands and conglomerates; Sandy and sand-marly units; Clayey and clayey-limestone units; Marly limestone units.
Pedology	Percentage of cover area of clay; Percentage of sand; Percentage of silt.
Land Cover	Percentage of cover area of Urban area; Agricultural area; Grassland; Forest; Natural area, Rivers and lakes.
Morphology	Distance to rivers; Distance to roads.
Climate	Mean annual maximum rainfall observed in 1 day; Mean annual rainfall.

Considered triggering factors were:

- *Seismic factor.* The dynamics of triggering landslides following an earthquake are well documented [33–35]. Earthquake-induced landslides are one of the worst natural hazards [36]. Specifically, there is a strong relationship between distance from faults and epicenters [37]. Furthermore, the factors triggering earthquake-induced landslides are also related to earthquake characteristics, such as ground acceleration [11].
- *Climate.* The climate affects landslides both directly and indirectly. For instance, intense rainfall is the most common triggering agent of landslides [38].

Finally, the following accelerating factors were considered:

- *Land cover/land use.* It influences the ability to eventually prevent and/or limit the extent and distribution of landslides. Forested and natural systems usually showed a statistically lower landslide occurrence, and they have an important role in preventing it. This is especially true when located in areas with critical topographical and lithological conditions [39].

- *Distance from road.* Roads are an important susceptibility factor to the triggering of landslides as their construction could modify the land topography and shape. As a matter of fact, roads represent an important driver of the soil profile and hillslope alteration [2].

Variable Selection

Predictors were selected starting from the original 47×9316 observation matrix dataset. In a multivariate statistical analysis, the selection process is one of the most important tasks [40]. In this work, the elimination of highly correlated predictors was carried out in the first variables selection according to previous works [41,42]. In this step, the predictors with Pearson's correlation coefficient >0.75 were removed. The final correlation matrix is reported in Table S1.

2.4. Logistic Regression

Logistic regression analyzes the relationship between multiple independent variables and a categorical dependent variable while estimating the occurrence probability of an event by fitting data to a logistic curve [43]. The logistic function is described as [31]:

$$P = \frac{1}{1 + (exp^{-Z})} \tag{1}$$

where P is the probability of landslide occurrence and Z a linear combination of casual X_i factors:

$$Z = \beta_0 + \beta_1 x_1 + \beta_2 x_2 + \beta_3 x_3 + \ldots + \beta_n x_n \tag{2}$$

where β_i are the landslide casual factor's coefficients.

The models were performed using R statistical software. In order to operate a predictor's selection, the regressions were performed using a stepwise model. Stepwise regression is a set of iterative search and model comparison procedures that identify which independent variables have the strongest association with the dependent one (Draper and Smith, 1981; Hauser, 1974). Using this approach, the model was performed for each type of landslide and for the total landslide.

For the four models, the Area under Curve (AUC), the Accuracy, and the Akaike information criterion (AIC) were calculated. In order to evaluate the predictive capacity of the logistic model, the McFadden pseudo R^2 area was calculated. This is intended as a logistic regression analog of R^2 and uses ordinary least-squares (OLS) regression [44].

3. Results

3.1. Main Factors Affecting Slope Stability

Table 3 reports the results of significant drivers for each model. In particular, stepwise logistic regression was performed using four different sub-datasets, from the whole landslide dataset, according to the following typology: (*i*) total landslides (TL); (*ii*) rotational/translational landslide (RTL); (*iii*) slow earth flow (SEF), and (*iv*) complex landslide (CL).

Table 3. List of coefficient drivers. Table S2 reports a complete list of factors with their significant metadata. The value of factors removed by stepwise logistic regression was not reported.

Factor	Total landslide (TL)			Rotational/Traslational Landslide (RTL)			Slow Earth Flow (SEF)			Complex Landslide (CL)		
	Coeff.	SE	*p*	Coeff.	SE	*p*	Coeff.	SE	*p*	Coeff.	SE	*p*
	−0.381	0.030	***	−0.704	0.037	***	−0.694	0.036	***	−0.388	0.031	***
T_TWI	−0.099	0.025	***	/	/	/	/	/	/	−0.207	0.023	***
T_VRM	0.086	0.023	***	0.076	0.027	*	/	/	/	0.050	0.026	.

Table 3. *Cont.*

Factor	Total landslide (TL)			Rotational/Traslational Landslide (RTL)			Slow Earth Flow (SEF)			Complex Landslide (CL)		
	Coeff.	SE	*p*	Coeff.	SE	*p*	Coeff.	SE	*p*	Coeff.	SE	*p*
T_DTM	/	/	/	0.112	0.026	***	−0.081	0.027	**	−0.040	0.026	.
T_F_acc	−0.042	0.025	.	−0.244	0.042	***	/	/	/	−0.223	0.040	***
T_Curv_pl	−0.053	0.022	**	0.100	0.025	***	−0.039	0.025	.	−0.064	0.023	**
T_Curv_st	0.106	0.022	***	0.146	0.024	***	−0.043	0.026	.	−0.037	0.020	.
T_N	/	/	/	0.054	0.025	*	/	/	/	−0.112	0.024	**
T_NE	/	/	/	−0.196	0.030	***	/	/	/	/	/	/
T_E	/	/	/	−0.172	0.029	***	/	/	/	/	/	/
T_SE	/	/	/	−0.040	0.025	.	/	/	/	/	/	/
T_S	0.117	0.021	***	/	/	/	0.061	0.025	.	/	/	/
T_SW	/	/	/	0.038	0.023	.	−0.052	0.025	.	0.060	0.021	**
T_W	−0.038	0.023	.	0.141	0.027	***	−0.088	0.025	***	−0.043	0.024	.
T_NW	0.067	0.022	**	/	/	/	/	/	/	0.089	0.021	.
T_FLAT	−0.060	0.035	.	−0.235	0.041	***	−0.089	0.034	**	/	/	/
T_Ls	0.182	0.024	***	0.115	0.029	***	0.175	0.027	***	0.132	0.022	***
T_TPI	0.112	0.022	***	0.060	0.022	**	0.253	0.022	***	−0.036	0.021	.
S_Epic	0.056	0.027	*	−0.252	0.028	***	0.286	0.031	***	−0.112	0.024	*
G_TA	−0.071	0.043	.	−0.098	0.046	*	−0.081	0.039	*	−0.141	0.041	***
G_SSC	0.137	0.018	***	−0.113	0.024	***	0.206	0.017	***	0.030	0.020	.
G_CM	−0.042	0.022	.	−0.132	0.025	***	−0.224	0.039	***	0.034	0.019	.
G_C	−0.236	0.031	***	−0.333	0.040	***	−0.260	0.031	***	−0.319	0.035	***
G_DDB	−0.262	0.037	***	−0.140	0.043	***	−0.244	0.038	***	−0.262	0.035	***
G_LG	/0.134	0.044	**	−0.077	0.038	*	−0.069	0.037	.	−0.068	0.037	.
G_SC	0.042	0.023	.	0.061	0.023	**	−0.280	0.033	***	/	/	/
G_SSM	0.045	0.019	*	−0.136	0.034	**	−0.119	0.036	**	0.101	0.010	***
G_CCL	0.099	0.022	***	−0.092	0.028	**	0.083	0.023	***	0.103	0.022	*
G_ML	/	/	/	0.157	0.027	**	/	/	/	−0.077	0.024	**
P_sand	−0.110	0.026	***	0.095	0.028	***	−0.241	0.030	***	/	/	/
P_silt	−0.169	0.023	***	−0.311	0.029	***	−0.161	0.024	***	0.106	0.028	***
LU_urb	0.068	0.013	***	0.130	0.011	***	−0.160	0.042	***	0.066	0.013	**
LU_grs	−0.072	0.024	**	−0.073	0.022	***	−0.257	0.037	***	−0.137	0.030	**
LU_for	/	/	/	0.079	0.023	***	−0.089	0.026	***	0.148	0.021	***
LU_nat	0.149	0.016	***	0.111	0.018	***	0.064	0.017	***	/	/	/
D_riv	−0.159	0.024	***	−0.041	0.022	.	−0.213	0.031	***	0.073	0.025	**
D_road	−0.338	0.026	**	−0.389	0.028	***	−0.303	0.027	***	−0.380	0.028	***
C_p_max	−0.039	0.023	.	/	/	/	−0.164	0.025	***	0.168	0.025	***

*** $p < 0.001$, ** $p < 0.01$, * $p < 0.05$, . $p < 0.1$.

3.2. Model Performance

In Table 4, performances of the model applied on four sub-datasets (*vide supra*) were reported.

Table 4. Models performance.

Landslide	AIC *	AUC **	Accuracy	McFadden's Pseudo R^2
All	8004.13	0.76	0.71	0.12
Rotational/Translational	7407.88	0.68	0.73	0.19
Slow Earth Flow	7200.94	0.80	0.77	0.21
Complex Landslide	7828.44	0.69	0.70	0.13

* AIC = Akaike Information Criterion. ** AUC = Area Under Curve.

Slow earth flow showed the best performance, with an accuracy >0.75, meaning that this can be considered an accurate model. Overall, all implemented models showed values > 0.70, thus considered suitable. The area under curve (AUC) values are comparable with previous similar studies [45]. To evaluate the model's performance, McFadden's

Pseudo R squared, instead of R squared, was applied. A McFadden's Pseudo R squared between 0.2–0.4 indicates a very good fit. The RTL and SEF landslide models showed the best fit with 0.19 and 0.21, respectively.

3.3. Susceptibility Mapping

According to the method performed by Elia et al. (2019), four susceptibility maps were created; each recorded landslide's detachment niche was georeferenced.

Implemented models and resulting maps show the landslide occurrence probability within the study area (Figure 3). Based on probability values, areas were classified in terms of susceptibility as: (i) high (>0.65, red areas); (ii) moderate (0.50–0.65, orange); (iii) moderate-low (0.30–0.50, yellow); (iv) low (20–30, light green); and (v) very low (0.00–0.20, green). In the total landslide map, areas affected by high-level susceptibility are distributed mainly in the northwest area. Conversely, in RTL maps, areas featured by high-level susceptibility are localized in the southern part of the region. In the slow heart flow model maps, the susceptibility distribution is closer to that observed for the total model; however, areas with high susceptibility are more concentrated in the extreme northwestern part. In the complex landslides map, high-level susceptibility spots were identified in several parts of the investigated area. Overall, in every implemented map, the low susceptibility is concentrated in the central–east part of the study area, in correspondence with the less hilly areas. In fact, the study area shows an elevation gradient from west to east, and the landslides distribution follows this gradient, especially in RTL and CL.

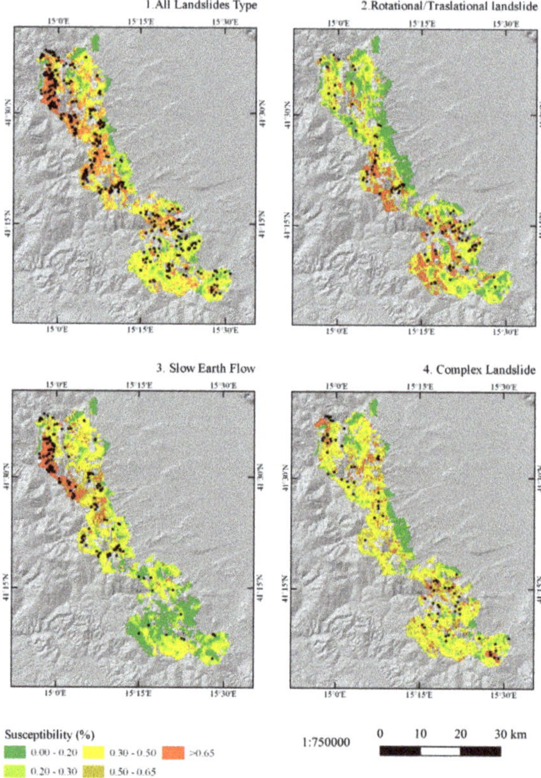

Figure 3. Landslide triggering probability distribution maps. The black dots represent the landslide positions.

4. Discussion

4.1. Logistic Regression Metrics

The assessment of areas likely to produce land movement phenomena was carried out by adopting a logistic regression model, one of the most used approaches in landslide susceptibility analysis. In fact, it allows predicting where landslides are more likely to occur based on the relationship between past events and ad hoc selected causative factors. The logistic regression was demonstrated to be highly suitable in assessing what the most correlated and the most predictive factors in estimating the landslide occurrence susceptibility are. This method provided consistent and significant results in the investigated area. By carrying out a cutoff of the probability of occurrence equal to 0.5, it appears that about 80% of the territory has been classified correctly, i.e., that for 80% of cases where no event has occurred, the probability is lower than 0.5; on the opposite, it is higher where events occurred.

According to the literature, AUC values of 0.5–0.7 are considered low accuracy, 0.7–0.9 suggest valuable applications, and values around 0.9 indicate high accuracy (Swets 1988). In our study, the AUC indicates 0.76, 0.68, 0.80, and 0.69 for TL, RTL, SEF, and CL, respectively, between predicted probabilities and observed outcomes. The findings are consistent with other models developed to estimate landslide susceptibility worldwide, especially given the high rate of human and natural variability across the study area [34,46,47].

4.2. Correlation Analysis

As shown in Table 2, the correlation analysis revealed that only 37 of the 47 causative factors were uncorrelated and were thus included in the subsequent logistic regression. These factors are related to topography (T_E, T_F_acc, T_FLAT, T_NW, T_TPI, T_S, T_TWI, T_VRM, T_W), seismicity (S_Epic), geolithology (G_TA, G_SSC, G_CM, G_C, G_DDB, G_LG, G_SC, G_SSM, G_CCL, G_ML), land cover (LU_urb, LU_grs, LU_for, LU_nat), morphology (D_riv, D_road), and climate (C_p_max). It is well known [48–51] that the slope length and steepness factor (T_Ls) is one of the main landslide predictors, being a parameter used to characterize the effects of topography and hydrology on soil loss [52,53]. In our study, the T_Ls showed the highest coefficient in predicting the totality of recorded events and one of the most predicting variables for the other landslide types, evidencing its influence in the evolution of landslide types. Similarly, Huangfu et al. (2021) [54] found that slope is one of the three most important predictors affecting landslide geo-hazard risk in China.

Generally speaking, landslides occurring on steep slopes are triggered by a reduction of the apparent cohesion of colluvium (soil and debris accumulated upon an impermeable bedrock), resulting from water infiltration into the soil. Concave shape also affects the development of rotational/translational mass movement, reflecting the importance of the hydrological regime in slope stability. In these areas, a convergence of surface and subsurface water streams saturates rapidly the soil, which becomes more prone to movements [51].

As expected, most of the other topographic variables showed a high predicting power in landslide occurrence. For example, T_TPI showed an elevated ability to predict the landslide, particularly the SEF. This finding is consistent with other studies [55,56], showing how the TPI is essential for slope stability analysis aiming at classifying priority zones in landslide occurrence [57]. However, the configuration of such a set of predictors is quite different among the four landslide types. Table 2 suggests that topographic and land cover variables are the most significant predictors for RTL type; seismic and geolithological variables had the highest predicting power for SEF, while the presence of forest and annual maximum rainfall significantly affects CL. These findings demonstrated that different landslide types are regulated by different predisposing or preparatory factors [58]. For example, distance to earthquake epicenter (S_Epic) had only a positive and significant relationship with the SEF landslide type. This is probably because mountainous landscapes, susceptible to landslides, are often characterized by moderate rates of earthquake events. Many authors similarly found this relationship [59–61], arguing that the presence of a fault damage zone is the primary control on the distribution of earth flows; this was primarily due to ground quaking and the subsequent expansion of superficial debris favoring accelerated water infiltration.

On the contrary, all types of landslides showed a negative relationship with the distance from the roads, while only two types with the distance from rivers. The more the distance from the road increases, the more the probability of a landslide decreases. This negative trend was observed by other authors worldwide [62–65], highlighting the importance of anthropogenic impact on slope stability. Additionally, other authors [26,66] reported that the landslide susceptibility gradually increases up to a proximity zone of 200 m from a road network.

4.3. Perspectives

Our findings in the Apulia region are of primary importance for a more broad assessment of landslide and seismic risk management. The proposed approach can be applied to a larger scale of analysis which constitutes one of the future perspectives. Indeed, we recognize the limits of our study. Firstly, to achieve a significant landslide susceptibility assessment, we used the most detailed available datasets on landslide occurrence and environmental predictors. However, although we employed the most harmonized and available dataset, we are aware that there are several limitations in its accuracy. Numerous factors such as the phenomenon's complexity, the high diversity of geomorphic landslide features, and the use of different mapping and sampling procedures undoubtedly affected the accuracy and precision of the applied landslide dataset [67–69]. Additionally, our model results from data collected in dynamic landscapes and ecosystems constantly subjected to changes. Therefore, starting from these findings, new and innovative approaches must be developed as new knowledge is obtained from a geological, hydrological, and geotechnical point of view.

5. Conclusions

Landslides are one of the most devastating natural disasters, causing human and economical losses. In order to mitigate landslide risk, it is fundamental to identify critical risk zones. In the current study, logistic regression was applied to a landslide-prone area in the Apulia Region (Italy) in order to identify the main causative factors and produce a landslide susceptibility map. The model has never been applied before in the study area.

In this case study, it was found that logistic regression achieved a good performance, with an AUC value higher than 70%. Thus, it could constitute a useful tool in identifying critical areas for landslide occurrence and defining a risk mitigation strategy and land use policy. For example, in high–moderate susceptibility areas, mitigation infrastructures, resilience building, and a relocation strategy for families living in those areas could be provided. In addition, as a result of this study, it was found that human infrastructures, such as road networks, are one of the main elements triggering landslides. Thus, peculiar attention must be paid to the choice of the route when designing the transport networks.

However, further detailed studies should be conducted to compare different models, widen the number of causative factors considered (i.e., geotechnical data), and assess landslide susceptibility individually for each type of landslide since the current study combined different types of landslides. Nonetheless, this study gives a comprehensive and preliminary understanding of the likely future insights for researchers and policy makers.

Supplementary Materials: The following supporting information can be downloaded at: https://www.mdpi.com/article/10.3390/su14148426/s1; Table S1: Correlation matrix; Table S2: Factors table; Figure S1: maps of some important driving factors.

Author Contributions: Conceptualization, A.G., M.E. and E.D.; methodology A.G. and M.E.; software, A.G. and E.D.; validation A.G., M.E., F.G. and G.S.; formal analysis, A.G.; investigation, A.G., M.E. and E.D.; resources, A.G., M.E., F.G. and G.S.; data curation, A.G., M.E. and E.D.; writing—original draft preparation, A.G., M.E., F.G., G.F.C., S.T. and G.S.; writing—review and editing, A.G., M.E., F.G., G.F.C., S.T. and G.S.; visualization A.G.; supervision, F.G. and G.S.; project administration, F.G. and G.S.; funding acquisition, F.G. and G.S. All authors have read and agreed to the published version of the manuscript.

Funding: This research has been developed within the project "Earthquake disasters management integrated system—ERMIS" (Code 5003751) by INTERREG V-A Greece-Italy (EL-IT) 2014–2020.

Institutional Review Board Statement: Not applicable.

Informed Consent Statement: Not applicable.

Conflicts of Interest: The authors declare no conflict of interest.

References

1. Chen, W.; Sun, Z.; Han, J. Landslide Susceptibility Modeling Using Integrated Ensemble Weights of Evidence with Logistic Regression and Random Forest Models. *Appl. Sci. Switz.* **2019**, *9*, 171. [CrossRef]
2. Nsengiyumva, J.B.; Luo, G.; Amanambu, A.C.; Mind'je, R.; Habiyaremye, G.; Karamage, F.; Ochege, F.U.; Mupenzi, C. Comparing Probabilistic and Statistical Methods in Landslide Susceptibility Modeling in Rwanda/Centre-Eastern Africa. *Sci. Total Environ.* **2019**, *659*, 1457–1472. [CrossRef] [PubMed]
3. Haque, U.; Blum, P.; da Silva, P.F.; Andersen, P.; Pilz, J.; Chalov, S.R.; Malet, J.-P.; Auflič, M.J.; Andres, N.; Poyiadji, E.; et al. Fatal Landslides in Europe. *Landslides* **2016**, *13*, 1545–1554. [CrossRef]
4. Trigila, A.; Iadanza, C.; Bussettin, M.; Lastoria, B. *Dissesto Idrogeologico in Italia: Pericolosità e Indicatori di Rischio*; Rapporto_Dissesto_Idrogeologico_ISPRA_287_2018_Web.Pdf; ISPRA: Milan, Italy, 2018.
5. Guzzetti, F.; Reichenbach, P.; Cardinali, M.; Galli, M.; Ardizzone, F. Probabilistic Landslide Hazard Assessment at the Basin Scale. *Geomorphology* **2005**, *72*, 272–299. [CrossRef]
6. Nepal, N.; Chen, J.; Chen, H.; Wang, X.; Pangali Sharma, T.P. Assessment of Landslide Susceptibility along the Araniko Highway in Poiqu/Bhote Koshi/Sun Koshi Watershed, Nepal Himalaya. *Prog. Disaster Sci.* **2019**, *3*, 100037. [CrossRef]
7. Lafortezza, R.; Tanentzap, A.J.; Elia, M.; John, R.; Sanesi, G.; Chen, J. Prioritizing fuel management in urban interfaces threatened by wildfires. *Ecol. Indic.* **2015**, *48*, 342–347. [CrossRef]
8. Elia, M.; D'Este, M.; Ascoli, D.; Giannico, V.; Spano, G.; Ganga, A.; Colangelo, G.; Lafortezza, R.; Sanesi, G. Estimating the Probability of Wildfire Occurrence in Mediterranean Landscapes Using Artificial Neural Networks. *Environ. Impact Assess. Rev.* **2020**, *85*, 106474. [CrossRef]
9. Soma, A.S.; Kubota, T.; Mizuno, H. Optimization of Causative Factors Using Logistic Regression and Artificial Neural Network Models for Landslide Susceptibility Assessment in Ujung Loe Watershed, South Sulawesi Indonesia. *J. Mt. Sci.* **2019**, *16*, 383–401. [CrossRef]
10. Meinhardt, M.; Fink, M.; Tünschel, H. Landslide Susceptibility Analysis in Central Vietnam Based on an Incomplete Landslide Inventory: Comparison of a New Method to Calculate Weighting Factors by Means of Bivariate Statistics. *Geomorphology* **2015**, *234*, 80–97. [CrossRef]
11. Dou, J.; Bui, D.T.; Yunus, A.P.; Jia, K.; Song, X.; Revhaug, I.; Xia, H.; Zhu, Z. Optimization of Causative Factors for Landslide Susceptibility Evaluation Using Remote Sensing and GIS Data in Parts of Niigata, Japan. *PLoS ONE* **2015**, *10*, e0133262. [CrossRef]
12. Elia, M.; Giannico, V.; Spano, G.; Lafortezza, R.; Sanesi, G. Likelihood and frequency of recurrent fire ignitions in highly urbanised Mediterranean landscapes. *Int. J. Wildland Fire* **2020**, *29*, 120. [CrossRef]
13. Merlini, S.; Mostardini, F. Appennino Centro-Meridionale: Sezioni Geologiche e Proposta Di Modello Strutturale. In Proceedings of the Geologia dell'Italia Centrale, Congresso Nazionale, Rome, Italy, 30 September–4 October 1986; Volume 73, pp. 147–149.
14. Patacca, E.; Scandone, P. Geology of the Southern Apennines. *Boll. Soc. Geol. Ital.* **2007**, *7*, 75–112.
15. Dazzaro, L.; Di Nocera, S.; Pescatore, T.; Rapisardi, L.; Romeo, M.; Russo, B.; Senatore, M.; Torre, M. Geologia Del Margine Della Catena Appenninica Tra Il F. Fortore e Il T. Calaggio (Appennino Meridionale). *Mem. Della Soc. Geol. Ital.* **1988**, *41*, 411–422.
16. Dazzaro, L.; Rapisardi, L. Integration of persistent scatterer interferometry and ground data for landslide monitoring: The Pianello landslide (Bovino, Southern Italy). *Mem. Soc. Geol. It.* **1996**, *16*, 143–147.
17. Rovida, A.; Locati, M.; Camassi, R.; Lolli, B.; Gasperini, P.; Antonucci, A. Italian Parametric Earthquake Catalogue (CPTI15), Version 3.0. Istituto Nazionale Di Geofisica e Vulcanologia (INGV). Available online: https://emidius.mi.ingv.it/CPTI15-DBMI1 5_v3.0/ (accessed on 1 June 2022).
18. Rovida, A.; Locati, M.; Camassi, R.; Lolli, B.; Gasperini, P. The Italian Earthquake Catalogue CPTI15. *Bull. Earthq. Eng.* **2020**, *18*, 2953–2984. [CrossRef]
19. Valensise, G.; Pantosti, D.; Basili, R. Seismology and Tectonic Setting of the 2002 Molise, Italy, Earthquake. *Earthq. Spectra* **2004**, *20*, 23–37. [CrossRef]
20. Miccolis, S.; Filippucci, M.; de Lorenzo, S.; Frepoli, A.; Pierri, P.; Tallarico, A. Seismogenic Structure Orientation and Stress Field of the Gargano Promontory (Southern Italy) From Microseismicity Analysis. *Front. Earth Sci.* **2021**, *9*, 179. [CrossRef]
21. Malagnini, L.; Scognamiglio, L.; Mercuri, A.; Akinci, A.; Mayeda, K. Strong Evidence for Non-Similar Earthquake Source Scaling in Central Italy. *Geophys. Res. Lett.* **2008**, *35*, 17. [CrossRef]
22. Soil Survey Staff. *Keys to Soil Taxonomy*, 12th ed.; USDA—Natural Resources Conservation Service: Washington, DC, USA, 2014.

23. Santini, M.; Grimaldi, S.; Nardi, F.; Petroselli, A.; Rulli, M.C. Pre-Processing Algorithms and Landslide Modelling on Remotely Sensed DEMs. *Geomorphology* **2009**, *113*, 110–125. [CrossRef]
24. Hamza, T.; Raghuvanshi, T.K. GIS Based Landslide Hazard Evaluation and Zonation – A Case from Jeldu District, Central Ethiopia. *J. King Saud Univ. Sci.* **2017**, *29*, 151–165. [CrossRef]
25. Kadavi, P.R.; Lee, C.-W.; Lee, S. Landslide-Susceptibility Mapping in Gangwon-Do, South Korea, Using Logistic Regression and Decision Tree Models. *Environ. Earth Sci.* **2019**, *78*, 116. [CrossRef]
26. Myronidis, D.; Papageorgiou, C.; Theophanous, S. Landslide Susceptibility Mapping Based on Landslide History and Analytic Hierarchy Process (AHP). *Nat. Hazards* **2016**, *81*, 245–263. [CrossRef]
27. Rozos, D.; Bathrellos, G.D.; Skillodimou, H.D. Comparison of the Implementation of Rock Engineering System and Analytic Hierarchy Process Methods, upon Landslide Susceptibility Mapping, Using GIS: A Case Study from the Eastern Achaia County of Peloponnesus, Greece. *Environ. Earth Sci.* **2011**, *63*, 49–63. [CrossRef]
28. Bednarik, M.; Magulová, B.; Matys, M.; Marschalko, M. Landslide Susceptibility Assessment of the Kraľovany–Liptovský Mikuláš Railway Case Study. *Phys. Chem. Earth Parts ABC* **2010**, *35*, 162–171. [CrossRef]
29. Youssef, A.M.; Al-Kathery, M.; Pradhan, B. Landslide Susceptibility Mapping at Al-Hasher Area, Jizan (Saudi Arabia) Using GIS-Based Frequency Ratio and Index of Entropy Models. *Geosci. J.* **2015**, *19*, 113–134. [CrossRef]
30. Kanungo, D.P.; Arora, M.K.; Sarkar, S.; Gupta, R.P. A Comparative Study of Conventional, ANN Black Box, Fuzzy and Combined Neural and Fuzzy Weighting Procedures for Landslide Susceptibility Zonation in Darjeeling Himalayas. *Eng. Geol.* **2006**, *85*, 347–366. [CrossRef]
31. Rasyid, A.R.; Bhandary, N.P.; Yatabe, R. Performance of frequency ratio and logistic regression model in creating GIS based landslides susceptibility map at Lompobattang Mountain, Indonesia. *GeoEnviron. Disasters* **2016**, *3*, 19. [CrossRef]
32. Meten, M.; PrakashBhandary, N.; Yatabe, R. Effect of Landslide Factor Combinations on the Prediction Accuracy of Landslide Susceptibility Maps in the Blue Nile Gorge of Central Ethiopia. *Geoenviron. Disasters* **2015**, *2*, 9. [CrossRef]
33. Rollo, F.; Rampello, S. Probabilistic assessment of seismic-induced slope displacements: An application in Italy. *Bull. Earthq. Eng.* **2021**, *19*, 4261–4288. [CrossRef]
34. Cao, J.; Zhang, Z.; Wang, C.; Liu, J.; Zhang, L. Susceptibility Assessment of Landslides Triggered by Earthquakes in the Western Sichuan Plateau. *Catena* **2019**, *175*, 63–76. [CrossRef]
35. Su, L.; Hu, K.; Zhang, W.; Wang, J.; Lei, Y.; Zhang, C.; Cui, P.; Pasuto, A.; Zheng, Q. Characteristics and Triggering Mechanism of Xinmo Landslide on 24 June 2017 in Sichuan, China. *J. Mt. Sci.* **2017**, *14*, 1689–1700. [CrossRef]
36. Tian, Y.; Xu, C.; Hong, H.; Zhou, Q.; Wang, D. Mapping Earthquake-Triggered Landslide Susceptibility by Use of Artificial Neural Network (ANN) Models: An Example of the 2013 Minxian (China) Mw 5.9 Event. *Geomat. Nat. Hazards Risk* **2019**, *10*, 1–25. [CrossRef]
37. Chen, X.L.; Yu, L.; Wang, M.M.; Li, J.Y. Brief Communication: Landslides Triggered by the $M_s = 7.0$ Lushan Earthquake, China. *Nat. Hazards Earth Syst. Sci.* **2014**, *14*, 1257–1267. [CrossRef]
38. Sidle, R.C. Using Weather and Climate Information for Landslide Prevention and Mitigation. In *Climate and Land Degradation*; Sivakumar, M.V.K., Ndiang'ui, N., Eds.; Environmental Science and Engineering; Springer: Berlin/Heidelberg, Germany, 2007; pp. 285–307. ISBN 978-3-540-72438-4.
39. Persichillo, M.G.; Bordoni, M.; Meisina, C. The Role of Land Use Changes in the Distribution of Shallow Landslides. *Sci. Total Environ.* **2017**, *574*, 924–937. [CrossRef]
40. Zellner, D.; Keller, F.; Zellner, G.E. Variable Selection in Logistic Regression Models. *Commun. Stat. Simul. Comput.* **2004**, *33*, 787–805. [CrossRef]
41. D'Ambrosio, E.; De Girolamo, A.M.; Barca, E.; Ielpo, P.; Rulli, M.C. Characterising the Hydrological Regime of an Ungauged Temporary River System: A Case Study. *Environ. Sci. Pollut. Res.* **2017**, *24*, 13950–13966. [CrossRef]
42. Elia, M.; Giannico, V.; Lafortezza, R.; Sanesi, G. Modeling Fire Ignition Patterns in Mediterranean Urban Interfaces. *Stoch. Environ. Res. Risk Assess.* **2019**, *33*, 169–181. [CrossRef]
43. Park, H.-A. An Introduction to Logistic Regression: From Basic Concepts to Interpretation with Particular Attention to Nursing Domain. *J. Korean Acad. Nurs.* **2013**, *43*, 154. [CrossRef]
44. Smith, T.J.; McKenna, C.M. A Comparison of Logistic Regression Pseudo R2 Indices. *Mult. Linear Regres. Viewp.* **2013**, *39*, 10.
45. Aditian, A.; Kubota, T.; Shinohara, Y. Comparison of GIS-Based Landslide Susceptibility Models Using Frequency Ratio, Logistic Regression, and Artificial Neural Network in a Tertiary Region of Ambon, Indonesia. *Geomorphology* **2018**, *318*, 101–111. [CrossRef]
46. Chen, W.; Yan, X.; Zhao, Z.; Hong, H.; Bui, D.T.; Pradhan, B. Spatial Prediction of Landslide Susceptibility Using Data Mining-Based Kernel Logistic Regression, Naive Bayes and RBFNetwork Models for the Long County Area (China). *Bull. Eng. Geol. Environ.* **2019**, *78*, 247–266. [CrossRef]
47. Murillo-García, F.G.; Steger, S.; Alcántara-Ayala, I. Landslide Susceptibility: A Statistically-Based Assessment on a Depositional Pyroclastic Ramp. *J. Mt. Sci.* **2019**, *16*, 561–580. [CrossRef]
48. Yang, X.; Yang, X. Digital Mapping of RUSLE Slope Length and Steepness Factor across New South Wales, Australia. *Soil Res.* **2015**, *53*, 216–225. [CrossRef]
49. Chen, W.; Pourghasemi, H.R.; Kornejady, A.; Zhang, N. Landslide Spatial Modeling: Introducing New Ensembles of ANN, MaxEnt, and SVM Machine Learning Techniques. *Geoderma* **2017**, *305*, 314–327. [CrossRef]
50. Çellek, S. Effect of the Slope Angle and Its Classification on Landslide. *Nat. Hazards Earth Syst. Sci. Discuss.* **2020**, 1–23. [CrossRef]

51. Akinci, H.; Zeybek, M.; Dogan, S. *Evaluation of Landslide Susceptibility of Şavşat District of Artvin Province (Turkey) Using Machine Learning Techniques*; IntechOpen: London, UK, 2021; ISBN 978-1-83969-024-2.
52. Hudson, N.W. Soil and Water Conservation (Second Edition). By F. R. Troeh, J. A. Hobbs and R. L. Donahue. Hemel Hempstead, Herts, UK: Prentice Hall, (1991), Pp. 530, £55.65, ISBN 13-832324-X. *Exp. Agric.* **1992**, *28*, 127. [CrossRef]
53. Troeh, F.R.; Hobbs, J.A.; Donahue, R.L.; Troeh, F.R. *Soil and Water Conservation*, 2nd ed.; Prentice Hall: Hoboken, NJ, USA, 1991.
54. Huangfu, W.; Wu, W.; Zhou, X.; Lin, Z.; Zhang, G.; Chen, R.; Song, Y.; Lang, T.; Qin, Y.; Ou, P.; et al. Landslide Geo-Hazard Risk Mapping Using Logistic Regression Modeling in Guixi, Jiangxi, China. *Sustainability* **2021**, *13*, 4830. [CrossRef]
55. Park, H.J.; Jang, J.Y.; Lee, J.H. Assessment of Rainfall-Induced Landslide Susceptibility at the Regional Scale Using a Physically Based Model and Fuzzy-Based Monte Carlo Simulation. *Landslides* **2019**, *16*, 695–713. [CrossRef]
56. Nseka, D.; Kakembo, V.; Bamutaze, Y.; Mugagga, F. Analysis of Topographic Parameters Underpinning Landslide Occurrence in Kigezi Highlands of Southwestern Uganda. *Nat. Hazards* **2019**, *99*, 973–989. [CrossRef]
57. Tağıl, Ş.; Jenness, J.; Tagil, S.; Jenness, J. GIS-Based Automated Landform Classification and Topographic, Landcover and Geologic Attributes of Landforms Around the Yazoren Polje, Turkey. *J. Appl. Sci.* **2008**, *8*, 910–921. [CrossRef]
58. Zêzere, J.L. Landslide Susceptibility Assessment Considering Landslide Typology. A Case Study in the Area North of Lisbon (Portugal). *Nat. Hazards Earth Syst. Sci.* **2002**, *2*, 73–82. [CrossRef]
59. Molnar, P.; Anderson, R.S.; Anderson, S.P. Tectonics, Fracturing of Rock, and Erosion. *J. Geophys. Res. Earth Surf.* **2007**, *112*. [CrossRef]
60. Clarke, B.A.; Burbank, D.W. Bedrock Fracturing, Threshold Hillslopes, and Limits to the Magnitude of Bedrock Landslides. *Earth Planet. Sci. Lett.* **2010**, *297*, 577–586. [CrossRef]
61. Scheingross, J.S.; Minchew, B.M.; Mackey, B.H.; Simons, M.; Lamb, M.P.; Hensley, S. Fault-Zone Controls on the Spatial Distribution of Slow-Moving Landslides. *GSA Bull.* **2013**, *125*, 473–489. [CrossRef]
62. Brenning, A.; Schwinn, M.; Ruiz-Páez, A.P.; Muenchow, J. Landslide Susceptibility near Highways Is Increased by 1 Order of Magnitude in the Andes of Southern Ecuador, Loja Province. *Nat. Hazards Earth Syst. Sci.* **2015**, *15*, 45–57. [CrossRef]
63. Pasang, S.; Kubíček, P. Landslide Susceptibility Mapping Using Statistical Methods along the Asian Highway, Bhutan. *Geosciences* **2020**, *10*, 430. [CrossRef]
64. Zhou, C.; Cao, Y.; Yin, K.; Wang, Y.; Shi, X.; Catani, F.; Ahmed, B. Landslide Characterization Applying Sentinel-1 Images and InSAR Technique: The Muyubao Landslide in the Three Gorges Reservoir Area, China. *Remote Sens.* **2020**, *12*, 3385. [CrossRef]
65. Ou, P.; Wu, W.; Qin, Y.; Zhou, X.; Huangfu, W.; Zhang, Y.; Xie, L.; Huang, X.; Fu, X.; Li, J.; et al. Assessment of Landslide Hazard in Jiangxi Using Geo-Information. *Front. Earth Sci. China* **2021**, *9*, 648342. [CrossRef]
66. Bathrellos, G.D.; Kalivas, D.P.; Skilodimou, H.D. GIS-Based Landslide Susceptibility Mapping Models Applied to Natural and Urban Planning in Trikala, Central Greece. *Estud. Geológicos* **2009**, *65*, 49–65. [CrossRef]
67. Guzzetti, F.; Carrara, A.; Cardinali, M.; Reichenbach, P. Landslide Hazard Evaluation: A Review of Current Techniques and Their Application in a Multi-Scale Study, Central Italy. *Geomorphology* **1999**, *31*, 181–216. [CrossRef]
68. Wu, X.; Chen, X.; Zhan, F.B.; Hong, S. Global Research Trends in Landslides during 1991–2014: A Bibliometric Analysis. *Landslides* **2015**, *12*, 1215–1226. [CrossRef]
69. Lima, P.; Steger, S.; Glade, T.; Tilch, N.; Schwarz, L.; Kociu, A. Landslide Susceptibility Mapping at National Scale: A First Attempt for Austria. In *Advancing Culture of Living with Landslides*; Mikos, M., Tiwari, B., Yin, Y., Sassa, K., Eds.; Springer International Publishing: Cham, Switzerland, 2017; pp. 943–951.

Article

Soil Order-Land Use Index Using Field-Satellite Spectroradiometry in the Ecuadorian Andean Territory for Modeling Soil Quality

Susana Arciniegas-Ortega [1,2,*], Iñigo Molina [1,*] and Cesar Garcia-Aranda [1,*]

[1] Department of Land Surveying and Cartography Engineering, Universidad Politécnica de Madrid, 28040 Madrid, Spain

[2] Department of Environmental Engineering, Universidad Central del Ecuador, Quito 170129, Ecuador

* Correspondence: srarciniegas@uce.edu.ec (S.A.-O.); inigo.molina@upm.es (I.M.); cesar.garciaa@upm.es (C.G.-A.)

Abstract: Land use conversion is the main cause for soil degradation, influencing the sustainability of agricultural activities in the Ecuadorian Andean region. The possibility to identify the quality based on the spectral properties allows remote sensing methods to offer an alternative form of monitoring the environment. This study used laboratory spectroscopy and multi-spectral images (Sentinel 2) with environmental covariates (physicochemical parameters) to find an affordable method that can be used to present spatial prediction models as a tool for the evaluation of the quality of Andean soils. The models were developed using statistical techniques of logistic regression and linear discriminant analysis to generate an index based on soil order and three indexes based on the combination of soil order and land use. This combined approach offers an effective method, relative to traditional laboratory methods, to derive estimates of the content and composition of soil constituents, such as electrical conductivity (CE), organic matter (OM), pH, and soil moisture (HU). For Mollisol index.3 with Páramo land use, a value of organic matter (OM) $\geq$8.6% was obtained, whereas for Mollisol index.4 with Shrub land use, OM was $\geq$6.1%. These results reveal good predictive (estimation) capabilities for these soil order–land use groups. This provides a new way to monitor soil quality using remote sensing techniques, opening promising prospects for operational applications in land use planning.

Keywords: remote sensing; soil quality; soil properties; indices; Andean region

1. Introduction

Soil is the main natural resource for food and energy production [1]. It controls the movement of water in the landscape and functions as a biological filter for the possible leaching of pollutants into environmental spheres [2]. However, soil can be degraded by chemical and physical processes, which reduces its ability to function as a base for the development of a healthy layer for vegetation. Therefore, acknowledging soil conditions by the effects on vegetation can represent site conditions [3,4].

Gholizadeh and Kopačková (2019) [1] considered that conventional methods of soil health evaluation in large areas involve several expensive and time-consuming variables such as collection of field data, chemical analysis in a laboratory, and geostatistical interpolation. Alternately, several studies have shown the possibility of characterizing soils and identifying their quality by correlating both physicochemical and spectral parameters [5]. Therefore, the use of remote sensing spectrometry products in environmental evaluation studies offers a complementary alternative to in situ monitoring procedures to aid in research, control, and monitoring of the soil component. The application field for these tools in the soil component is extensive [6] through the study of soil characteristics such

Citation: Arciniegas-Ortega, S.; Molina, I.; Garcia-Aranda, C. Soil Order-Land Use Index Using Field-Satellite Spectroradiometry in the Ecuadorian Andean Territory for Modeling Soil Quality. *Sustainability* **2022**, *14*, 7426. https://doi.org/10.3390/su14127426

Academic Editors: Mario Elia, Antonio Ganga and Blaž Repe

Received: 16 May 2022
Accepted: 15 June 2022
Published: 17 June 2022

Publisher's Note: MDPI stays neutral with regard to jurisdictional claims in published maps and institutional affiliations.

as reflectance, degradation, and possible polluting agents with the processing of satellite images permitting the inspection and monitoring of large areas in a fixed time and place [7].

In Ecuador, the properties and pedogenetic processes of soil have been studied in terms of rock type, geomorphology, taxonomic classification, and soil order. An increasing breach between the available information on the main soils and their quality [8] leads to understanding the condition of the soil to allow for the planification of healthy and sustainable territories, as determined by goals 12 and 15 of the Sustainable Development Goals (SDG) [9]. These goals correspond to responsible consumption and production, and life on land. Therefore, it is necessary to determine the quality of soil to develop fast, feasible, and affordable estimation methods for monitoring and assessing areas.

In the study area, the predominate soils are Andisol and Mollisol, which originate from weathering of volcanic material (ash) [10]. These relatively young soils can convey high agricultural potential [11]. Andisols, also known as *páramos*, are clay loam soils capable of retaining enormous amounts of water; on the contrary, Mollisols are fertile soils with a high organic matter content that cover approximately 70% of the Cayambe canton, a political–administrative unit of Ecuador, where the research's basin is located. However, the lack of land use and occupation policies has caused the expansion of agricultural activity boundaries [12], causing the loss of the *páramo*. The main goal is to understand whether there is any relationship between the spectra measured in the samples collected in field with the corresponding bands measured by satellite, combining physicochemical analysis to quantify and model the quality of Andean soils caused by agricultural activity. This will be achieved by (i) compiling and analyzing the physicochemical parameters of soil based on quality standards for agricultural activities, obtaining indices that classify the soils based on their order (Andisol and Mollisol) and use, and (ii) determining whether the soil is associated with some of the physicochemical qualities considered, validating land use and order models based on field reflectance data, satellite reflectance, and physicochemical qualities.

There are several studies based on national models capable of predicting spectra limited in an infrared laboratory with statistical algorithm analysis [13,14]. In this research, with the use of these spectroscopic methods for the evaluation of soil quality, models were developed for estimating indicators based on the combination of soil order, land use, and physicochemical characteristics, using logistic regression analysis, linear discriminant analysis, and regression trees. The approach offers a method that derives the estimates using the ratio of laboratory/satellite spectra when the soil is well represented by the calibration samples used to build the predictive models [13]. Therefore, the performance of these local models can be used in other geographic spaces by incorporating the spectra into a dataset for that area [15].

Consequently, the combination of laboratory spectroscopy and multispectral images with environmental covariates is an adequate methodological alternative to obtain models that are adjusted for the prediction of the quality of Andean soils, independently of other methodological approaches that have been used [16–19].

2. Materials and Methods

2.1. Study Area

This study was done at Río Blanco basin, located in the Cayambe canton, Pichincha province, Ecuador (Figure 1), where agricultural activities are mainly related to the dairy–floricultural corridor via the cultivation of exportation products, such as roses and summer flowers (17.32 km^2), livestock for milk production (399.201 km^2), and, in smaller quantities, agricultural activity, especially high Andean crops (10.984 km^2) [12]. Unfortunately, in Río Blanco basin, no studies have been aimed towards soil quality. The few studies available are related to land use and occupation, risks associated with the presence of the Cayambe volcano, agriculture [20], soil rehabilitation with Cangahua [21], and watershed management [22].

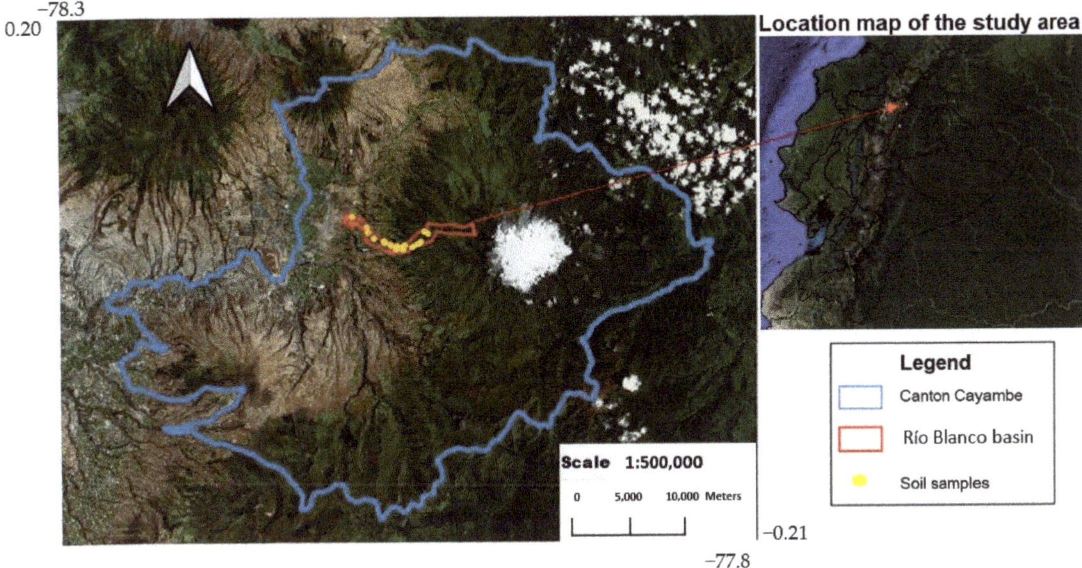

Figure 1. Geographical location of the study area and location of the soil sampling points (Google Maps).

The existing black and brown soils in the basin can retain high amounts of water and organic matter content, known as Mollisol and Andisol [10]. The soils have been influenced by the volcanic activity of the Cayambe during its genesis, causing slopes ranging from gentle to steep [7]. Vegetation coverage is characterized by the presence of cultivated grass (27%), herbaceous and shrubby moorland (45.6%), flowers and short-cycle crops (16.5%), and eucalyptus trees (8.1%) [10].

2.2. Sentinel-2 Satellite Images

The satellite imagery used for this research was acquired by the Sentinel-2 (S2) satellite constellation (2A and 2B Earth Observing Missions) launched in 2015 and 2016, which has been used extensively for monitoring land cover and vegetation [23,24]. The S2 satellites are identical and operate in a sun-synchronous orbit at a mean altitude of 786 km. The main S2 payload is a multi-spectral instrument (MSI), which is a push-broom sensor that registers the radiation reflected from the Earth passing through the atmosphere in 13 spectral bands distributed in four bands at 10 m, six bands at 20 m, and three bands at 60 m spatial resolution. Figure 2 depicts the range and spectral response functions of the S2A/S2B MSI instruments for these bands [25]. The S2 satellites have a swath width of 290 km. The visible bands (VIS) B1, B2, B3, B4 at 10 m resolution; near-infrared bands (NIR) B5, B6, B7, B8A at 20 m resolution and B8 at 10 m; and shortwave infrared (SWIR) B11 and B12 are most useful for retrieving geophysical surface parameters [26].

Meanwhile, the 60 m resolution bands are used for atmospheric corrections, which are of crucial importance for most EOS applications and permit the development and evaluation of robust atmospheric correction algorithms such as Sen2Cor. In the study area, located inside a swath overlap, the revisit frequency of each satellite is four to five days in an 11° forward-looking view angle; the presence of two identical satellites allows a geometric revisit time between two and three days, supporting near-continuous monitoring of vegetation and land surface processes [26].

The imagery used in this research was acquired on 16 July 2018 by the Sentinel 2B platform (see Supplementary Materials). The Level-2A product was used after the processing carried out in the Level-1C product available in the Open Access Hub of the European Space Agency (ESA) DataHUB server [27]. The identifier of the product is referenced as S2B_MSIL1C_20180716T153619_N0206_R068_T17NRA_20180716T202613. Furthermore, in the text, it will be referred to as S2BL2A.

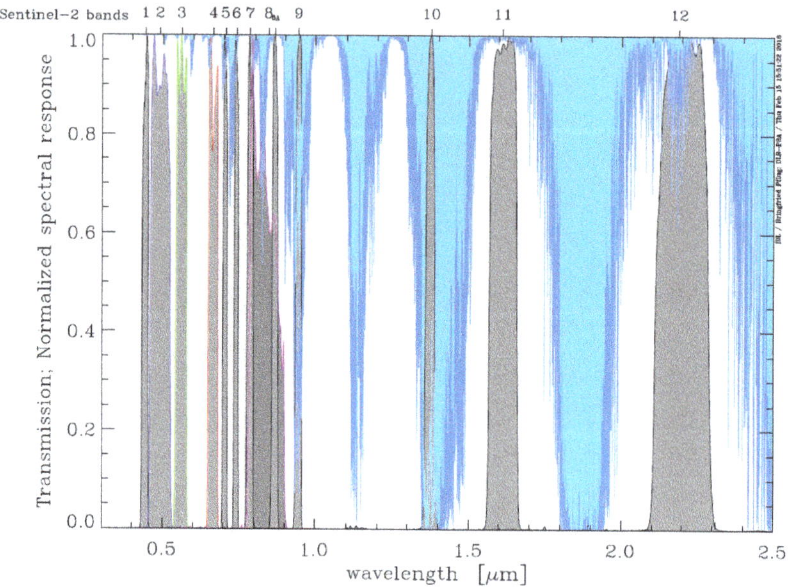

Figure 2. S2A and S2B-MSI spectral response functions (SRF) [28].

2.3. Soil Sampling

Soil samples were collected during four field trips during the months of June, July, and August 2018. The samples were placed in hermetically sealed sleeves for physicochemical analysis (Figure 3a,b), and in two bulk density cylinders (Figure 3c) for spectroradiometry analysis. The cylinder lids were covered with geomembranes to keep the samples unaltered, as explained by Yánez and Arciniegas (2019) [7], to comply with the provisions of Section 2.5.

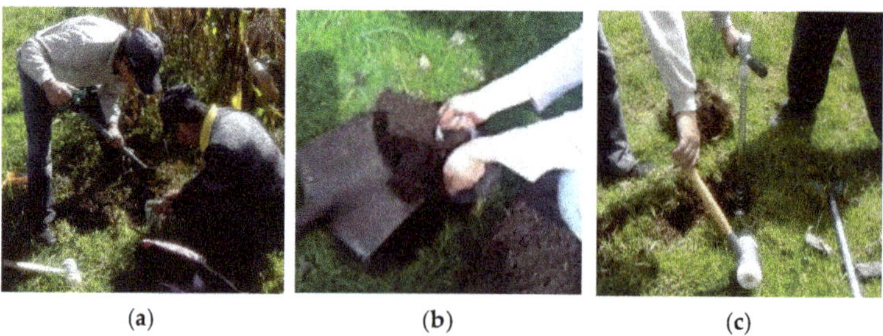

(a) (b) (c)

Figure 3. Soil sample collected in the field. (a) Shows site preparation. (b) Shows sample put in sealed sleeves for physicochemical analysis (c) Shows bulk density cylinders for spectroradiometry analysis.

A total of 36 surface soil samples (0–10 cm) were collected from the study site using core drilling, cylinders, a stainless-steel shovel, and a GARMIN GPSMAP 62sc handheld navigator (accurate to within ±4 m). After removing vegetation from the soil surface in a quadrant of approximately 30 × 30 cm, 1 kg of soil sample was collected from each sampling point. The number of samples was calculated by the type of composite sample according to Ecuadorian environmental standards (Book VI, Annex2) [29], establishing two samples per homogeneous zone, and one sample for zones ID02 and ID08, which presented collection problems.

2.4. Physicochemical Parameter Measurements

To determine the quality of soil, the physical, chemical, and biological components of the soil and their interactions must be considered; despite the different measurements possible, not all parameters are relevant for the soil in a particular scenario [30]. In this study, the physicochemical components were selected considering two criteria: The first was based on the reflectivity of the soils that are conditioned to organic matter, that interfere with the spectral curves [31], and the second according to Friedman et al. 2001 [32], who set a minimum number of indicators for agricultural activities, due to human use and management. The parameters measured were soil moisture, pH, electrical conductivity (CE), and organic matter (OM). Other parameters, such as heavy metals, could not be measured for economic reasons.

For the 36 soil samples, the analysis was carried out as follows: For soil moisture, pH, and conductivity, the methods established in NOM-021-RECNAT-2000 (Mexican official standard that establishes the specifications of fertility, salinity, and soil classification) were used [33]. The AS-05 gravimetric method was applied to soil moisture, the AS-02 electrometric method was applied to pH, and the AS-20 method was applied to electrical measure conductivity.

For the determination of OM, the soil samples were sent to the certified soil, foliar, and water laboratory of the Agency for the Regulation and Control of Phytosanitary and Zoosanitary (Agrocalidad) of Ecuador, where the volumetric PEE/SFA 09 method (Walkley Black's analytical method consisting of wet oxidation of the soil sample) was applied [34].

2.5. Spectral Measurements in Laboratory and Satellite Image Sentinel-2

Spectral analysis of the soil samples was carried out in the Ecuadorian Space Institute (IEE) laboratory, which facilitated the use of the ASD FieldSpec4 spectroradiometer (Analytical Spectral Devices). The equipment has a spectral resolution of 3 nm in the range of 350–1000 nm and 10 nm in the range of 1001–2500 nm. In this regard, the spectral bands of the MSI sensor onboard satellite S2 are within the spectral range of the ASD instrument.

In the laboratory, the soil spectra were collected using a small spectralon placed in the HiBrite MudLight device with an optical fiber connected to generate an artificial light source (Figure 4). The ASD FieldSpec4 spectroradiometer was used to obtain soil spectral data in the laboratory. The measurement protocol followed the methodology described in Figure 5 [35].

Figure 4. Location of cylinder with respect to HiBrite MudLight device on support.

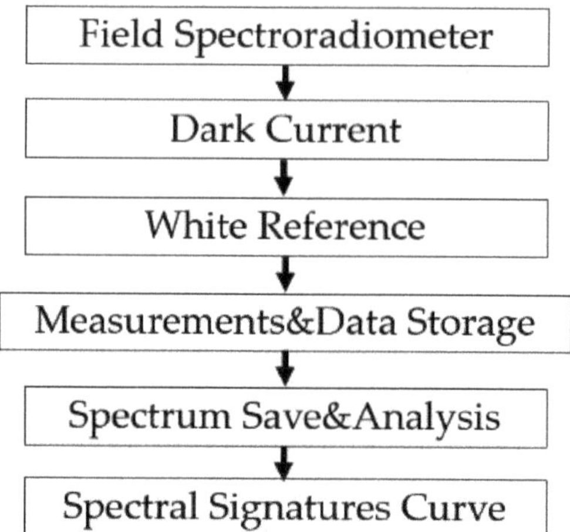

Figure 5. Methodology for the generation of spectral signatures [35,36].

After the common configuration and control settings, i.e., output folders, connecting optical fiber, etc., the appropriate integration time was set given the lighting conditions to optimize the measurements. Subsequently, the dark current was also recorded. Then, a reference target or white reference (spectralon) was measured until a horizontal line with a reflectance value of 1 was presented. Laboratory soil spectra measurements followed these pre-operation phases, and the resulting measured spectra were processed with the corresponding software (ViewSpec Pro). An important issue that also needed to be addressed was that of the signal-to-noise ratio, or SNR, during measurements, which is related to the signal component. The spectra measurement procedure was performed using the principle of a continuous fiber optic cable, as specified in [36]. This technique has the advantage of significantly degrading the SNR, as it avoids interactions with other media between the recording device and the sample. Soil measurements were carried out by controlling the direction of the optical fiber to always point towards the target in order to avoid anisotropy effects. Subsequently, an average of all measurements was made to obtain the spectrum. Finally, the displays of the spectra that were selected were shown.

The S2BL2A product specified in 2.2 was used for the satellite image, with bottom-of-atmosphere (BOA) reflectance, in which only the bands matching the S2 product were considered. The centroid of each pixel was determined to obtain the spectral values per band on a 20 × 20 m grid.

These laboratory spectroradiometry measurements and satellite images were used to establish indices (the table in Section 2.8.1) that compare the physicochemical parameters of the soil with the spectral bands of the S2-MSI sensor. Since it was agreed to only use the equipment supplied in the Laboratory of the Ecuadorian Space Institute, in-field spectroradiometry measurements were not possible. The measurements of the spectra were carried out only in the laboratory, as in [37–39], generating a different model to those already known to evaluate soil quality [40,41].

2.6. Land Use/Soil Order Dataset

Before the processing specified in Section 2.8, a dataset was built based on the in-field soil samples (spectra and physicochemical parameters) and reflectance per homogeneous area in the S2BL2A satellite image. These data consisted 345,408 observations as a product of the combination of two datasets and constituted the geographic population of the area under study. The data were then treated based on the combination of the variables "land order" and "land use", whose total set of observations was as follows (Table 1). The combinations shown in Table 1 allowed different models to be established based on soil order and land use.

Table 1. Soil samples classified according to the order of the soil and by land use.

Land Use	Order of the Soil		Total
	Andisol	**Mollisol**	
Agricultural	0	4284	4284
Shrub	968	13,496	14,464
Forest	40	18,476	18,516
Páramo *	188,760	38,072	226,832
Pasture	9892	71,420	81,312
Total	199,660	145,748	345,408

* A *páramo* is a fragile neotropical high mountain ecosystem. In Ecuador it has an average height of 3300 m above sea level.

Model 1 was established from the dataset by applying a simple random method of 5% of the population to compare what was provided by each sample until one dataset was left as a result of the information provided between one model and another being the same. The data were divided into training and testing groups. The training dataset consisted of 70% of the total number of observations in the sample, and the test dataset consisted of 30% of the total number of observations in the sample. With the dataset the model was elaborated and later executed.

Model 2 was developed from the dataset by selecting those corresponding to the Andisol soil order (Table 1), with 199,660 observations. Three categories were selected: Shrub, Páramo, and Pasture, leaving Forest out of the analysis due to its low frequency. Composite samples of 70% of each land use category were randomly selected, thus leaving a composite sample of 139,734 random observations, which were part of the training dataset. Therefore, the test dataset contained the remaining 30% of the 59,886 observations.

For the Mollisol order, considering the data's trend, based on the geographical distribution of the different land uses (Figure 6), it was observed that some of the soils had a particular use owing to their nature and population growth, among other characteristics of the area.

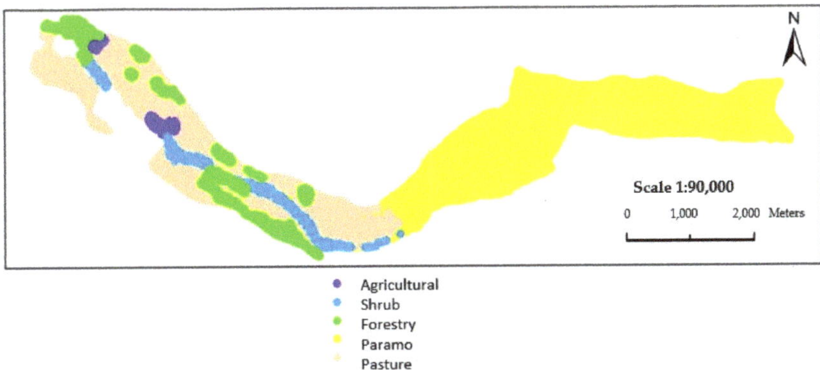

Figure 6. Geographical distribution of land uses.

The dataset corresponding to the Mollisol soil order (Table 1), with 145,748 observations, was considered to form Model 3 from a stratified random sample of 70% of the total. This allowed us to consider two subsets based on land use, resulting in two more models. The first subset (Model 3), called "Mollisol 1", was made up of Forest, Páramo, and Pasture; the second subset (Model 4), "Mollisol 2", was composed of Agriculture and Shrub. Each training dataset represented a 70% stratified random sample based on the categories of Mollisol 1 (Table 2) and Mollisol 2 (Table 3). Consequently, the test dataset included the remaining 30% of the data. For Mollisol 1 it was 38,391, and for Mollisol 2, 5335.

Table 2. Training sample to obtain the discriminant function in the classification of land use of the Mollisol 1 order.

Land Use	Forest	Páramo	Pasture	Total
Samples	12,933	26,650	49,994	89,577
Ratio	0.1444	0.2975	0.5581	1.00

Table 3. Training sample to obtain the discriminant function in the classification of land use of the Mollisol 2 order.

Land Use	Agriculture	Shrub	Total
Samples	2998	9447	12,445
Ratio	0.2409	0.7591	1.00

2.7. Homogeneous Zones

The area of the Rio Blanco basin was extracted from the hydrological database of the Cayambe canton. Land use was extracted from the national productive systems database [10], and slope data and soil order data were obtained from a geopedology base map [10]. The multi-criteria technique was applied to create homogeneous zones [7] considering the following criteria: soil order and land use. This made it possible to establish a spatial analysis unit for correlation evaluation tests, determining the number of samples to be taken in the field. In total, within the Rio Blanco basin, 31 zones with similar characteristics were identified in previous work [7], of which 19 were analyzed for easy access in the study area (Figure 7).

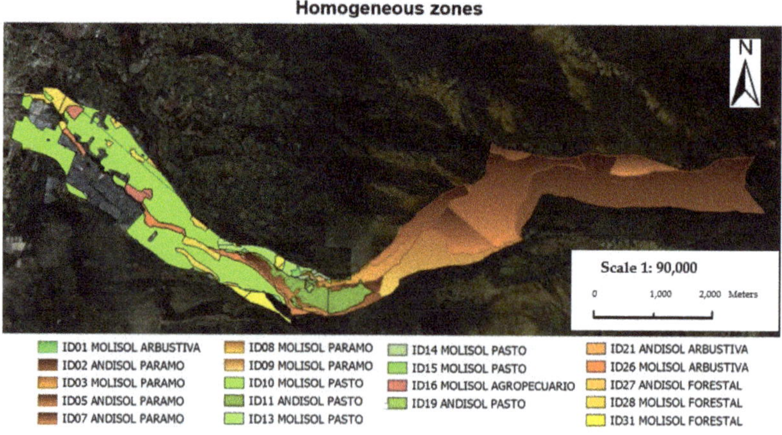

Figure 7. Homogeneous areas analyzed in the study.

Using the physicochemical laboratory analysis results, the Thiessen polygons were calculated in each homogeneous zone from the soil-sampling points. Likewise, the topology of spatial relations (intersection and inside) was applied [42] with the spectral values per satellite band to identify the spatial relationship of the satellite spectral values concerning the physicochemical characteristics of each homogeneous zone [43].

The Mexican (NOM-021-RECNAT-2000) [33] and Ecuadorian (Book VI, Annex 2) environmental standards [29] were also considered to establish homogeneous areas as a unit of spatial analysis. This allowed sampling points to be defined based on the criteria of slope, soil order, and land use, resulting in 19 homogeneous areas (Figure 7).

2.8. Regression and Spatial Analysis

To identify the relationship between the spectral behavior of soil in the laboratory and satellite images associated with the physicochemical qualities of soils, we proceeded to apply regression techniques based on the theory of machine learning, which makes them useful for predicting outcomes, identifying patterns, and making decisions with minimal human intervention [44]. Subsequently, discriminant analysis techniques, geostatistics, and non-parametric models (regression trees) were applied.

This study started by creating a logistic regression model (Figure 8) with the dependent variable "soil order" (Andisol, Mollisol), and as explanatory variables, the reflectance levels of the spectral behavior of soil in laboratory and satellite images, called Model 1, to later select those variables that were statistically significant, from which this model generated the soil index required (Table 4). The basic form of the logistic regression model is as follows (Equation (1)) [45]:

$$\text{logit (yy)} = \beta_0 + \beta_1 x_1 + \beta_2 x_2 + \cdots + \beta_i x_i \tag{1}$$

where

y: soil order dependent variables Andisol and Mollisol;
xi: independent variables (predictors): wavelengths of the spectroradiometer convoluted by the spectral response of soils in the laboratory (B04c, B05c, B06c, B07c, B08c, B08Ac) and bands corresponding to the MSI sensor of satellite S2 (B04s, B05s, B06s, B07s, B08s, B08As);
βo: constant (intercept);
βi: coefficients of predictor variables.

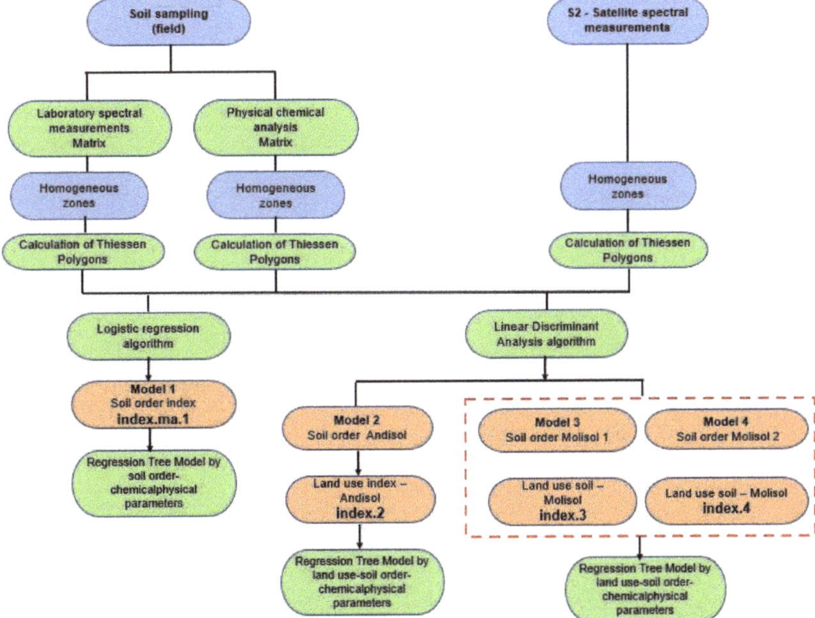

Figure 8. Flowchart indicating the methodological steps implemented to create the model, indexes, and predictions of physicochemical parameters.

Table 4. Models and indices by soil order and soil order–land use.

Model	Index	Description	Dependent Variable	Independent Variable
Model 1	index.ma.1	Separates soil order: Mollisol, Andisol	Mollisol Andisol	Field bands, Satellite bands,
Model 2	index.2	Separates land use of the Andisol order	Andisol	Field bands, Satellite bands, Shrub, Páramo, Pasture
Model 3	index.3	Separates land use of the Mollisol order	Mollisol (Mollisol 1 group) for land use: Forest, Páramo, Pasture	Field bands, Satellite bands, Forest, Páramo, Pasture
Model 4	index.4	Separates land use of the Mollisol order	Mollisol (Mollisol 2 group) for land use: Agriculture, Shrub	Field bands, Satellite bands, Agriculture, Shrub

Several logistic regression trials with the dependent variable of soil order were tested. The statistically significant variables ($p \leq 0.05$) were verified until the most highly significant set of variables was obtained. The statistically significant model consisted of a combination of independent variables related to the reflectance levels of the soil spectra in the laboratory and satellite image.

The land use variable was related to the taxonomic classification of soil, so when considering only this variable, there was no way to separate the different land uses and generate an index that allowed us to estimate the physicochemical characteristics based on the land use variable of soil. To resolve this difficulty, the soil order variable was considered, and within each of these categories—Andisol and Mollisol—the land use variable that yielded the best classification levels by soil order was analyzed, as discussed in Section 2.6. Considering land use as the dependent variable, a statistical methodology known as linear discriminant analysis was applied (Figure 8) [46], which allowed a linear combination of variables to be identified that could be used to determine the group to which each individual belonged. In this case, the individuals were identified as homogeneous areas in which different soil samples were collected to develop Model 2, Model 3, and Model 4 (Table 4).

2.8.1. Obtaining Index

The standardization of the coefficients from the logistic regression model (Model 1) followed a pattern where each coefficient was divided by the sum of its coefficients, in such a way that the sum of the coefficients of the index was equal to 1 (Equation (2)).

$$\text{index.ma.1} = k + cso[i] * xi + \ldots \ldots \tag{2}$$

where

k: constant;
cso[i] = standardized soil order coefficients;
xi: independent variables (predictors): wavelengths of the spectroradiometer convoluted by the spectral response of soils in the laboratory (B04c, B05c, B06c, B07c, B08c, B08Ac) and bands corresponding to the MSI sensor of satellite S2 (B04s, B05s, B06s, B07s, B08s, B08As).

From the linear discriminant function, the coefficients of the models were standardized by land use by soil order of the models: Model 2, Model 3, and Model 4, where each coefficient of the model was divided by the sum of its coefficients in such a way that the sum of the coefficients of the indices (Index 2, Index 3, Index 4) was equal to 1 Equation (3).

$$\text{index} = cso[1] * x1 + \ldots . + cso[i] * xi \tag{3}$$

where

cso[i] = standardized coefficients of land use by soil order;

xi: independent variables (predictors): wavelengths of the spectroradiometer convoluted by the spectral response of soils in the laboratory (B02c, B03c, B04c, B05c, B06c, B07c, B08c, B08Ac, B011c, B012c) and bands corresponding to the MSI sensor of satellite S2 (B02s, B03s, B04s, B05s, B06s, B07s, B08s, B08As, B011c, B012c).

In this case, the data were treated based on the combination of soil order and land use, and the entire set of observations is shown in Table 1.

An association analysis was applied between each physicochemical parameter and the indices, which determined no linear trend between the different pairs of variables that were compared. The geostatistical surface was created using the inverse distance weighted (IDW) method, which assumes that closer objects are similar to those far apart. Therefore, any unknown location will probably have an equal value to the nearest known locations [47]. It was possible to determine how the order of soil and land use were spatially distributed in terms of probability and to establish predictions of each physicochemical parameter through a set of non-parametric models known as decision trees [48], where the dependent variable can be categorical or numeric. The regression tree model was applied since the dependent variable was each of the calculated indices and was numerical, and the explanatory variables (covariate) were the order of the soil and the order–land use, together with each physicochemical parameter, considered as independent variables, as shown examples in Table 5, which are explained in Section 3.5. The space defined by the regression tree models, as part of a non-parametric analysis, consisted of dividing the predictor space into boxes (regions) [49]. For example, the areas were a function of Index 1 and the order of soil to estimate the physicochemical organic matter (OM) parameter to make the prediction. Consequently, a regression tree model was generated to describe the association between each index and each physicochemical parameter, as shown in Figure 8.

Table 5. Regression trees.

Regression Trees	Dependent Variable	Covariates (Soil Order and/or Land Use)	Physicochemical Parameter Covariate (Examples)
ARMAH	index.ma.1	Soil order: Andisol—Mollisol	Soil moisture
ARUSAMO	index.2	Soil order: Andisol Land use: Shrub, Páramo, Pasture	OM
ARUSM3MO	index.3	Soil order: Mollisol 1 Land use: Forest, Páramo, Pasture	OM
ARUSM4MO	index.4	Soil order: Mollisol 2 Land use: Shrub, Agriculture	OM

For a better understanding, the general methodological framework was divided into three parts, as shown in Figure 8. All statistical and graphical calculations were performed using RStudio software [50].

2.8.2. Validation

The validity of Model 1 was tested by calculating the confusion matrix [49] to determine the classification error of the samples and the accuracy, together with the Kappa statistic to indicate the degree of agreement between the measured data and the predicted value by the model, concerning their order in Andisol or Mollisol, as well as the sensitivity and specificity, which indicate the probability of correctly classifying the soil samples from the model. The Kappa coefficient must be equal to zero. When the Kappa coefficient differs from 0, it means that the data obtained from the validation model, as predicted data, agree with the measured data used to generate the model. Sensitivity refers to the ability of a model to identify the order of soils. In contrast, specificity indicates the ability of the model to identify soil samples that do not correspond to the order of the soil to be classified.

In Model 2, among the different land uses of the order Andisol with each of the two sets of test data selected randomly, the classification errors were evaluated using the confusion matrix and the accuracy metric.

For Model 3, among land uses such as Forest, Páramo, and Pasture of soil order Mollisol, those that did not participate in the elaboration of the model needed to be classified based on Model 3 and have their corresponding confusion matrices calculated in order to determine the classification error and accuracy metric.

The same criteria applied for Model 4, between the agricultural and shrubland uses of the soil order Mollisol.

Once each model was validated, the effects of soil order and land use on each physicochemical parameter were determined; one-way analysis of variance (ANOVA) was used [49], with a dependent variable for each physicochemical parameter and an independent variable for the order and land use of the soil. Several ANOVA models were tested for soil order and other land use models within each order. For each case, the null hypothesis was that the mean of each physicochemical parameter is the same in each order of soil, or the mean of each physicochemical parameter is the same for each land use within each order, versus the alternate hypotheses indicating that at least one pair of mean values is different. The statistical decision criteria are based on a significance level of 5% ($\alpha = 0.05$); thus, if the p-value is less than or equal to 0.05 the null hypothesis is rejected, and it can be concluded that the effect of the independent variable is significant in relation to the mean of the dependent variable. Otherwise, if the p-value is greater than 0.05, then the available data do not yield enough information to conclude that the independent variable is significant in relation to some variation of the dependent variable.

3. Results

3.1. Physicochemical Analysis

The different soil samples were analyzed using standardized physicochemical methods. The soil moisture results were in a range from 12.23% to 74.99%. The areas that predominated with the highest percentage of soil moisture were moors, and those with the lowest soil moisture were forest and shrub areas. The lowest water content was observed in Mollisol.

Regarding pH, the most acidic soils corresponded to undisturbed moors, whereas the least acidic soils corresponded to cultivated pastures. The range was 4.55 to 5.76, which, according to Mexican regulations, ranges from moderately acidic to strongly acidic, and according to Ecuadorian regulations, it would be out of range.

The electrical conductivity of the soils was within the limits established by both the Ecuadorian and Mexican regulations, <200 µS/cm and <1 dS/m, respectively. The areas with the lowest electrical conductivity were moors, whereas the highest electrical conductivity was observed in grasses.

For OM, the values ranged from 2.78% to 16.06%. According to Mexican standards, the zones range from very low to very high levels for soils of volcanic origin. The zone with the lowest OM percentage was forest, followed by passage areas, and the zone with the highest OM content was *páramo* areas.

3.2. Analysis of Spectral Signatures

In this section the behavior of each band was determined with respect to the intensity of reflectance of the soil samples. For each cylinder, two spectral measurements were made in opposite sections of the same tube, and for each soil sample 10 spectral measurements per section were averaged for a single representative spectrum per homogeneous area, which resulted in graphs (Figure 9a,b) as a function of reflectance and wavelength per sample in each homogeneous area.

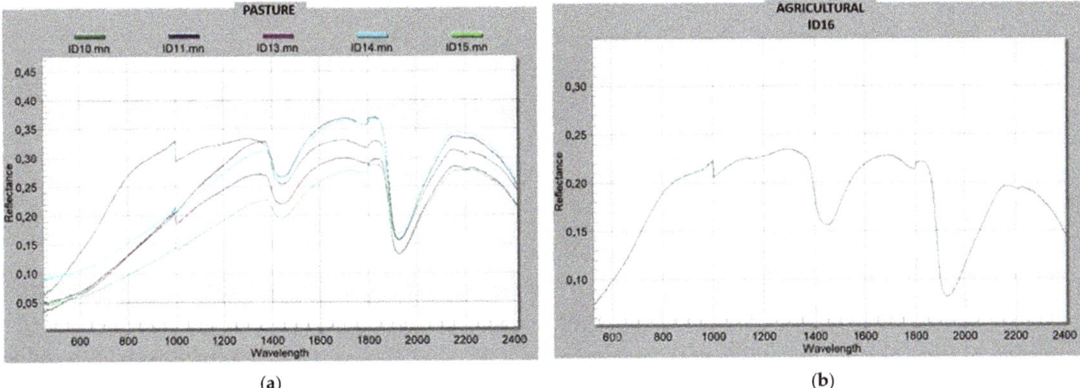

(a)

(b)

Figure 9. Spectral measurements in the laboratory according to land use. (**a**) Shows the curves by homogeneous zones ID10, ID11, ID13, ID1, ID15. (**b**) Shows the curve by homogeneous zone ID16.

The reflectance of the spectra was graphically analyzed in the laboratory to determine the behavior of the soils related to their spectral signatures of the Andisol and Mollisol orders. The spectral signatures obtained in the laboratory presented a pattern related to the typical spectral signature of soils, ranging from the visible range (VNIR) to near-infrared (NIR) to short-wave infrared (SWIR).

The graph in Figure 9a shows the pasture curves, where sample ID11 of the Andisol soil order presented the same intensity of reflectance as sample ID14 of the Mollisol order, which were the highest compared to the other samples. Sample ID10 of the Mollisol order had a medium intensity of reflectance, unlike sample ID15 of the Mollisol order and sample ID13 of the Mollisol order with lower values of reflectance intensity. This variation in the curves is related to the properties and state of these soils [51], considering the variation of each of the land uses. Thus, in the graph in Figure 9b, sample ID16 of the Mollisol order, with agricultural land use, may indicate changes in the characteristics and status of agricultural use in the months of June, July, and August.

It can be said that the graphs made a difference in the behavior of the soil order based on the associated land use.

This could be related to the reflectance records of the Sentinel-2 satellite images (Figure 10a,b) to improve spectral differences by calculating soil order indices based on land use and physicochemical parameters, as explained in the next section.

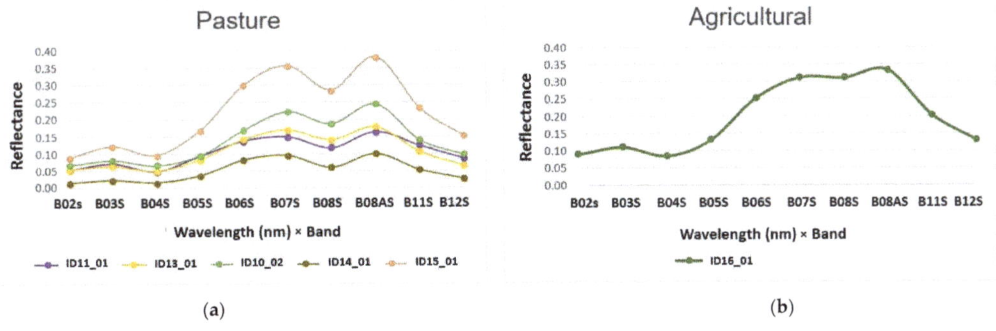

(a)

(b)

Figure 10. Spectral signatures of Sentinel 2B. (**a**) Shows the spectral signatures of Pasture by homogeneous zones. (**b**) Shows the average spectral signature of Agriculture obtained in the homogeneous zone 16 of Mollisol soil.

3.3. Development and Validation of Models Based on Soil Reflectance Levels in Laboratory and Satellite Image

3.3.1. Model 1, by the Orders of Andisol and Mollisol Soils

From the logistic regression calculation, Model 1 was obtained, whose structure is shown in Table 6.

Table 6. Logistic regression of Model 1 based on the reflectance levels of the soil spectra in the laboratory and satellite image.

Variables	Estimated Coefficients	Error Std	Value z	Pr(Z > |z|)	Decision
(Intercept)	-6.125×10^0	8.011×10^2	-76.459	$<2.0 \times 10^{-16}$	$p < 0.0001$
B04c	-7.596×10^3	5.784×10^1	-131.317	$<2.0 \times 10^{-16}$	$p < 0.0001$
B05c	9.382×10^3	1.080×10^2	86.872	$<2.0 \times 10^{-16}$	$p < 0.0001$
B06c	9.149×10^3	1.284×10^2	71.263	$<2.0 \times 10^{-16}$	$p < 0.0001$
B07c	-1.395×10^4	1.105×10^2	-126.278	$<2.0 \times 10^{-16}$	$p < 0.0001$
B08c	-3.630×10^0	4.775×10^0	-0.760	0.447	$p > 0.1000$
B08Ac	3.021×10^3	2.114×10^1	142.912	$<2.0 \times 10^{-16}$	$p < 0.0001$
B04s	5.655×10^0	1.151×10^0	4.913	8.96×10^{-7}	$p < 0.0001$
B05s	4.412×10^1	1.417×10^0	31.131	$<2.0 \times 10^{-16}$	$p < 0.0001$
B06s	-5.892×10^1	1.968×10^0	-29.941	$<2.0 \times 10^{-16}$	$p < 0.0001$
B07s	-5.889×10^1	2.164×10^0	-27.216	$<2.0 \times 10^{-16}$	$p < 0.0001$
B08s	8.076×10^0	5.844×10^1	13.819	$<2.0 \times 10^{-16}$	$p < 0.0001$
B08As	8.902×10^1	1.933×10^0	46.059	$<2.0 \times 10^{-16}$	$p < 0.0001$

p-value of the model: $p < 0.0001$.

The coefficients of Model 1 were both positive and negative. This model was composed of explanatory variables, consisting of a combination of the spectral behavior of the soil in the laboratory and satellite image, with the particularity that there are reflectance levels related to the characteristics of red, red border, and near-infrared. Classical vegetation indices were composed, but in this case, the objective was to classify the order of the soil in Andisol and Mollisol. Furthermore, one of the independent variables was not significant (B08c), which did not influence the global significance of this logistic regression model ($p < 0.0001$).

Based on the training dataset, we obtained a confusion matrix (Table 7).

Table 7. Training dataset from Model 1 based on reflectance levels of soil spectra in laboratory and satellite image.

Predicted Data	Data Real		Total
	Andisol	Mollisol	
Andisol	137,147	6006	143,153
Mollisol	2615	96,018	98,633
Total	139,762	102,024	241,786

The confusion matrix indicated a training error of 3.5%, which means that the model was good for classifying soils in relation to their order in Andisol or Mollisol based on the spectral behavior of the soil in laboratory and satellite image, and satellite related to the red reflectance of any modality. The false-positive and false-negative coefficients were relatively low (Table 7), at 1.9% (2615/139,762) and 5.89% (6006/102,024), respectively. In other words, 2615 soil samples of the Andisol order were classified as Mollisol, and 6006 Mollisol soil samples were classified as Andisol. We then evaluated the model using a test dataset to describe the validation process.

Model 1 Validation

The diagnostic evaluation of Model 1, from the diagnostic statistics using the test dataset, was generally good because the accuracy, sensitivity, and specificity were above 95%. On the other hand, the p-value of the Kappa statistic (Table 8) was more significant than 0.05 ($p > 0.05$), indicating that the null hypothesis that the measurements obtained through Model 1 were equivalent to the real data is not rejected.

Table 8. Diagnostic statistics for the validation of Model 1.

Statistical	Value	Statistical	Value	Statistical	Value
Success	0.9644	Kappa	0.9268	Sensitivity	0.9811
CI 95%	0.9633–0.9656			Specificity	0.9416

3.3.2. Model 2 for Variable Land Use of the Andisol Order

As explained in Section 2.8.1, to classify land use according to the Andisol soil order, linear discriminant analysis was applied (Figure 8). The following results were obtained (Table 9).

Table 9. Median reflectance levels of soil spectra in laboratory and satellite image, according to the group defined by land use for the Andisol order.

	Laboratory (B0ic) Satellite Image (B0is)						
Land Use	B02c	B03c	B04c	B05c	B06c	B07c	B08c
Shrub	0.0679	0.0765	0.0980	0.1108	0.1229	0.1381	0.1644
Páramo	0.0733	0.0855	0.1183	0.1363	0.1514	0.1693	0.1995
Pasture	0.0554	0.0650	0.0878	0.1012	0.1137	0.1290	0.1461
Land Use	B08Ac	B11c	B12c	B02s	B03s	B04s	B05s
Shrub	0.1680	0.3096	0.2812	0.0090	0.0363	0.0228	0.0698
Páramo	0.2006	0.3537	0.2843	0.0147	0.0400	0.0400	0.0785
Pasture	0.1600	0.3409	0.3161	0.0166	0.0481	0.0354	0.0848
Land Use	B06s	B07s	B08s	B08As	B11s	B12s	
Shrub	0.2134	0.2502	0.2791	0.2710	0.1296	0.0624	
Páramo	0.1591	0.1844	0.2083	0.2065	0.1594	0.0902	
Pasture	0.2413	0.2940	0.3272	0.3208	0.1791	0.1006	

The spectral values of the soil in the laboratory had greater weight in the classification of the different land uses than in the satellite image as a function of spectrometry in the laboratory and satellite image. Regardless of the sign, the coefficients of the soil spectral values in the laboratory were greater than the coefficients of the values in the satellite image. Consequently, the first component of this linear discriminant function explains that 96.5% of the total variability of the three different land uses had lower coefficients; even though the reflectance values in the satellite image were lower, these variables were important for the classification of land use as a function of the Andisol soil order. The first and second discriminant components were the linear combinations of the variables that best discriminate between the three land uses of the Andisol order, which in this case corresponded to the entire spectrum of soil in the laboratory and satellite image, respectively. Figure 11 shows the results of the soil classification based on the linear discriminant function model (Model 2).

The numbers 1, 2, and 3 represent the mean of each dataset. The means were quite separate, which implies a good classification of the land use of the Andisol order. In addition, based on the first linear discriminator, better discrimination was observed between the soils of Pasture and Páramo or between soils of Shrub and Páramo use than between the soils of Shrub and Pasture use. This situation could be because these land uses, in some cases, have relatively small neighboring units. Based on the training dataset, a confusion matrix was obtained (Table 10). In Figure 11 and Table 10, a good classification of the land uses of the Andisol order was observed, with a classification error of 0.51%.

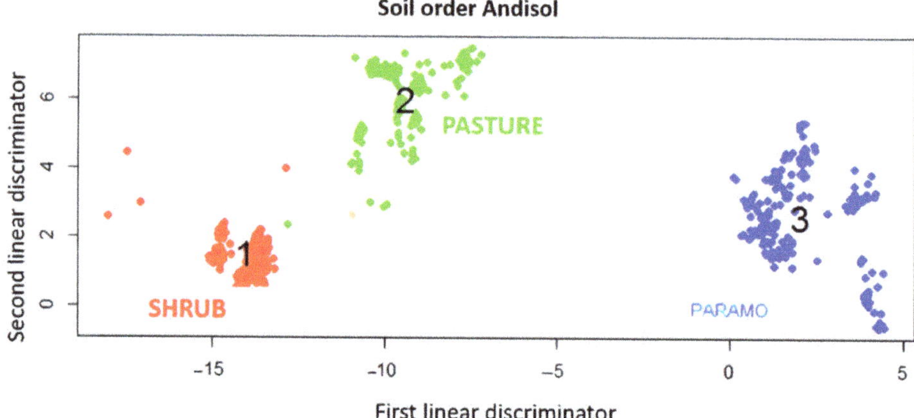

Figure 11. Classification of land uses of the Andisol order.

Table 10. Classification of land uses of the Andisol order based on Model 2 for the training set.

Real Land Use	Land Use Predicted by Model 2			Total
	Shrub	Páramo	Pasture	
Shrub	671	0	0	671
Páramo	0	132,132	0	132,132
Pasture	693	12	6220	6925
Total	1364	132,144	6220	139,728

Model 2 Validation

From the first group data for Shrub, Páramo, and Pasture land uses of the Andisol order, corresponding to the 30% that were not part of the model calculation, a confusion matrix was obtained (Table 11) from which a good classification of the land uses of the Andisol order was obtained, whose classification error was only 0.50% and accuracy 99.5%. Similar results were obtained for the second set of randomly selected data.

Table 11. Classification of land uses of the Andisol order based on Model 2 for a first test set.

Real Land Use	Land Use Predicted by Model 2			Total
	Shrub	Páramo	Pasture	
Shrub	289	0	2	291
Páramo	0	56,628	0	56,628
Pasture	295	5	2667	2967
Total	584	56,633	2669	59,886

3.3.3. Model for Variable Land Use of the Mollisol Order

(a) Model 3 for the variable of land use of the Mollisol order 1 from all wavelengths of the soil spectrum in laboratory and satellite image

The results were obtained from the application of LDA (Table 12).

Table 12. Coefficients of the linear discriminant function with dependent variable of the land use of the Mollisol 1 order and independent variables of reflectance levels of soil spectral behavior in laboratory and satellite image.

Field Bands	Coefficients LD1	Satellite Bands	Coefficients LD1	Field Bands	Coefficients LD2	Satellite Bands	Coefficients LD2
B02c	−708.68	B02s	−5.48	B02c	−410.48	B02s	13.59
B03c	2125.64	B03s	20.24	B03c	466.47	B03s	−33.49
B04c	−3693.53	B04s	−13.93	B04c	−883.87	B04s	17.64
B05c	887.38	B05s	−5.09	B05c	451.14	B05s	9.81
B06c	2377.94	B06s	3.01	B06c	105.54	B06s	−12.59
B07c	−805.64	B07s	3.55	B07c	327.71	B07s	−7.35
B08c	180.12	B08s	−0.92	B08c	−280.28	B08s	3.01
B08Ac	−388.23	B08As	−5.63	B08Ac	468.29	B08As	16.89
B11c	−32.67	B11s	4.67	B11c	−25.013	B11s	−4.61
B12c	50.53	B12s	1.16	B12c	6.73	B12s	−1.11

For the Mollisol 1 order, observing the coefficients of the linear discriminant function (Table 12), it resulted that the spectral behavior of soils measured in the laboratory exhibited a higher contribution to discriminate land uses Forest, Páramo, and Pasture for the Mollisol 1 order, compared to the coefficients derived from the satellite image. Consequently, the first component of this linear discriminant function (LD1) explained 70.9% of the total variability of the three different land uses, implying that although the reflectance values in the satellite image had lower coefficients, these variables were important for the classification of land use from the Mollisol 1 soil order. The first and second discriminant components were the linear combinations of the variables that best discriminated between the three Mollisol 1 land use types. Figure 12 shows a representation of the linear discriminant function for this particular case of Mollisol 1 land use, with a minimum overlap between Páramo and Pasture, with a classification error of only 0.47% and an accuracy of 99.52%.

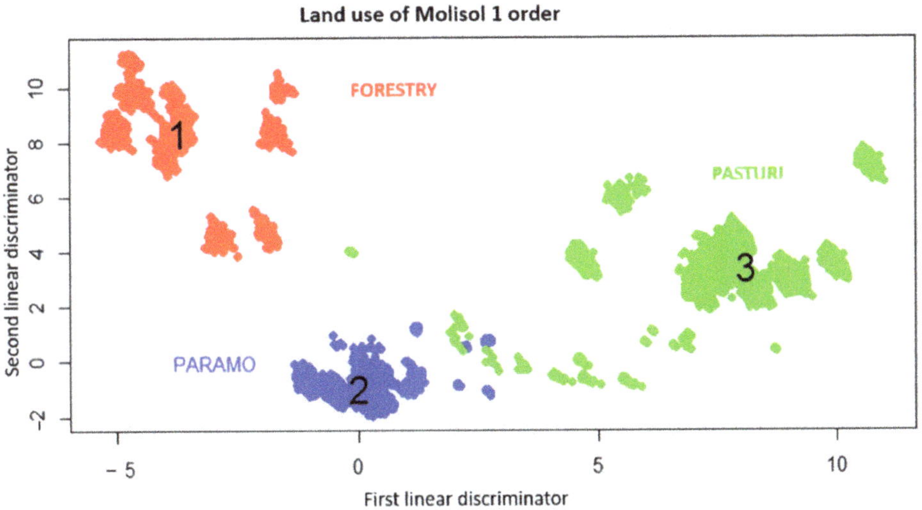

Figure 12. Representation of the linear discriminant function: land use—Mollisol 1.

(b)   Model 4 for the variable of land use of the Mollisol 2 order from all wavelengths of the soil spectra in laboratory and satellite image

The following results were obtained from the application of LDA (Table 13):

Table 13. Coefficients of the linear discriminant function with dependent variable of the use of soil of the Mollisol 2 order and independent variables of the reflectance levels of the soil spectra in laboratory and satellite image.

Ground Bands	Ground Coefficients LD	Satellite Bands	Satellite Coefficients LD
B02c	2796.26	B02s	−2.27
B03c	−3969.19	B03s	−1.87
B04c	−6701.17	B04s	4.90
B05c	18,157.90	B05s	2.94
B06c	−10,820.44	B06s	3.71
B07c	−311.00	B07s	−2.37
B08c	−1444.39	B08s	0.21
B08Ac	2194.74	B08As	−1.85
B11c	35.04	B11s	1.42
B12c	−36.25	B12s	−8.34

In relation to the coefficients of the linear discriminant function (Table 13), it was found that the soil spectral values measured in the laboratory had a higher contribution for the classification of the considered land uses compared to those of the satellite image. Regardless of the sign, the coefficients of the soil spectrum in the laboratory were higher than the coefficients of the spectrum in the satellite image. Consequently, the only component of this linear discriminant function explained 100% of the total variability of the two different land uses, which implies that even though the spectral values of the satellite image had lower coefficients, these variables were important for the classification of these land uses as a function of the Mollisol 2 soil order. Figure 13 represents a good classification of the uses of soils order Mollisol 2.

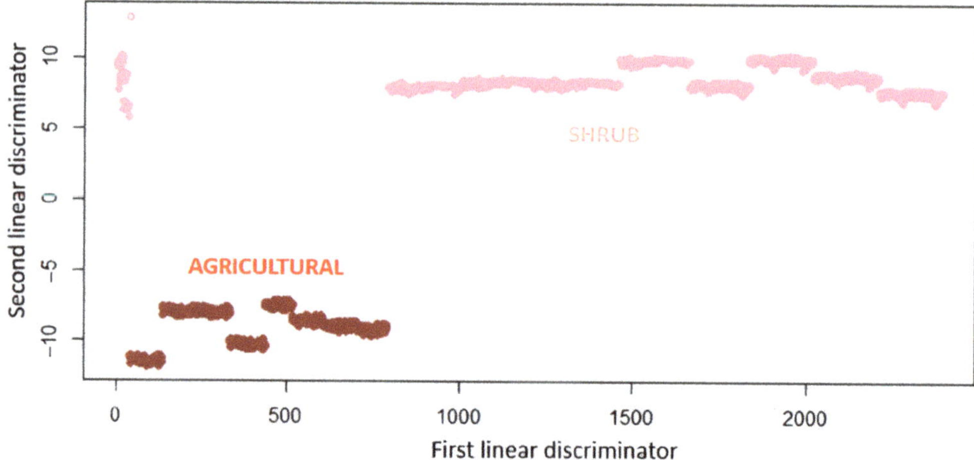

Figure 13. Representation of the linear discriminant function: land use—Mollisol 2.

Model 4 Validation

For the validation of Model 4, we tested 30% of the remaining data, called the test dataset, to classify the soils based on Model 4, and obtained the following confusion matrix, shown in Table 14.

Table 14. Classification of land use of the Mollisol 2 order based on Model 4 for the test dataset.

Real Land Use	Land Use Predicted by Model 4		Total
	Agriculture	Shrub	
Agriculture	1286	0	1286
Shrub	0	4049	4049
Total	1286	4049	5335

In Table 14, for the remaining 30%, a good classification of the land use was observed in Agriculture and Shrub for the Mollisol 2 order, considering the same behavior indicated in the training data.

3.4. Index Development

3.4.1. Index for Andisol and Mollisol Soil Orders from Model 1

According to the methodological process indicated in Section 2.8.1, the index.ma.1 (Mollisol Andisol Index) was obtained (Equation (4)):

$$\text{index.ma.1} = -\,0.21 - 272.64B04c + 336.76B05c + 328.37B06c$$

$$- 500.64B08c - 0.13B08c + 108.45B08Ac + 0.20B04s \tag{4}$$

$$+ 1.58B05s - 2.11B06S - 2.11B07s + 0.29B08s + 3.20B08As$$

The index.ma.1 separates the soils according to their order into Andisol and Mollisol. If the index values are positive, they correspond to soils of the Andisol order; if index.ma.1 takes negative values, they correspond to soils of the Mollisol order. The descriptive statistics of index.ma.1 are shown in Table 15.

Table 15. Descriptive statistics of the index.ma.1 according to the order of the soil.

Orden	Descriptive Statistics						
	Minimum	Q1	Median	Mean	S	Q3	Maximum
Andisol	−0.6407	0.0936	0.1908	0.2377	0.1792	0.3310	1.3658
Mollisol	−3.6134	−2.1854	−1.9723	−1.4961	0.5688	−0.1383	0.4836
Global	−3.6134	−0.3100	0.0669	−0.1340	0.5868	0.2290	1.3658

For the soils of the Andisol order, the mean level of the index was 0.2377 with a relatively low level of variability, equal to 0.1792. For soils of the Mollisol order, the mean level of the index was lower, −1.4961, presenting a higher level of variability equal to 0.5688, which can also be classified as a high level of variability. In this way, index.ma.1 classifies soils according to their order.

3.4.2. Indices Depending on the Variable of Land Use of the Andisol and Mollisol Orders

(a)   Index for the Land Use of the Andisol Order from Model 2

The index obtained from the discriminant function of Model 2 was expressed as follows (Equation (5)):

$$\text{index.2} = -\,22.24B02c - 67.95B03c - 149.75B04c + 402.09B05c$$

$$+ 438.95B06c - 705.75B07c + 108.45B08Ac + 0.20B04s$$

$$+ 29.83B11x - 23.95B12c - 2.67B02s - 2.49B03s \tag{5}$$

$$+ 2.90B04s + 1.13B05s + 1.56B06 - 0.99B07s$$

$$+ 0.18B08s - 0.89B08As + 0.67B11s - 1.55B12s$$

The descriptive statistics of Index 2 are displayed in Table 16.

Table 16. Descriptive statistics of Index 2 for land use of the Andisol order.

Land Use	Descriptive Statistics						
	Minimum	Q1	Median	Mean	S	Q3	Maximum
Shrub	−1.98	−1.29	−1.14	−1.04	0.33	−1.04	−0.43
Páramo	0.25	0.49	0.53	0.54	0.06	0.58	1.14
Pasture	−1.46	−1.30	−1.18	−1.12	0.22	−0.99	−0.28
General	−1.98	0.48	0.52	0.45	0.38	0.58	1.14

For land use of the Andisol order, the mean level of Index 2 was higher in Páramo (0.54), with a level of variability equal to 0.06 (the table in Section 3.5.2). For the land use of the Pasture type of the Andisol order, the average level of Index 2 was −1.12, with a level of variability equal to 0.22 (the table in Section 3.5.2). In the use of bushland of the Andisol order, the mean level of Index 2 was −1.04, with a level of variability of 0.33. The level of variability of the groups defined according to the land use of the Andisol order was very different, representing the natural behavior of these variables.

(b)   Index for the Land Use of the Mollisol 1 Order from Model 3.

The index obtained from the discriminant function of Model 3 is expressed as follows (Equation (6)).

$$index.3 = 127.77B02c - 383.23B03c + 665.89B04c - 159.98B05c$$
$$- 428.71B06c + 145.25B07c - 32.47B08c + 69.99B08Ac$$
$$+ 5.89B11c - 9.11B12c + 0.99B02s - 3.65B03s \tag{6}$$
$$+ 2.51B04s + 0.92B05s - 0.54B06s - 0.64B07s$$
$$+ 0.16B08s + 1.014B08As - 0.84B11s - 0.21B12s$$

For land uses of the Mollisol 1 order, the mean level of Index 3 was higher in forest land use, at 0.79, with the highest level of variability, equal to 0.22 (the table in Section 3.5.3). For the use of páramo land of the Mollisol 1 order, the mean level of Index 5 was −0.06, with the lowest level of variability, equal to 0.12 (Table 17). The level of variability of the groups defined as a function of the land use of the Mollisol order was different, representing the natural behavior of these variables.

Table 17. Descriptive statistics of Index 3 for land use of the Mollisol 1 order.

Land Use	Descriptive Statistics						
	Mín	Q1	Median	Mean	S	Q3	Max
Forest	0.32	0.56	0.85	0.79	0.22	0.98	1.15
Páramo	−0.11	−0.08	−0.07	−0.06	0.12	0.03	0.19
Pasture	−2.19	−1.36	−1.23	−1.25	0.19	−1.18	0.082

(c)   Index for the land Use of the Mollisol 2 Order from Model 4

The fourth index obtained from the discriminant function (Model 4) was obtained by standardizing the coefficients of this model. Each coefficient of Model 4 was divided by the sum of its coefficients in such a way that the sum of the coefficients of Index 4 was equal to 1, obtaining (Equation (7)):

$$\begin{aligned}
index.4 = &- 27.40B02c + 38.90B03c + 65.66B04c - 177.93B05c \\
&+ 106.03B06c + 3.05B07c + 14.15B08c - 21.51B08Ac \\
&- 0.34B11c + 0.35B12c + 0.02B02s + 0.02B03s \\
&- 0.05B04s - 0.03B05s - 0.04B06s + 0.02B07s \\
&- 0.002B08s + 0.02B08As - 0.01B11s + 0.08B12s
\end{aligned} \tag{7}$$

For land uses of the Mollisol 2 order, the mean Index 4 was higher in agricultural land use, at 0.12, with the highest level of variability, equal to 0.02 (Table 18). For the use of shrub soil of the Mollisol 2 order, the mean level of Index 6 was −0.07, with the lowest level of variability, equal to 0.01 (Table 18). The level of variability of the groups defined as a function of the land use of the Mollisol order was different, representing the natural behavior of these variables.

Table 18. Descriptive statistics of Index 4 for the land use of the Mollisol 2 order.

Land Use	Descriptive Statistics						
	Mín	Q1	Median	Mean	S	Q3	Max
Agriculture	0.09	0.10	0.11	0.12	0.02	0.12	0.15
Shrub	−0.11	−0.08	−0.07	−0.07	0.01	−0.07	−0.05

3.5. Regression Tree Models to Define Association between Indices and Physicochemical Parameters

3.5.1. Regression Tree Models of Physicochemical Parameters as a Function of Soil Order through Model 1 (index.ma.1)

The first model predicted the values of index.ma.1. as a function of the covariates, soil order, and physicochemical parameters. An example with soil moisture is presented, where for soils of the Andisol order, the mean of the index.ma.1 (Figure 14) is equal to −0.134. If the soil moisture (HU) is greater than or equal to 36.7 for soils of the Andisol order, the predicted value of the index.ma.1 is on average equal to 0.109 (i = 0.109) for n = 74,000 soil samples. However, if the soil moisture is less than 36.7, the predicted value of index.ma.1 is on average equal to 0.382 (i = 0.382), for n = 65,700 soil samples. The same interpretation for soils of Mollisol order.

Likewise, if the value of the index is positive, it corresponds to an Andisol soil order and negative to a Mollisol soil order. The predicted moisture value is at least 36.7.

3.5.2. Regression Tree Models of Physicochemical Parameters as a Function of Land Use of the Andisol Order through Model 2 (index.2)

As shown in Table 19, it was possible to obtain the effect of land use of the Andisol order on the physicochemical parameters.

Table 19. Application of one-way ANOVA to determine the effect of land use of the Andisol order on PFQ.

PFQ	F-Value	p-Value	Decision
OM	21,289	$<2 \times 10^{-16}$	Significant
CE	82,125	$<2 \times 10^{-16}$	Significant
pH	19,808	$<2 \times 10^{-16}$	Significant
HU	5213	$<2 \times 10^{-16}$	Significant

An example of the non-parametric regression tree model is presented below (Figure 15), with the dependent variable as index.2 and the independent variables as soil use and organic matter (OM) for soils of the Andisol order (Table 20) (ARUSAMO).

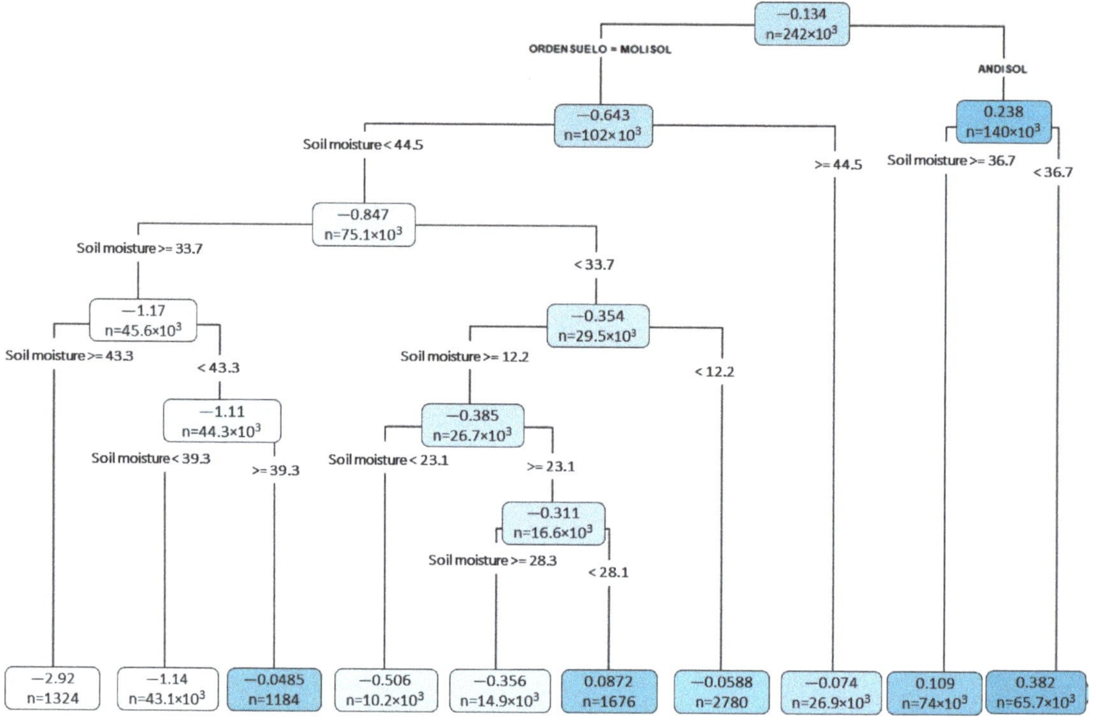

Figure 14. Moisture Regression Tree Model for Soil Order Andisol-Mollisol (ARMAH).

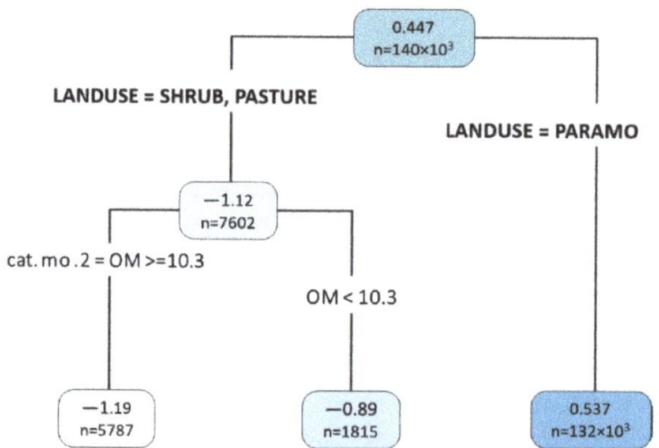

Figure 15. Regression tree for Index 2, Andisol soil, and OM.

Table 20. Classification of land uses of the Andisol order according to the mean of organic matter.

Land Use	Mean of OM		Total
	OM < 10.3%	OM $\geq$ 10.3%	
Shrub	677	0	677
Páramo	122,254	9878	132,132
Pasture	1138	5787	6925
Total	**124,069**	**15,665**	**139,734**

3.5.3. Regression Tree Models of Physicochemical Parameters as a Function of Land Use of the Mollisol Order through Model 3 (index.3)

In Figure 16, the regression tree model of index.3 can be observed as an example in the function of land use of soil order Mollisol 1 and organic matter, which are related based on Table 21 (ARUSM3MO).

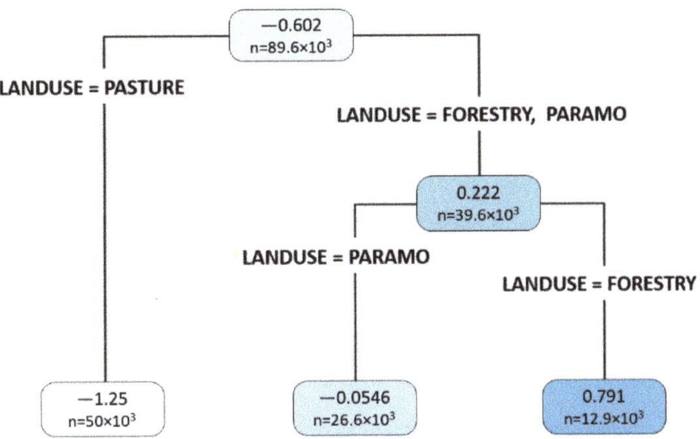

Figure 16. Regression tree for index 3: land use of Mollisol 1 and OM.

Table 21. Classification of the soil samples of the Mollisol 1 order according to the use of the soil by the mean of OM.

Land Use	Mean of OM		Total
	OM < 8.6%	OM ≥ 8.6%	
Forest	10,119	2814	12,933
Páramo	0	26,650	26,650
Pasture	41,775	8219	49,994
Total	51,894	37,683	89,577

OM was greater in Páramo ($\geq$8.6%) than Forest and Pasture, with a misclassification of 22% for Forest and 17% for Pasture (Table 21).

3.5.4. Regression Tree Models of Physicochemical Parameters as a Function of Land Use of the Mollisol Order through Model 4 (index.4)

The land use regression tree model for soil order Mollisol 2 and OM, which are related based on Table 22 (ARUSM4MO), are shown in Figure 17.

Table 22. Classification of the soil samples of the Mollisol 2 order according to the use of the soil by the mean of OM.

Land Use	Mean of OM		Total
	OM < 6.1%	OM ≥ 6.1%	
Agriculture	2998	0	2998
Shrub	94	9353	9447
Total	2979	9466	12,445

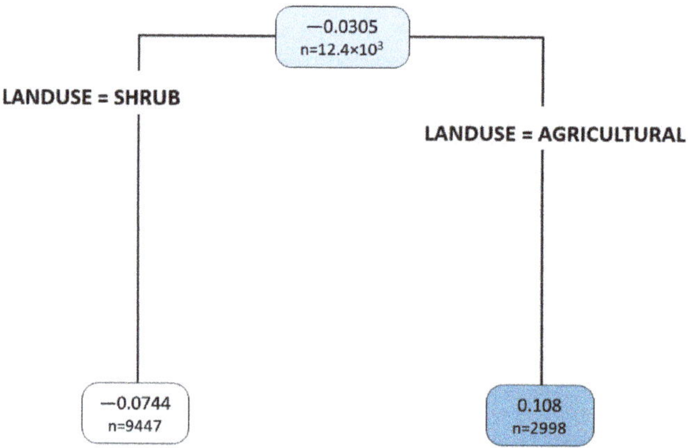

Figure 17. Regression tree for Index.4: land use of Mollisol 2 and OM.

OM was greater in Shrub ($\geq$6.1%) than Agriculture (<6.1%), with a misclassification of 1% for Shrub (Table 22).

4. Discussion

The results of this study allow for a description of the correlation between the physicochemical parameters with index.2 (Andisol), index.3 (Mollisol 1), index.4 (Mollisol 2), according to the soil order–land use homogeneous zones defined in Section 2.7 and based on the criteria of Zebrowski 1997 [52]. In the case of the Mollisol order soil, the prediction values for Páramo were $\geq$8.6% (Table 21), which maintained the characteristic behavior of the Ecuadorian Andean zone, as cited by Podwojewski (1999) [53]. On the contrary, for Forest and Pasture, the prediction models presented a behavior with a lower value in organic matter (<8.6%) (Table 21), a less acidic pH and lower soil moisture percentage, and a higher electrical conductivity [10]. This behavior shows the effects of the impact of human activity, with a lesser value of OM in the Agriculture land use (<6.1%) (Table 22).

The results presented in this study differ from other studies that compared different classification techniques using Sentinel-2 images [18,54], or considered the capacity of satellite observations to monitor and determine the state of the vegetation due to environmental stress factors by evaluating vegetation and chlorophyll indices [1].

Unlike other methodological approaches [17,55–57], this study demonstrates that the combination of laboratory spectroscopy and multispectral images with environmental covariates is an adequate alternative to establish spatial analysis models to predict the quality of Andean soils in terms of physicochemical variables such as CE, OM, pH, and HU. For this purpose, performing soil order–land use associations was revealed to be an important possible tool for assessing the accomplished predictive models.

(1)   Performance of the Models

Equation (4) shows the distributions of the logistic regression coefficients in the R-NIR spectral range for soil order, with low false-positive (1.9%) and false-negative (5.89%) coefficients. For the Andisol soil order, the mean level of the index (index.ma.1) was 0.2377, with a level of variability equal to 0.1792 (Table 15). For the Mollisol soil order, the mean level of the index (index.ma.1) was −1.4961, with a level of variability equal to 0.5688 (Table 15). The index values ranged from approximately −6 to 2, with some outliers below −4 and a very low frequency of occurrence. This corroborates the potential of promoting soil studies based on laboratory spectral data and remote sensors, such as Ali Aldabaa et al. (2014) [19], who evaluated the feasibility of the methods for the prediction of soil surface salinity by visible near-infrared diffuse reflectance spectroscopy (VisNIR) and remote sensing (RS). Equations (5)–(7) show the distributions of the coefficients of the linear discriminant function in the VIS-NIR-SWIR spectral range for the order–land use both in laboratory and S2 configurations. For land uses of the Andisol order, the level of variability of the defined groups was very different, which represents the natural behavior of these variables (Shrub, Páramo, Pasture), which, unlike previous research on the VIS-NIR, presented greater sturdiness considering SWIR.

(2)   Predictions of Physicochemical Parameters

The prediction performance of the R-NIR model, based on the Student t-test with $p < 0.0001$ for OM, CE, pH, and soil moisture, shows that the mean of each parameter in the Andisol and Mollisol soil order were different, concluding that the mean of each one was lower for the Mollisol soil order, unlike CE, where its mean was higher in Mollisol.

Regarding the results obtained from the VIS-NIR-SWIR or full spectrum model, using non-parametric regression tree models, excellent results were obtained for OM, pH, CE, and soil moisture as explanatory variables of order–land use [57]. For Mollisol 1, the 95% confidence intervals for the difference in means for the set of physicochemical parameters (CE, OM, pH, HU) were negative for Pasture and Páramo, and for Forest. This means that on average the given set of parameters had higher values in Forest. For the Mollisol 2 soil type, the 95% confidence intervals for the difference in means for the considered set of physicochemical parameters were negative for Shrub and positive for Agriculture. For Andisol-type soils, the 95% confidence intervals for the difference in means for OM in Páramo are higher than the average OM in Shrub. Similarly, but in an opposite direction, when comparing the mean OM in Pasture and Páramo land uses ($p = 0.0000 < 0.05$), the 95% confidence interval for the mean difference was negative, which implies that the average OM in soils used for pasture was less than the average OM in soils used for Páramo. Very similar results were obtained in relation to the pH physicochemical parameter, and in relation to CE and HU in all pairs of established comparisons there were statistically significant differences.

In the methodological process, the nonparametric regression tree method was successfully applied to predict the values of the model covariates by soil order or land use order (Figure 8). This statistical analysis methodology differed from those applied to date, like Adeline (2017) [41], Bao (2017) [40], Soriano-Disla (2014) [17], and Ali Aldabaa (2014) [19], where it was established that soil properties were derived from reflectance spectra that can be applied from various sources of spectral measurements, such as measurements in the laboratory, in the field, or from remote sensing systems.

These regression tree models were more flexible than those presented by Hill (2011) [55], because they did not consider non-compliance with statistical assumptions such as normality or collinearity problems between predictor variables. The regression tree models allowed for approximate estimates supported by 95% confidence intervals as a measure of the variation range of each physicochemical parameter, allowing for a reading of this from the top to the final nodes and vice versa, which was not possible in other applied models (Figures 14–17; Tables 21 and 22) [19,55,56].

Soil quality and soil degradation are crucial to develop sustainable agricultural activities [58]. The usual methods for environmental soil monitoring are very labor intensive and costly to cover large areas of land [1,13]. Satellite data in this field open up new research opportunities with great applications, as large areas of land can be analyzed and soil quality can be assessed in areas that are difficult to access [6,51].

Finally, this study is very valuable for the Ecuadorian Andean region for soil sustainability. Additionally, the results obtained in this study could be adapted in future research to other geographical regions after reviewing the soil order and land use that allow the relationships observed in the proposed model indices to be confirmed.

5. Conclusions

Soil quality is an important factor in sustainable land management. Its evaluation allows for the development and implementation of sustainable agriculture management techniques. Thus, in this study, an alternative method for the prediction of the parameters

OM, CE, pH, and soil moisture based on the R-NIR and VIS-NIR-SWIR models is presented to demonstrate its applicability in the Ecuadorian Andean region. For this purpose, logistic regression analysis and linear discriminant function analysis were used. This required the establishment of homogeneous zones defined by soil order and land use combinations to design and implement soil-sampling strategies and field–satellite spectral measurements. The findings of this study suggest that soil + RS spectroscopy is a useful technique to predict soil properties, presenting good potential as an impetus towards future soil studies.

According to the results of this study:

(1) The logistic regression function made it possible to predict the values as a soil order function and each of the physicochemical parameters described above.

(2) The linear discriminant function made it possible to treat the data based on the linear combination of the Andisol soil order variables by land use (Shrub, Páramo, Pasture), Mollisol soil order by land use (group 1: Forest, Páramo and Pasture; group 2: Agriculture and Shrub).

(3) Non-parametric models had the advantage of predicting the values of the independent variables OM, CE, pH, and soil moisture (soil properties).

Therefore, because of the achieved results, the proposed methodology might be applied to other regions and adapted to predict soil properties as a function of the site-specific soil order and land use properties. Future research should explore the variability of soil quality parameters geographically with the aim of building regional models.

Supplementary Materials: The Sentinel 2 dataset used in this study can be downloaded from https://scihub.copernicus.eu/dhus/ (accessed on 20 August 2018).

Author Contributions: S.A.-O.: conceptualization, methodology, fieldwork, software, validation, formal analysis, visualization, writing and original draft. I.M.: conceptualization, writing, review, and editing. C.G.-A.: writing, review, and editing. All authors have read and agreed to the published version of the manuscript.

Funding: This research did not receive any specific grant from funding agencies in the public, commercial, or not-for-profit sectors.

Institutional Review Board Statement: Not applicable.

Informed Consent Statement: Not applicable.

Data Availability Statement: Not applicable.

Acknowledgments: The first author thanks Maria Eugenia Jacome for the logistical support for taking soil samples in the field.

Conflicts of Interest: The authors declare no conflict of interest.

References

1. Gholizadeh, A.; Kopačková, V. Detecting vegetation stress as a soil contamination proxy: A review of optical proximal and remote sensing techniques. *Int. J. Environ. Sci. Technol.* **2019**, *16*, 2511–2524. [CrossRef]

2. Stenberg, B.; Viscarra Rossel, R.; Mouazen, A.M.; Wetterlind, J. Visible and Near Infrared Spectroscopy in Soil Science. *Adv. Agron.* **2010**, *107*, 163–215. [CrossRef]

3. Chaerle, L.; van Der Straeten, D. Seeing is believing: Imaging techniques to monitor plant health. *Biochim. Biophys. Acta-Gene Struct. Expr.* **2001**, *1519*, 153–166. [CrossRef]

4. Fedotov, Y.; Bullo, O.; Belov, M.; Gorodnichev, V. Experimental research of reliability of plant stress state detection by laser-induced fluorescence method. *Int. J. Opt.* **2016**, *2016*, 4543094. [CrossRef]

5. Correa, R.; de Souza, C.; Álves, L. Espectroscopia de reflectancia aplicada a la caracterización espectral de suelos contaminados. In Proceedings of the Latinoamerican Remote Sensing Week—LARS, Santiago, Chile, 23–25 October 2013; pp. 1–7. (In Spanish)

6. Pérez González, E.; García Rodríguez, P. Aplicaciones de la teledetección en degradación de suelos. *Bol. Asoc. Geógr. Esp.* **2013**, *61*, 285–308. (In Spanish) [CrossRef]

7. Yánez, G.; Arciniegas, S. Caracterización Físico-Química y Espectral de Suelos con Actividad Agropecuaria en la Microcuenca del Río Blanco—Cayambe. Bachelor's Thesis, Universidad Central del Ecuador, Quito, Ecuador, 2019. (In Spanish)

8. Cruzatty, L.C.G.; Vollmann, J.E.S. Caracterización de suelos a lo largo de un gradiente altitudinal en Ecuador. *Rev. Bras. Cienc. Agrar.* **2012**, *7*, 456–464. (In Spanish) [CrossRef]

9. UNDP Ecuador. Available online: https://www.ec.undp.org/content/ecuador/es/home/sustainable-development-goals.html (accessed on 29 December 2021).

10. Instituto Espacial Ecuatoriano. *Memoria Técnica: Cantón Cayambe. Generación de Geoinformación para la Gestión del Territorio a Nivel Nacional Escala 1:25,000*; Ministerio de Defensa Nacional, SENPLADES, IEE, GADPP: Quito, Ecuador, 2013. (In Spanish)

11. Franco, W.; Peñafiel, M.; Cerón, C.; Freire, E. Biodiversidad productiva y asociada en el valle interandino norte del Ecuador. *Bioagro* **2016**, *28*, 181–192. (In Spanish)

12. GADIP Cayambe. *Actualización del Plan de Desarrollo y Ordenamiento Territorial del Cantón Cayambe 2015–2025*; GADIP Cayambe: Cayambe, Ecuador, 2015. (In Spanish)

13. Baldock, J.; Hawke, B.; Sanderman, J.; MacDonald, L. Predicting contents of carbon and its component fractions in Australian soils from diffuse reflectance mid-infrared spectra. *Soil Res.* **2013**, *51*, 577–595. [CrossRef]

14. Stevens, A.; Nocita, M.; Tóth, G.; Montanarella, L.; van Wesemael, B. Prediction of soil organic carbon at the European scale by visible and near infrared reflectance spectroscopy. *PLoS ONE* **2013**, *8*, e66409. [CrossRef] [PubMed]

15. Bayat, B.; van der Tol, C.; Verhoef, W. Remote sensing of grass response to drought stress using spectroscopic techniques and canopy reflectance model inversion. *Remote Sens.* **2016**, *8*, 557. [CrossRef]

16. Bellon-Maurel, V.; McBratney, A. Near-infrared (NIR) and mid-infrared (MIR) spectroscopic techniques for assessing the amount of carbon stock in soils—Critical review and research perspectives. *Soil Biol. Biochem.* **2011**, *43*, 1398–1410. [CrossRef]

17. Soriano-Disla, J.M.; Janik, L.J.; Viscarra Rossel, R.A.; Macdonald, L.M.; McLaughlin, M.J. The performance of visible, near-, and mid-infrared reflectance spectroscopy for prediction of soil physical, chemical, and biological properties. *Appl. Spectrosc. Rev.* **2014**, *49*, 139–186. [CrossRef]

18. Borras, J.; Delegido, J.; Pezzola, A.; Pereira, M.; Morassi, G.; Camps-Valls, G. Clasificación de usos del suelo a partir de imágenes Sentinel-2. *Teledetección* **2017**, *48*, 55–66. (In Spanish) [CrossRef]

19. Ali Aldabaa, A.; Weindorf, D.; Chakraborty, S.; Sharma, A.; Li, B. Combination of proximal and remote sensing methods for rapid soil salinity quantification. *Geoderma* **2014**, *239–240*, 34–46. [CrossRef]

20. Calvache, M.; Ballesteros, H.; Andino, J. Caracterizacion del potencial natural de los suelos dedicados a la ganaderia lechera en la sierra ecuatoriana: Caso Cayambe. In Proceedings of the VIII Congreso Ecuatoriano de la Ciencia del Suelo, Manabí, Ecuador, 26 September 2002. (In Spanish)

21. Prat, C.; Moreno, J.; Hidrobo, J.; Trujillo, G.; Ortega, C.; Etcheveres, J.; Hidalgo, C.; Baéz, A.; Gallardo, J. *Understanding Mountain Soils: A Contribution from Mountain Areas to the International Year of Soils*; FAO: Rome, Italy, 2015; pp. 111–115.

22. Tucci, C. *Plan de Manejo Integrado de los Recursos Hídricos en la Cuenca Alta del Río Guayllabamba*; BID Banco Interamericano de Desarrollo Económico, FONAG Fondo Para La Protección Del Agua: Quito, Ecuador, 2009. (In Spanish)

23. Clevers, J.; Kooistra, L.; van den Brande, M. Using Sentinel-2 data for retrieving LAI and leaf and canopy chlorophyll content of a potato crop. *Remote Sens.* **2017**, *9*, 405. [CrossRef]

24. Immitzer, M.; Vuolo, F.; Atzberger, C. First experience with Sentinel-2 data for crop and tree species classifications in central Europe. *Remote Sens.* **2016**, *8*, 166. [CrossRef]

25. European Space Agency. *Sentinel 2 Document Library*; ESA Technical Document, ref. COPE-GSEG-EOPG-TN-15-0007; ESA: Paris, France, 2017.

26. ESA. Available online: https://sentinels.copernicus.eu/web/sentinel/missions/sentinel-2/instrument-payload (accessed on 28 January 2021).

27. ESA Copernicus. Available online: https://scihub.copernicus.eu/dhus/#/home (accessed on 16 July 2018).

28. ESA Sentinel Online. Available online: https://sentinel.esa.int/web/sentinel/technical-guides/sentinel-2-msi/msi-instrument (accessed on 28 January 2021).

29. Ministry of the Environment and Water. *Book VI, Annex 2, Norma de Calidad Ambiental del Recurso Suelo y Criterios de Remediación para Suelos Contaminados*; Registro Oficial del Ecuador: Quito, Ecuador, 2015.

30. USDA. *Guía para la Evaluación de la Calidad y Salud del Suelo*; USDA: Washington, DC, USA, 1999. (In Spanish)

31. Baumgardner, M.F.; Silva, L.R.F.; Biehl, L.L.; Stoner, E.R. Reflectance properties of soils. In *Advances in Agronomy*; Academic Press: Cambridge, MA, USA, 1986; Volume 38, pp. 1–44. [CrossRef]

32. Friedman, D.; Hubbs, M.; Tugel, A.; Seybold, C.; Sucik, M. *Guidelines for Soil Quality Assessment in Conservation Planning*; USDA: Washington, DC, USA, 2001.

33. Secretaría de Medio Ambiente y Recursos Naturales de México. *Norma Oficial Mexicana NOM-021-SERMANAT-2000 que Establece las Especificaciones de Fertilidad, Salinidad y Clasificación de Suelos, Estudio, Muestreo y Análisis*; Diario Oficial de la Federación: Mexico City, Mexico, 2002. (In Spanish)

34. Hidrobo, J.R.; Mendez, E.G. Validación del Método Analítico Walkley y Black de Materia Orgánica en Suelos Arcillosos, Francos y Arenosos del Ecuador. Bachelor's Thesis, Universidad Central del Ecuador, Quito, Ecuador, 2016. (In Spanish)

35. Borana, S.; Yadav, S.; Parihar, S. Hyperspectral Data Analysis for Desertic Vegetation of Jodhpur Area. In Proceedings of the 3rd International Conference on Emerging Technologies in Computer Engineering: Machine Learning and Internet of Things (ICETCE), Jaipur, India, 7–8 February 2020; pp. 171–175. [CrossRef]

36. Field Spectroradiometer Signal-to-Noise Ratio—The ASD Advantage. Available online: https://www.materials-talks.com/field-spectroradiometer-signal-to-noise-ratio-the-asd-advantage/ (accessed on 29 May 2022).

37. Orjuela, I.; Camacho, J.; Rubiano, Y. Uso de espectrómetros de infrarrojo medio (MIRS) para la determinación de carbono del suelo en oxisoles. In Proceedings of the XII Congreso Latinoamericano y Del Caribe de Ingeniería Agrícola, Bogotá, Colombia, 23 May 2016. (In Spanish)

38. Estrella, S.; Jácome, N. Cuantificación del Contenido de Arcilla en los Suelos del Estado de Santa Catarina, Brasil a Través de los Datos Espectrales Obtenidos Emdiante Fieldspec Pro 3. Bachelor's Thesis, Universidad Distrital Francisco José de Caldas, Bogotá, Colombia, 2016. (In Spanish)

39. Bonett, J.; Camacho, J.; Ramirez, L. Análisis de respuestas espectrales en el infrarrojo medio de suelos de Colombia. In Proceedings of the XX Congreso Latinoamericano y XVI Congreso Peruano de la Ciencia del Suelo, Cusco, Peru, 9–15 November 2014. (In Spanish) [CrossRef]

40. Bao, N.; Wu, L.; Ye, B.; Yang, K.; Zhou, W. Assessing soil organic matter of reclaimed soil from a large surface coal mine using a field spectroradiometer in laboratory. *Geoderma* **2017**, *288*, 47–55. [CrossRef]

41. Adeline, K.R.M.; Gomez, C.; Gorretta, N.; Roger, J.M. Predictive ability of soil properties to spectral degradation from laboratory Vis-NIR spectroscopy data. *Geoderma* **2017**, *288*, 143–153. [CrossRef]

42. Mark, D.; Egenhofer, M. Topology of Prototypical Spatial Relations between Lines and Regions in English and Spanish. In Proceedings of the Third International Symposium on Spatial Data Handling, Sydney, Australia, 17–19 August 1988.

43. Poveda, K.; Arciniegas-Ortega, S. Análisis del Comportamiento Espectral del Suelo a Partir de sus Características Físico-químicas en la Microcuenca del río Blanco, Cayambe. Bachelor's Thesis, Universidad Central del Ecuador, Quito, Ecuador, 2020. (In Spanish)

44. Amat, J. Estadística con R. Git Hub. 2020. Available online: https://rpubs.com/Joaquin_AR/229736 (accessed on 20 May 2020). (In Spanish)

45. Dalgaard, P. *Introductory Statistics with R*; Springer: New York, NY, USA, 2008; pp. 227–248. [CrossRef]

46. James, G.; Witten, D.; Hastie, T.; Tibshirani, R. *An Introduction to Statistical Learning: With Applications in R*; Springer: New York, NY, USA, 2013; pp. 130–163, 303–330.

47. Witold, F.; Heather, S. *Interpolate Temperatures Using the Geostatistical Wizard*; Esri: Redlands, CA, USA, 2019.

48. Rojo, J. *Árboles de Clasificación y Regresión*; Consej Superior de Investigaciones Científicas Laboratorio de Estadística: Madrid, Spain, 2006. (In Spanish)

49. Gareth, J.; Witten, D.; Hastie, T.; Tibshirani, R. *An Introduction to Statistical Learning*; Springer: New York, NY, USA, 2021. [CrossRef]

50. The R Project for Statistical Computing. Available online: https://www.r-project.org/ (accessed on 18 April 2020).

51. Chuvieco, E. *Teledetección Ambiental: La Observación de la Tierra Desde el Espacio*, 3rd ed.; Ariel Ciencias: Barcelona, Spain, 2008. (In Spanish)

52. Zebrowski, C. Los suelos con cangahua en el Ecuador. In *Suelos Volcánicos Endurecidos*; IRD: Quito, Ecuador, 1997; pp. 128–137. (In Spanish)

53. Podwojewski, P. *Los Suelos de las Altas Tierras Andinas: Los Páramos del Ecuador*; Sociedad Ecuatoriana de la Ciencia del Suelo, IRD: Quito, Ecuador, 1999. (In Spanish)

54. Peng, Y.; Kheir, R.B.; Adhikari, K.; Malinowski, R.; Greve, M.B.; Knadel, M.; Greve, M.H. Digital Mapping of Toxic Metals in Qatari Soils Using Remote Sensing and Ancillary Data. *Remote Sens.* **2016**, *8*, 1003. [CrossRef]

55. Hill, J.; Udelhoven, T.; Vohland, M. The Use of Laboratory Spectroscopy and Optical Remote Sensing Soil Properties. In *Precision Crop Protection—The Challenge and Use of Heterogenety*; Springer: Dordrecht, The Netherlands, 2010; pp. 67–85. [CrossRef]

56. Yufeng, G.; Thomasson, A.; Ruixiu, S. Remote sensing of soil properties in precision agriculture. *Front. Earth Sci.* **2011**, *5*, 229–238. [CrossRef]

57. Rosero-Vlasova, O.; Pérez-Cabello, F.; Montorio, R.; Vlassova, L. Assessment of Laboratory VIS-NIR-SWIR Setups with Different Spectroscopy Accessories for Characterisation of Soils from Wildfire Burns. *Biosyst. Eng.* **2016**, *152*, 51–67. [CrossRef]

58. Baroudy, A.; Ali, A.; Mohamed, F.; Shokr, M.; Savin, I.; Poddubsky, A.; Ding, Z.; Kheir, A.; Aldosari, A.; Elfadaly, A.; et al. Modeling Land Suitability for Rice Crop Using Remote Sensing and Soil Quality. *Sustainability* **2020**, *12*, 9653. [CrossRef]

Review

Current Status and Development Trend of Soil Salinity Monitoring Research in China

Yingxuan Ma [1,2,3] and Nigara Tashpolat [1,2,3,*]

1   College of Geography and Remote Sensing Sciences, Xinjiang University, Urumqi 830046, China;
    mayx_1201@163.com
2   Xinjiang Key Laboratory of Oasis Ecology, Xinjiang University, Urumqi 830046, China
3   MNR Technology Innovation Center for Central Asia Geo-Information Exploitation and Utilization,
    Urumqi 830046, China
*   Correspondence: ngr.t@xju.edu.cn

Abstract: Soil salinization is a resource and ecological problem that currently exists on a large scale in all countries of the world. This problem is seriously restricting the development of agricultural production, the sustainable use of land resources, and the stability of the ecological environment. Salinized soils in China are characterized by extensive land area, complex saline species, and prominent salinization problems. Therefore, strengthening the management and utilization of salinized soils, monitoring and identifying accurate salinization information, and mastering the degree of regional salinization are important goals that researchers have been trying to explore and overcome. Based on a large amount of soil salinization research, this paper reviews the developmental history of saline soil management research in China, discusses the research progress of soil salinization monitoring, and summarizes the main modeling methods for remote sensing monitoring of saline soils. Additionally, this paper also proposes and analyzes the limitations of China's soil salinity monitoring research and its future development trend, taking into account the real needs and frontier hotspots of the country in related research. This is of great practical significance to comprehensively grasp the current situation of salinization research, further clarify and sort out research ideas of salinization monitoring, enrich the remote sensing monitoring methods of saline soils, and solve practical problems of soil salinization in China.

Keywords: soil salinization; saline soil treatment; remote sensing monitoring; model construction

Citation: Ma, Y.; Tashpolat, N. Current Status and Development Trend of Soil Salinity Monitoring Research in China. *Sustainability* **2023**, *15*, 5874. https://doi.org/10.3390/su15075874

Academic Editors: Blaž Repe, Mario Elia and Antonio Ganga

Received: 23 February 2023
Revised: 22 March 2023
Accepted: 24 March 2023
Published: 28 March 2023

1. Introduction

Soil salinization refers to the accumulation of soluble salts in soil caused by certain natural factors such as climate, hydrology, and topography or caused by the combination of destructive human factors and fragile ecological environments, thus leading to the deterioration of soil quality to form saline soils [1]. Soil salinization, as a resource and ecological problem that currently exists on a large scale in all countries of the world, is one of the main types of land desertification and soil degradation [2]. It seriously restricts the production and development of the agricultural industry, the sustainable use of land resources, and the security and stability of the ecological environment. Saline soils are a collective term for all types of soils that are negatively affected by saline components. The unique physicochemical-biological properties of saline soils have a variety of negative impacts. These include reduction in soil fertility and productivity levels, reduction in crop yields and harvests [3,4], waste of agricultural resources, destabilization of the ecological environment, and other secondary hazards [5]. Therefore, strengthening the management and utilization of salinized soils, monitoring and identifying accurate salinization information, and mastering the salinization level of regional arable farmland have been important goals for scientists to research and overcome.

The total area containing saline soils worldwide is currently about 1.1×10^9 hm^2, which is widely distributed in more than a hundred countries and regions around the world, and the global soil salinization level is still showing a rising trend [6]. The total area of saline soils in China has reached 3.69×10^7 hm^2, which is close to 4.88% of the available land area in China [7]. It is mainly distributed in arid and semi-arid regions and coastal areas with arid climate and little rainfall, high soil evaporation, a high groundwater table, and more soluble salts [8–10]. Examples include semi-humid regions such as the Yellow River Basin in North China, the plains in northeast China, and arid and semi-arid regions in northwest China such as Gansu, Ningxia, and Xinjiang. Therefore, study of the spatial distribution and management of soil salinity prevention as well as improvement of monitoring accuracy and early warning capability are gradually becoming a hot spot of concern in the field of saline soil research today.

In order to explore the current research status and current research hotspots of soil salinity monitoring, this paper searched in the China National Knowledge Infrastructure (CNKI) and the Web of Science (WOS) databases using "soil salinity monitoring" as the key search term. CiteSpace software was used to perform keyword co-occurrence analysis on the large number of highly relevant literature datasets obtained from the search (Figures 1 and 2). By extracting the frequency distribution of keywords that express the core content of the literature, a co-word matrix is thus generated based on the keyword matrix. The co-word matrix was visualized as a network to study the development trends and research hotspots in the field of salinity monitoring. The core nodes in the figure can fully reflect the focus and branches of research in the field in recent years. The size of the node represents the frequency of the keyword; the larger the node, the more frequently the keyword appears and the higher the relevance to the topic. Among them, Figure 1 shows the keyword co-occurrence analysis graph based on the relevant research articles in the CNKI database. The analysis shows that the nodal framework consisting of "saline soil", "remote sensing monitoring", "hyperspectral", "arid zone", "multisource remote sensing", and "salinity index" appears more frequently and has stronger correlations among the research articles published in Chinese database. Figure 2 shows a keyword co-occurrence analysis graph based on the relevant international research articles in the WOS database. It can be seen that the nodal framework consisting of "soil salinity", "model", "spatial distribution", "change detection", "prediction", and "remote sensing" appears more frequently and has stronger correlations among the research articles published in international databases. These keywords provide important information for us to analyze the progress of research on soil salinity monitoring in China, and they are the focus of our attention.

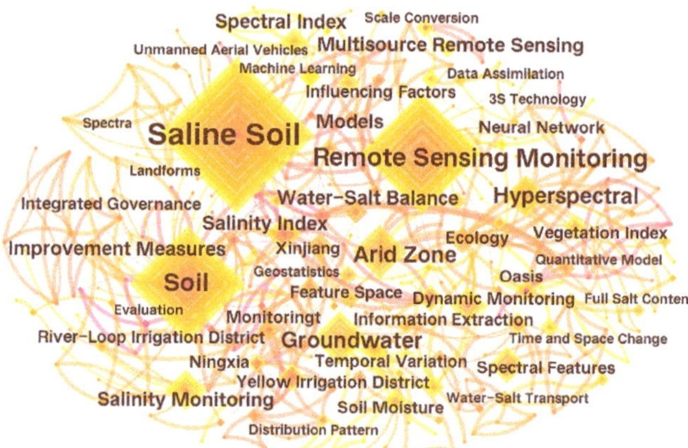

Figure 1. Keyword co-occurrence mapping based on the literature related to soil salinity monitoring in the China National Knowledge Infrastructure (CNKI) database.

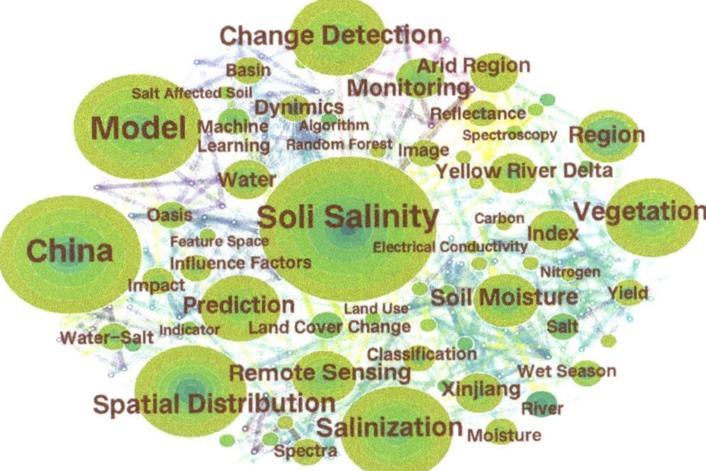

Figure 2. Keyword co-occurrence mapping based on the literature related to soil salinity monitoring in the Web of Science (WOS) database.

This paper reviews the developmental history of saline soil management research in China, discusses the research progress of soil salinity monitoring, and summarizes the main modeling methods for remote sensing monitoring of saline soils. Based on a large amount of domestic and foreign soil salinization research, this paper combines the real needs and frontier hotspots of the country in related research, proposes limitations of soil salinization monitoring research in China, and analyzes the future developmental trend of soil salinization monitoring research in China. This is of great practical significance to comprehensively grasp the current situation of salinization research, further clarify and sort out the research ideas of salinization monitoring, enrich the remote sensing monitoring methods of saline soils, and solve practical problems of soil salinization in China.

2. Research History and Importance of Saline Soil Management in China

As an important actual and potential arable land resource in China, saline soils have strong development and utilization value. Different types of saline soils can be managed and improved in terms of their physicochemical and biological properties by using various types of effective soil improvement tools and other comprehensive measures, thus improving soil quality and productivity levels [11]. Theoretical and technological research on saline soil management in China has been steadily developing. The country attaches great importance to the treatment and utilization of saline soils, policy research, and technological innovation. In the 1950s, the State organized a lot of research on saline resources and mastered the situation of many different regions and different types of saline lands [12–14]. In the 1990s, researchers started to study the regional water and salt movement of saline soils and its regulation and management on the basis of the regional water table, water quality, and a soil water and salt co-forecasting model, and this helped facilitate the improvement of saline soils [15,16]. Thus, during the 20th century, Chinese research in the field of soil salinization focused on research and classification, investigation of causes, and improvement and prevention. Several provinces have conducted focused analyses for regional saline soils [17,18] and have conducted in-depth research while gaining a basic understanding of saline soils, laying the foundation for future monitoring and management of saline soils.

In the 21st century, during the 11th Five-Year Plan, the Chinese Academy of Sciences, together with the relevant domestic forces, organized and implemented the "Research and Demonstration of Supporting Technologies for the Efficient Utilization of Saline Soil in Agriculture", which is a public welfare industry special project for the whole country, and carried out comprehensive research on saline soil management [19,20]. During the 12th Five-Year Plan period, the report of the Chinese Academy of Sciences, "The National Demonstration of Saline Land Management Technology", was received by the national leaders [21–24]. During the 13th Five-Year Plan period, China deployed the national key project of "Typical Fragile Ecological Restoration and Protection Research", which has strongly driven enthusiasm and encouraged researchers to publish related research articles [25–27]. Internationally, the theme of the 8th World Soil Day (WSD) in 2021 is "Preventing Soil Salinization and Improving Soil Productivity" [28]. This theme aims to raise awareness of soils, strengthen national research capacity, and work together to preserve the Earth's environmental carrying capacity. Summarizing the research history of soil salinization management in China since the beginning of the 21st century, it can be seen that during this period, China started to focus on advancing the theory and technology of salinization prevention and control. Comprehensive measures have been applied to manage salinized soils and improve their physicochemical and biological properties. The potential of saline soils as an important arable resource has been fully exploited.

At present, with policy support and implementation of various departments at the national level, research and development of saline soil management and monitoring technology in China has achieved certain results. The research not only covers the direction and content of international saline soil research but also highlights domestic characteristics, with a richer connotation, broader coverage, and more common interdisciplinary association [29]. This further indicates that the nation has recognized the importance and necessity of saline soil treatment to ensure food security and promote ecological stability.

3. Advances in Soil Salinity Monitoring Research

The literature in CNKI and WOS databases was searched based on the similarity of literature keywords. A keyword clustering analysis was performed on the retrieved literature datasets related to soil salinity monitoring (Figures 3 and 4). Keyword clustering analysis is the process of analyzing the set of keywords extracted from the literature into multiple categories consisting of similar objects. There are many different algorithms for clustering analysis, and this study chose Logarithmic Likelihood Ratio (LLR) as the base calculation for the analysis. The labels of each cluster are the key keywords in the co-occurrence network. Based on this, closely related keywords are clustered. The higher the ranking of the cluster number, the more keywords are included in the cluster. Conversely, the more backward the ordinal number, the fewer keywords are contained in that cluster. The modularity value of the clustering metric Q ranges 0–1. The larger the value, the better the clustering effect. Usually, when Q is less than 0.3, it indicates that the literature data set analyzed by this clustering is not well structured. In Figure 3, a value of Q = 0.8293 was obtained from cluster analysis of the literature in the CNKI database. In Figure 4, a value of Q = 0.7128 was obtained from cluster analysis of the literature in the WOS database. This indicates that the data collected in the Chinese literature database and the international literature database were reliable and the keyword clustering analysis structure was significant.

A total of 16 clustering tags were obtained from the keyword clustering analysis based on the CNKI database (Figure 3). Among the top seven clusters, three of them are related to remote sensing monitoring, namely "quantitative remote sensing", "remote sensing monitoring", and "monitoring models". A total of 10 clustering tags were obtained from the keyword clustering analysis based on the WOS database (Figure 4). Among them, both "machine learning" and "feature space" are specific remote sensing monitoring modeling methods. Based on these methods, the researcher explores the "change detection", "spatial distribution", and "evolution trend" of saline soils in China, making full use of the advanced remote sensing information technology. This demonstrates that to achieve the purpose of saline soil management and utilization on a large scale, it is an important prerequisite to use scientific means to quickly and accurately grasp the information of saline soil distribution and to clarify the spatial and temporal variability characteristics of salinization [30]. With the decades of continuous exploration and practice of soil salinization research in China, there are more and more methods and means of soil salinization monitoring. In summary, they can be divided into two main categories: (1) traditional field investigation and experimental methods and (2) modern remote sensing information technology monitoring methods.

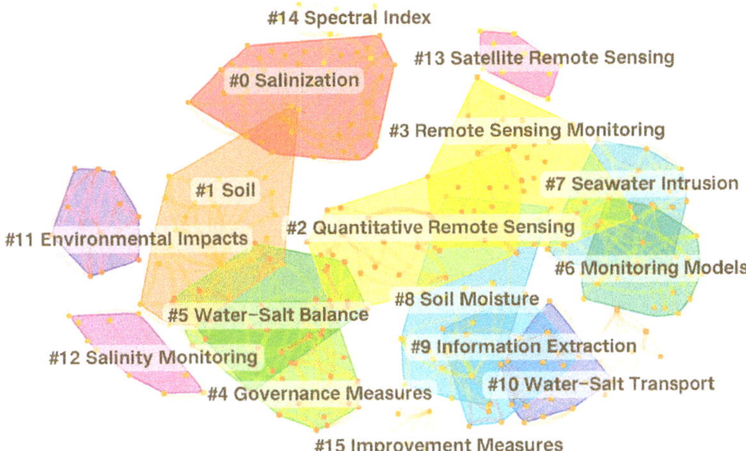

Figure 3. Keyword clustering mapping based on the literature related to soil salinity monitoring in the CNKI database.

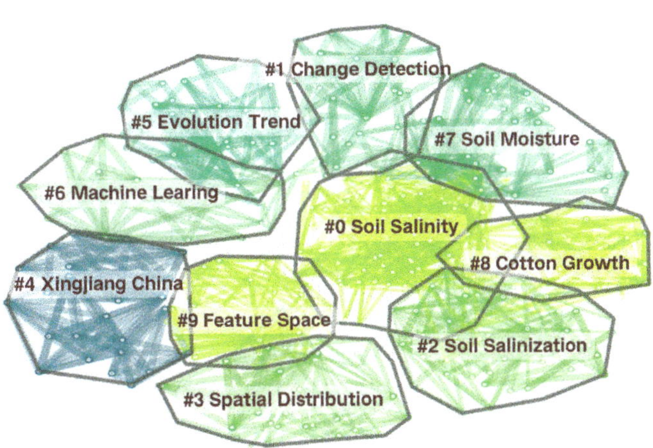

Figure 4. Keyword clustering mapping based on the literature related to soil salinity monitoring in the WOS database.

3.1. Field Investigations and Experiments

The field survey is to select test areas in the field where salinity characterization exists and to obtain visual information on soil salinity from soil samples to provide an accurate and reliable data source for the study. During the sampling process, the soil sample points should be evenly distributed throughout the work area. The sample points should also be divided into typical sampling areas containing four different landscapes, vegetation cover, soil types, etc. that are more representative. For soil samples collected at different depths, soil conductivity can be measured using the EM38 Geodetic Conductivity Probe [31]. The content of each ion in the soil can also be detected, and the corresponding total soil salt content can be calculated [32]. We can also use dry weight and wet weight to determine soil water content as well as soil evapotranspiration [33], and we can use a portable spectrometer to obtain spectral curve data from different sampling points [34]. In this way, we can achieve the purpose of extracting information on soil salinization.

The field survey method of collecting soil samples in the field and then analyzing them has become a basic monitoring tool for accurate information on soil salinization [35]. Kai Deng et al. [36] established a linear mixed model based on the linear relationship between magnetic susceptibility apparent conductivity and actual measured soil salinity to assess the spatial distribution of salinity in the soil profile. Wenping Xie et al. [37] constructed a multiple regression model between magnetic susceptibility geodetic conductivity and soil multiple regression models between geodetic conductivity and soil salinity to quantitatively assess the spatial and temporal evolution of soil salinity in the estuary over the past decade. Yasenjiang Kahaer et al. [38] performed indoor hyperspectral measurements and conductivity measurements on soil samples obtained from fieldwork and established a hyperspectral estimation model of soil conductivity after screening parameters. Finally, effective monitoring of soil salinity was achieved.

By analyzing the results of scientists' investigations, we can find that field surveys and experimental methods can extract saline soil information very precisely. However, this method is mainly based on manual point-by-point examination, which is less efficient and difficult to obtain the salinity variation characteristics of large areas at a macroscopic scale [39–41]. Especially in the study area where the vegetation cover is complex and the natural environment is harsh, the number of monitoring stations is not sufficient, and the difficulty of field investigation is increased. All these problems render the traditional monitoring method of field surveys insufficient to meet research needs [42,43].

3.2. Remote Sensing Information Technology Monitoring

Among the many soil physicochemical parameters, soil salinity content is the primary parameter for measuring salinization. The higher the soil salinity, the higher the risk of soil salinization. When the accumulation of salts in the soil exceeds a certain level, it leads to weakening of the bond between soil particles, loosening of soil structure, reduction of soil fertility, and even directly affects the survival of vegetation [2,44]. At the same time, there is a close relationship between soil conductivity and soil salinity. Soil conductivity is a measure of the ability of ions in the soil to conduct electricity, and it reflects the amount of dissolved ions in the soil, including salt ions and other dissolved substances. Similar to soil salinity, conductivity can be an important factor in measuring salinity [45,46]. Moisture in the soil can also largely reflect the salinity of the soil. Salt moves with water, and soil salts are prone to shift with changes in moisture. Under strong evaporation, salts in groundwater and deep soil rise to the surface along soil capillaries and accumulate, resulting in salinization of regional soils [47–49]. Additionally, the heavy metal content in soil also interacts with soil salinization and constrains it. Some heavy metal elements, such as cadmium and lead, can affect the growth of soil microorganisms and form insoluble complexes when combined with salts. They can reduce the activity of salts in the soil, thus affecting the process of conversion and circulation of salts in the soil and aggravating the degree of soil salinization [50,51].

Different levels of soil salinity, water content, and some heavy metals can be distinguished by using the different spectral reflectance of remote sensing images for different features. The multidimensional combination of different bands in spectral images can also construct a variety of model indices that can monitor soil salinization. Therefore, the use of remote sensing to track physicochemical parameters such as soil salinity, conductivity, water content, and heavy metals can all be effective in monitoring saline soils. In recent years, remote sensing information technology has been innovating, and the spectral resolution and spatial resolution of remote sensing images have been increasing. Efficient, convenient, and large-scale means of monitoring soil salinity have been rapidly developed. The method of monitoring salinity based on remote sensing images is gradually becoming common and is rapidly developing into an important tool for studies such as soil salinity information extraction, monitoring, and forecasting [52–54].

A keyword timeline analysis was performed on the literature datasets related to soil salinity monitoring from 1981 to 2022 and 2002 to 2022 retrieved from the CNKI and the WOS databases, respectively (Figures 5 and 6). The keywords in the figure are spread out in the clusters they belong to according to the chronological order in which they appear in the corresponding years, showing the development of keywords in each cluster. The size of the keyword node represents the frequency of the keyword occurrence, and the warm and cold colors of the node periphery represent the emergence and the duration of the keyword. The larger the keyword node, the warmer the color of the edge of the node, the more frequently the hotspot appears, and the longer it lasts. Conversely, the smaller the node, the cooler the color of the node edge, the less frequently the hotspot appears, and the shorter the duration of the hotspot. The analysis of the development of soil salinity monitoring showed that in the mid-1990s, results for "dynamic monitoring" clustering began to appear in the Chinese literature database, and attention was focused on "remote sensing monitoring" (Figure 5). In the 21st century, "machine learning" and "feature space" methods for monitoring "soil salinity" and "soil moisture" have begun to appear in the international literature database (Figure 6). While a large number of studies on soil salinity monitoring based on remote sensing have gradually emerged, keyword nodes such as "remote sensing technology", "hyperspectral", "radar remote sensing", "feature space", "inversion models", "random forest", "classification", and "index" have also begun to appear in other clusters one after another. There are a large number of keyword links between them, both within and across clusters. These remote sensing monitoring-related research terms link the whole clustering trend and become important research hotspots in domestic and international literature databases.

Based on data analysis and processing of multiple remote sensing images, Yanhua Li et al. [55] established a soil salinity monitoring index model using Landsat-TM multispectral remote sensing image data to invert and estimate soil salinity in the Weigan River-Kuche River basin. Also using field collection samples and Landsat8 image data, Mingkuan Wang et al. [56] collected soil samples in the key study area in the Yellow River Delta region in the field and acquired simultaneous phase Landsat8 image data. They constructed multiple models for remote sensing inversion of soil salinity and inversion of the spatial distribution of soil salinity in the study area based on the optimal model. The results showed that the relationship between reflectance of remote sensing images and soil salinity content is not purely linear, and the constructed salt estimation model can better simulate the relationship between soil salinity and spectral data. Yanling Li et al. [39] established a machine learning and statistical regression model based on the fusion of multispectral and hyperspectral images, which significantly improved the accuracy of salt inversion. Similarly, Wumuti Aishanjiang et al. [57] matched field-measured hyperspectral data with WorldView-2 remote sensing images to improve the prediction accuracy and mapping accuracy of soil salinity. The quantitative inversion model developed in this paper considering vegetation and moisture does not require complex parameters. To a certain extent, it meets the needs of salinity monitoring in arid and semi-arid regions. This can promote further applications of high-spatial-resolution satellites such as WorldView-2 in salinity monitoring. In the meantime, some scientists have also used radar remote sensing data combined with soil moisture and pH factors to achieve predictive inversion of soil salinity. In terms of the variability and correlation used to reveal soil properties, M. Samiee et al. [58] used geostatistical methods to simulate the spatial correlation of soil salinity, and the results were used to predict the spatial distribution of soil properties by spatial interpolation methods. In addition, more and more kinds of remote sensing images have been applied to salinity monitoring. For example, Junying Chen et al. [59] used unmanned aerial vehicle (UAV) aerial imagery and high-resolution remote sensing imagery to construct a salinity inversion model and used an improved scale conversion method to achieve soil salinity monitoring at the ascending scale. Predictive inversion of soil salinity was achieved by using radar remote sensing data combined with soil moisture and pH factors by Zaytungul Yakup et al. [60]. The monitoring model developed in this paper does not need to consider complex dielectric constants. This can meet the needs of soil salinity monitoring to a certain extent and promotes the application of Phased Array type L-band Synthetic Aperture Radar data in soil salinity monitoring. Yumei Li et al. [61] focused on the current status of the application of lidar three-dimensional remote sensing observation technology in the three-dimensional dynamic monitoring of various natural resources. The paper provides a comprehensive analysis of the potential and limitations of lidar applications in natural resource surveys. These authors are also considering how to combine multi-source, multi-scale, and multi-platform remote sensing data with artificial intelligence. The goal is to build an integrated "sky-air-ground" natural resources survey and monitoring technology system. This is the developmental direction of future methods for three-dimensional dynamic monitoring of natural resources.

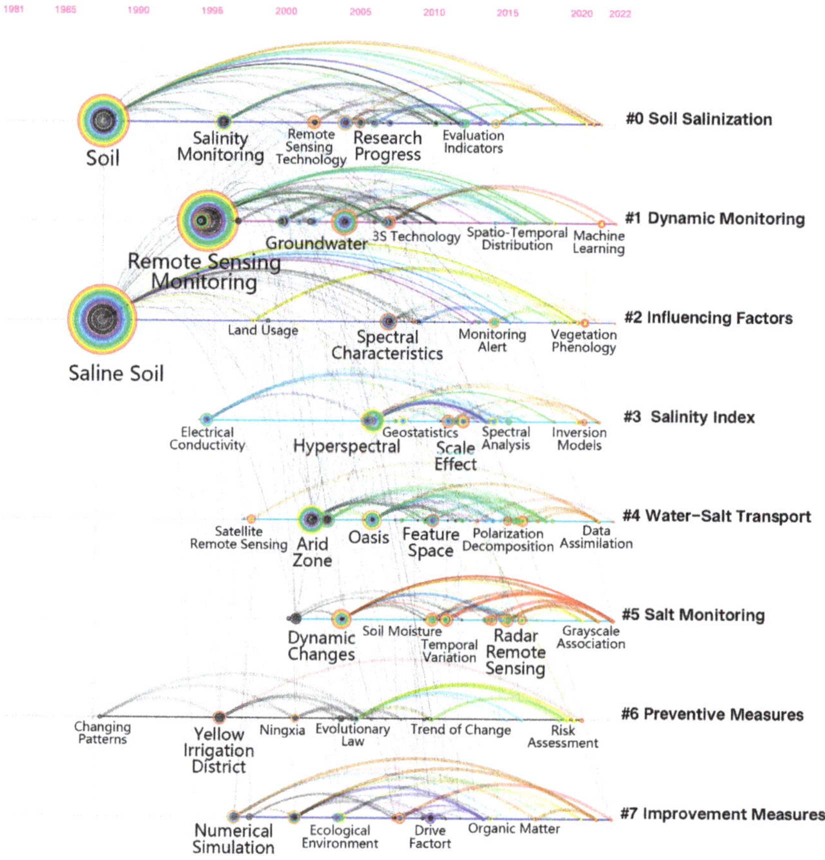

Figure 5. Keyword timeline mapping based on the literature related to soil salinity monitoring in the CNKI database from 1981 to 2022.

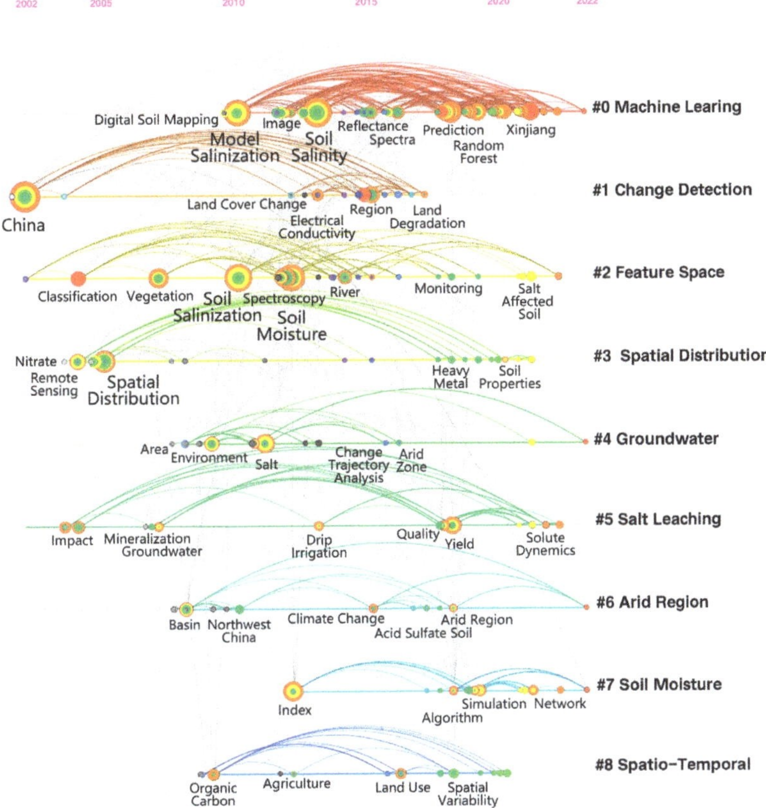

Figure 6. Keyword timeline mapping based on the literature related to soil salinity monitoring in the WOS database from 2002 to 2022.

New theories and technologies in image processing [62,63], machine learning [64], remote sensing monitoring [65], and other research fields are gradually developing and innovating. By summarizing a large number of domestic and foreign researchers' studies in recent years, it is clear that research using satellite remote sensing data to extract salinity information by classification or soil salinity inversion by multiple spectral indices has further deepened [66,67]. At the same time, more researchers are also focusing on the application of integration between different observation elements, different scales, and different data. We try to improve the accuracy of inversion of soil salinization information in terms of classification algorithms and model construction. This will provide reliable information for the development of saline soil management, ecological maintenance, and sustainable agricultural development in China.

4. Main Modeling Approaches for Soil Salinity Remote Sensing Monitoring

In recent years, countless domestic and international researchers have created many additional soil salinity monitoring techniques by tirelessly decoding an enormous amount of sensing image data. Keyword burst detection analysis was performed in CiteSpace software using the retrieved literature dataset related to soil salinity monitoring from 1981 to 2022 obtained by searching in the CNKI and the WOS databases (Figure 7). This analysis can be used to detect large changes in the number of citations of keywords in the literature during a certain time period. Thus, it helps to obtain the latest information on the frontiers of research in the field. The keyword burst detection mapping shows that since the information extraction and dynamic monitoring of salinized soil, many researchers have started to focus on the construction of models for soil salinization information by remote sensing. For example, "classification" appeared in 2012, "feature space" appeared in 2013, "quantitative models" appeared in 2015, "neural networks" and "random forest" appeared in 2016, "methods plsr" appeared in 2018, and "model", "machine learning", and "monitoring models" appeared in 2020. Among the 35 main keywords obtained from the analysis, a variety of keywords related to remote sensing monitoring modeling studies have begun to appear frequently in the last decade. These modeling studies are updated quickly and span a long period of time, and all of them have gradually achieved valuable research results while broadening the methods of soil salinity monitoring. This indicates that modeling studies have become a hot method in remote sensing monitoring of soil salinity.

Top 35 Keywords with the Strongest Citation Bursts

Keywords	Year	Strength	Begin	End
Water–Salt Movement	1988	2.33	1988	2003
Water–Salt Balance	1988	2.16	1988	2005
Yellow Irrigation District	1996	1.98	1996	2011
Numerical Simulation	1997	2.17	1997	2005
Governance	1998	1.92	1998	2008
Seawater Intrusion	2007	2.73	2007	2012
Saline Soils	1988	2.21	2007	2008
Information Extraction	2007	2.24	2007	2008
Dynamic Changes	2001	3.56	2008	2012
Dynamic Monitoring	1995	1.78	2009	2012
Ecosystem Services	2009	1.90	2009	2016
Oasis	2003	2.78	2010	2011
Improvement Measures	2010	2.43	2010	2013
Evaluation	2010	2.18	2010	2011
Spectral Characteristics	2007	1.78	2012	2014
Classification	2003	3.48	2012	2019
Soil Moisture	2012	2.62	2012	2014
Electrical Conductivity	2013	2.08	2013	2015
Feature Space	2013	2.97	2013	2018
Genesis	2008	1.97	2013	2014
Plain Reservoir	2014	2.50	2014	2016
Hyperspectral	2006	4.08	2014	2020
Quantitative Models	2015	1.95	2015	2018
Salt Affected Soils	2015	4.89	2015	2016
Neural Networks	2016	2.13	2016	2018
Random Forest	2016	1.68	2016	2022
Methods PLSR	2018	3.17	2018	2019
Indicator	2013	1.99	2018	2019
Regression Analysis	2018	2.17	2018	2022
Scale Conversion	2016	1.82	2019	2022
Landsat 8 OLI	2020	2.09	2020	2022
Model	2010	2.36	2020	2022
Machine Learning	2020	4.91	2020	2022
Monitoring Models	2020	2.17	2020	2022
Unmanned Aerial Vehicles	2020	1.76	2020	2022

Figure 7. Keyword burst detection mapping based on the literature related to soil salinity monitoring from 1981 to 2022 in the CNKI and WOS databases. Note: In the Figure, the blue lines correspond to the year in which the keywords first started to appear; the red lines correspond to the year in which the keywords started and ended as research frontiers.

The core method of quantitative assessment of soil salinity using remote sensing data is to explore the correlation between the content of relevant salinity indicators and remote sensing data [68]. Among them, salinity indexes can be obtained by bringing soil samples collected from field surveys back to the laboratory and performing measurement experiments for analysis. Pre-processed remote sensing image data are rich in spectral information of features, which are contained in the reflectance of different wavelength bands of remote sensing data. The spectral information contained in a single band is limited. Therefore, after extracting the spectral reflectance of the bands of the remotely sensed image, a combination operation between different bands can be performed. These indices are usually a priori formulas derived from existing studies (Tables 1 and 2).

Table 1. Salinity indexes and related calculation formulas.

Salinity Index	Calculation Formula	Note	Reference
Salinity Index (SI)	$\sqrt{B \times R}$	B is the blue band R is the red band	[69]
Salinity Index 1 (SI1)	$\sqrt{G \times R}$	G is the green band R is the red band	[70]
Salinity Index 2 (SI2)	$\sqrt{G^2 + R^2 + NIR^2}$	G is the green band R is the red band NIR is the near infrared band	[71]
Salinity Index 7 (SI7)	$\frac{NIR \times R}{G}$	NIR is the near infrared band R is the red band G is the green band	[72]
Salinity Index-T (SI-T)	$\frac{R}{NIR} \times 100$	R is the red band NIR is the near infrared band	[73]
Salinity Index (S)	$\frac{R}{NIR}$	R is the red band NIR is the near infrared band	[74]
Salinity Index (S1)	$\frac{B}{R}$	B is the blue band R is the red band	[74]
Salinity Index (S2)	$\frac{B-R}{B+R}$	B is the blue band R is the red band	[74]
Salinity Index (S3)	$\frac{G \times R}{B}$	G is the green band R is the red band B is the blue band	[74]
Salinity Index (S5)	$\frac{B \times R}{G}$	B is the blue band R is the red band G is the green band	[74]
Brightness Index (BI)	$\sqrt{R^2 + NIR^2}$	R is the red band NIR is the near infrared band	[75]
Brightness Index (BRI)	$\sqrt{G^2 + R^2}$	G is the green band R is the red band	[76]
Intensity Index 1 (Int1)	$\frac{G+R}{2}$	G is the green band R is the red band	[77]
Intensity Index 2 (Int2)	$\frac{G+R+NIR}{2}$	G is the green band R is the red band NIR is the near infrared band	[77]
Salinity Ratio Index (SRI)	$(R - NIR) \times (G + NIR)$	R is the red band NIR is the near infrared band G is the green band	[78]
Normalized Difference Salinity Index (NDSI)	$\frac{R-NIR}{R+NIR}$	R is the red band NIR is the near infrared band	[79]
Normalized Difference Water Index (NDWI)	$\frac{G-NIR}{G+NIR}$	G is the green band NIR is the near infrared band	[79]
Canopy Response Salinity Index (CRSI)	$\sqrt{\frac{(NIR \times R)-(G \times B)}{(NIR \times R)+(G \times B)}}$	NIR is the near infrared band R is the red band G is the green band B is the blue band	[80]
Clay Index (CLEX)	$\frac{SWIR1}{SWIR2}$	SWIR1 and SWIR2 are the short-wave infrared bands	[81]
Carbonate Index (CAEX)	$\frac{R}{G}$	R is the red band G is the green band	[82]

Table 2. Vegetation indexes and related calculation formulas.

Vegetation Index	Calculation Formula	Note	Reference
Ratio Vegetation Index (RVI)	$\frac{NIR}{R}$	NIR is the near infrared band R is the red band	[83]
Enhanced Ratio Vegetation Index (ERVI)	$\frac{NIR+SWIR2}{R}$	NIR is the near infrared band SWIR2 is the short-wave infrared band R is the red band	[83]
Green Ratio Vegetation Index (GRVI)	$\frac{NIR}{G}$	NIR is the near infrared band G is the green band	[84]
Difference Vegetation Index (DVI)	$NIR - R$	NIR is the near infrared band R is the red band	[85]
Enhanced Difference Vegetation Index (EDVI)	$NIR + SWIR1 - R$	NIR is the near infrared band SWIR1 is the short-wave infrared band R is the red band	[86]
Renormalized Difference Vegetation Index (RDVI)	$\frac{NIR-R}{\sqrt{NIR+R}}$	NIR is the near infrared band R is the red band	[87]
Generalized Difference Vegetation Index (GDVI)	$\frac{NIR^2-R^2}{NIR^2+R^2}$	NIR is the near infrared band R is the red band	[88]
Normalized Difference Vegetation Index (NDVI)	$\frac{NIR-R}{NIR+R}$	NIR is the near infrared band R is the red band	[89]
Green Normalized Difference Vegetation Index (GNDVI)	$\frac{NIR-G}{NIR+G}$	NIR is the near infrared band G is the green band	[90]
Extended Normalized Difference Vegetation Index (ENDVI)	$\frac{NIR+SWIR2-R}{NIR+SWIR2+R}$	NIR is the near infrared band SWIR2 is the short-wave infrared band R is the red band	[91]
Enhanced Vegetation Index (EVI)	$2.5 \times \frac{NIR-R}{NIR+6R-7.5B+1}$	NIR is the near infrared band R is the red band B is the blue band	[92]
Two Band Enhanced Vegetation Index (EVI2)	$2.5 \times \frac{NIR-R}{NIR+2.4R+1}$	NIR is the near infrared band R is the red band	[93]
Chlorophyll Index Green (CIgreen)	$\frac{NIR}{G} - 1$	NIR is the near infrared band G is the green band	[94]
Simple Ratio Index (SRI)	$\frac{NIR}{R}$	NIR is the near infrared band R is the red band	[95]
Nonlinear Vegetation Index (NLI)	$\frac{NIR^2-R}{NIR^2+R}$	NIR is the near infrared band R is the red band	[96]
Modified Nonlinear Vegetation Index (MNLI)	$\frac{1.5(NIR^2-R)}{NIR^2+R+0.5}$	NIR is the near infrared band R is the red band	[96]
Modified Simple Ratio (MSR)	$\frac{NIR-B}{R+B}$	NIR is the near infrared band B is the blue band R is the red band	[97]
Triangular Vegetation Index (TVI)	$0.5[120(NIR - G) - 200(R - G)]$	NIR is the near infrared band G is the green band R is the red band	[98]
Modified Triangular Vegetation Index (MTVI)	$1.2[1.2(NIR - G) - 200(R - G)]$	NIR is the near infrared band G is the green band R is the red band	[99]
Soil-Adjusted Vegetation Index (SAVI)	$\frac{(1+L)(NIR-R)}{NIR+R+L}$	L is the soil adjustment coefficient, which is generally close to 0.5 NIR is the near infrared band R is the red band	[100]
Green Soil-Adjusted Vegetation Index (GSAVI)	$(1 + L)\frac{NIR-G}{NIR+G+L}$	NIR is the near infrared band G is the green band L is the soil adjustment coefficient, which is generally close to 0.5	[101]
Optimization Soil-Adjusted Vegetation Index (OSAVI)	$\frac{NIR-R}{NIR+R+0.16}$	NIR is the near infrared band R is the red band	[102]
Green Optimization Soil-Adjusted Vegetation Index (GOSAVI)	$\frac{NIR-G}{NIR+G+0.16}$	NIR is the near infrared band G is the green band	[103]
Composite Spectral Response Index (COSRI)	$\frac{B+G}{R+NIR} \times \frac{NIR-R}{NIR+R}$	B is the blue band G is the green band NIR is the near infrared band R is the red band	[104]
Visible Atmospherically Resistant Index (VARI)	$\frac{G-R}{G+R-B}$	G is the green band R is the red band B is the blue band	[105]
Atmospherically Resistant Vegetation Index (ARVI)	$\frac{NIR-(2R-B)}{NIR+(2R+B)}$	NIR is the near infrared band R is the red band B is the blue band	[106]

A remote sensing monitoring model was constructed with the aim of establishing the relationship between soil salinity content and the abovementioned modeling factors. Therefore, the research hotspot of remote sensing monitoring of soil salinity mainly lies in the construction of an efficient and reliable remote sensing monitoring model [107,108]. Various models are used to extract salinity information mainly by remote sensing technology to obtain more accurate inversion results and to assess the regional soil salinity distribution. Among them, the most commonly used remote sensing models for salinity monitoring mainly include the following.

4.1. Partial Least Squares Regression

Partial least squares regression (PLSR) is a new multivariate statistical data analysis algorithm that combines the advantages of multiple stepwise linear regression models, principal component models, and simple linear regression models in one [109,110]. Partial least squares regression implements a combination of simplified data, structural regression modeling, and the analysis of correlations between two groups of variables. It also implements cross-validity validation of component extraction in the calculation process, reorganization and screening of information in the variable system, and selection of several new components with the best systematic explanatory power for regression modeling. This provides a great convenience for statistical analysis of multivariate data [111]. Therefore, partial least squares regression, which can automatically filter variables based on correlation and ensure model stability, has been widely used in recent years to model spectral data and achieve high-accuracy predictions.

Viscarra Rossel et al. [112] showed that the partial least squares regression method has better soil nutrient prediction capability. This method allows modeling when the number of sample points is smaller than the number of variables and provides excellent processing and analysis capabilities for complex problems. Numerous researchers have also used this model to invert soil organic matter and to filter the most important organic matter content variables from spectral data [113], resulting in better robustness of the established models. For example, Yumiti Maiming et al. [114] set the independent variable as the original spectral reflectance and the dependent variable as the soil organic matter content to construct an estimation model based on methods such as partial least squares regression. It was shown that the inverse logarithmic first-order differential transformation could help to improve the accuracy of the partial least squares regression estimation model.

4.2. Support Vector Machines

Support vector machines (SVM) are a method that better implements the idea of structural risk minimization. It can better solve practical problems such as small samples, nonlinearity, and high-dimensional data. It can be extended to other machine learning problems such as function fitting [115]. Support vector machines are a kind of machine learning method based on statistical theory, which has a strong mathematical foundation and is intuitive, stable, accurate, and efficient. Compared with traditional statistical methods, it has the advantages of strong functional expression, good generalization ability, and high learning efficiency. It is widely used in the fields of soil salinity information extraction because it is easy to combine with multiple sources of information and has relatively high accuracy [116]. However, at the same time, support vector machines are also too sensitive to the selection of parameters and functions and have shortcomings in solving multi-classification problems [117].

Using the spectral information of Landsat ETM+ images, Fei Zhang et al. [118] found that the support vector machines regression method based on texture features had an improved effect on the monitoring accuracy of soil salinity information. Similarly, Yiliyas Jiang [119] referred to the salinity classification system and used support vector machines to determine the optimal parameters, thus obtaining more accurate soil salinity classification information. In addition, Xi Wang et al. [120] selected support vector machines as a modeling method for machine learning based on the small sample size and practical situation of the study and performed remote sensing inversion of salinized soil organic matter.

4.3. BP Neural Network

BP neural network (BPNN) is a multilayer feedforward neural network trained according to the error back-propagation algorithm. It mainly contains two processes: forward propagation of signals and backward propagation of errors. BP neural networks can identify nonlinear relationships between input and output data sets in complex systems without constructing mathematical equations. It can eliminate the influence of specific values to a certain extent, has strong self-learning ability, strong adaptability, and anti-interference ability, and it is widely used by researchers in the inversion monitoring of salinity [121].

Wenzhe Feng et al. [122] simultaneously introduced BP neural network, support vector machines, and extreme learning machines to build a soil salinity monitoring model so as to improve the monitoring accuracy of soil salinity by satellite remote sensing. Among them, a three-layer BP neural network was used to build a soil salinity dynamics model according to the principle of relatively small error in training results. The results showed that the BP neural network model based on multi-source remote sensing images is more accurate. Xueli Feng et al. [123] found that the stability and prediction accuracy of the BP neural network model combining the second-order derivatives of spectral feature bands, radar backscattering characteristics, and combined surface roughness were higher than the rest of the models. This shows that the BP neural network model combined with multi-source remote sensing data can quickly and accurately monitor the distribution of soil salinization, which provides important guidance for the prevention and control of soil degradation.

4.4. Random Forest

Random forest (RF) is a new classification and prediction model that uses multiple decision trees for training and prediction of samples. The model randomly selects the training sample set with split attribute set, which is a combination of multiple decision trees and is less sensitive to outliers [124]. This gives random forest models high noise immunity and nonlinear mining ability, and the distribution of data does not need to conform to any assumptions, which performs well in the randomization training phase of samples and variables. Random forest models have been increasingly used for classification and regression in recent years because of their high accuracy in predicting results, the importance of computable variables, and the ability to model complex interactions among a large number of predictor variables [125,126].

Jie Hu [127] compared partial least squares regression and random forest regression methods using hyperspectral first-order differentiation, broad-band spectral indices, and narrow-band spectral indices as independent variables. The results showed that the random forest regression model could better predict soil salinity using spectral data, and the model for the bare soil area had the highest prediction accuracy compared to the model for the other sample areas. Meanwhile, in order to monitor the spatial variability of soil salinity as accurately as possible on a large scale, Lina Meng [128] used ordinary kriging, geographically weighted regression, and a random forest model combined with several environmental auxiliary variables to map the distribution of surface soil salinity in the study area. The results show that the random forest model has the highest prediction accuracy among the various prediction methods, indicating that the model is effective in quantitatively estimating soil salinity at the regional scale.

4.5. Feature Space

The modeling method based on the spectral feature space can also be applied to construct a soil remote sensing monitoring model [129]. The distance to a certain feature point in the two-dimensional or three-dimensional feature space of soil salinization covariates can reflect different degrees of salinity and determine the change trend between different covariates. We can analyze the scattered spatial map with practical experience so that we can use spatial characteristic covariates of the scattered map to build the corresponding model [130]. Among them, two-dimensional feature space can clearly express the distribution pattern of factors affecting the soil salinization process. For example, the two-dimensional scatter diagram between vegetation and soil salinity has been elaborated and discussed by domestic and foreign researchers. In view of the future development path of soil salinity monitoring, two-dimensional feature space can no longer satisfy the analysis and display of multiple factors involved in salinization at the same time. With the development of remote sensing technology, the emergence of three-dimensional technology can help promote the development of soil salinity feature space research to three-dimensional or multi-dimensional space.

On the one hand, in the study of two-dimensional feature space application, Jianli Ding et al. [131] constructed a two-dimensional feature space based on a modified soil-adjusted vegetation index and moisture index to derive a remote sensing monitoring index model for soil salinization. The results showed that the two-dimensional feature space correlated well with the soil surface salinity of the oasis in the arid zone. Fei Wang et al. [132] constructed a quantitative relationship between surface-based feature vectors and the occurrence of the salinization process. It was found that the feature space model could invert the soil salinity distribution of the delta oasis in the study area more accurately. In addition, Lingling Bian et al. [133] quantitatively explored the patterns between soil salinity and surface biophysical parameters and also constructed a soil salinity eigenspace model. The results showed that the salinity index and albedo characteristic spatial model has a strong predictive ability and is most suitable for the inversion of salinization in coastal areas.

On the other hand, the use of three-dimensional feature spaces is increasingly developed. The relationship between soil salinization, albedo, and the modified type of soil adjusting the vegetation index was used by Juan Feng et al. to construct the feature space [134] The monitoring model constructed by surface albedo and soil-adjusted vegetation index was found to have a high correlation with soil salinity, which can better quantify and monitor the degree of soil salinization in the study area. Not only that, XuePing Ha et al. [135] used the relationship between salinity index, surface albedo, and soil salinity based on three-dimensional feature space to efficiently and accurately extract salinization information of the oasis in the study area.

5. Discussion

The extraction of soil salinization information by remote sensing monitoring technology has recently become a hot spot in the field of remote sensing research. Many researchers domestic and abroad are gradually developing new technical means and research methods to expand research in the field of high-precision monitoring of regional soil salinization.

Various models including those summarized above have been widely used for remote sensing monitoring of soil salinity. Many of these improved and optimized mathematical models have been continuously applied to the study of the spatial distribution of soil salinity (Table 3). In addition, the back trajectory model can be used to track the trajectory of saline sands and salt dusts. The vertical distribution characteristics of sands and dusts in the atmosphere are detected by using satellite remote sensing data or UAV remote sensing data combined with the back trajectory model for the purpose of monitoring the time-series trajectory of saline soils [136,137]. Many of these models, which have achieved high accuracy, provide a favorable basis for spatial and temporal analysis and prediction of regional salinization. At the same time, however, this paper argues that there are still some shortcomings in the current research on soil salinity monitoring in China that need to be further discussed and addressed by researchers.

Table 3. Multiple modeling approaches for remote sensing to monitor soil salinity.

Authors	Monitoring Model	Reference
Jie Wang et al.	Multiple Linear Regression (MLR)	[138]
Jianwen Wang et al.	Multiple Stepwise Linear Regression (MSLR)	[139]
Yasenjiang Kahaer et al.	Nonlinear Regression (NR)	[140]
Haifeng Wang et al.	Quadratic Polynomial Regression (QPR)	[141]
Elia Scudiero et al.	Ordinary Least Square (OLS)	[142]
Shengmin Peng et al.	Partial Least Squares Regression (PLSR)	[143]
Lornbardo et al.	Quantile Regression (QR)	[144]
Pingping Jia et al.	Poisson Regression (PR)	[145]
Richard H. Anderson et al.	Ridge Regression (RR)	[146]
Glen Fox et al.	Principal Component Regression (PCR)	[147]
Yan Shen et al.	Stepwise Regression (SR)	[148]
Haorui Chen et al.	Multiple Mixed Regression (MMR)	[149]
Ayetiguli Sidike et al.	Stepwise Multiple Regression (SMR)	[150]
Nurmemet Erkin et al.	Multiple Adaptive Regression Spline (MARS)	[151]
Akshar Tripathi et al.	Decision Tree Algorithm (DTA)	[152]
Yinyin Wang et al.	Random Forest (RF)	[153]
Jinjie Wang et al.	Classification and Regression Tree (CART)	[154]
Xiaoyan Guan et al.	Support Vector Machine (SVM)	[115]
Xiaoping Wang et al.	Grid Search Support Vector Machine (GSSVM)	[155]
Utpal Barman et al.	Differential Evolutionary Support Vector Machine (DESVM)	[156]
Zheng Wang et al.	Particle Swarm Optimization Support Vector Machine (PSOSVM)	[157]
Zhongyi Qu et al.	BP Neural Network (BPNN)	[158]
Christina Corbane et al.	Convolutional Neural Network (CNN)	[159]
Sedaghat A. et al.	Artificial Neural Network (ANN)	[160]
Dawei Hu et al.	BP Artificial Neural Network (BPANN)	[161]
Xiaoping Wang et al.	Bootstrap-BP Neural Network (Bootstrap-BPNN)	[162]
Gopal Ramdas Mahajan et al.	Ordinary Krieger (OK)	[163]
Jialin Zhang et al.	Universal Kriging (UK)	[164]

Table 3. *Cont.*

Authors	Monitoring Model	Reference
Eldeiry A. A. et al.	Modified Residual Kriging (MRK)	[165]
Ku Wang et al.	Residual Universal Kriging (RUK)	[166]
Ting Du et al.	Two-Dimensional Feature Space (2DFS)	[167]
Bing Guo et al.	Three-Dimensional Feature Space (3DFS)	[168]
Yueru Wu et al.	Dobson Model (DM)	[169]
Ya Liu et al.	Structural Equation Model (SEM)	[170]
Purandara B. K. et al.	Solute Transport Model (STM)	[171]
Suchithra M. S. et al.	Extreme Learning Machine (ELM)	[172]
Ya Liu et al.	Spectral Index Regression (SIR)	[173]
Elia Scudiero et al.	Spatial Autoregressive Model (SAM)	[174]
Zhen Li et al.	Geographically Weighted Regression (GWR)	[175]

First of all, the most important shortcoming is that the universal applicability of monitoring models needs to be further improved. Saline soil resources in China are distributed in different climatic zones, and there are large differences in soil types, water and heat conditions, and planting methods. Therefore, the physical and chemical properties of soils in the current study area, temporal and climatic conditions, relevant environmental factors, and model application preferences considered by researchers in constructing monitoring models can lead to a generally low applicability of the optimal models derived from the final experiments. However, this can invert the soil salinization information of the current study area with high accuracy. Nevertheless, it is difficult for it to be applied to other seasons in the same study area or to other study areas with different environmental states.

Secondly, sometimes there are more types and a greater abundance of salt-tolerant vegetation cover in the study area. The soil salinity monitoring model constructed in the study area needs to be adjusted because of the large differences in spectral characteristics between different vegetation types and bare soil. Therefore, the model can be applied to areas with different vegetation growth stages and areas with different arable, grassland, and woodland coverage to eliminate the influence of vegetation on monitoring results as much as possible. This not only increases the difficulty of designing and executing the distribution of soil sampling points during fieldwork but also incorporates more variables into the construction of the model. This will lead to a decrease in the accuracy of monitoring models to invert salinity. Therefore, how to construct a salinity monitoring model that can efficiently and accurately reflect the salinity of vegetation cover areas will be a major problem to be faced in this field in the future.

6. Research Perspectives on Soil Salinity Monitoring in China

A synthesis of previous studies carried out on soil salinization shows that most monitoring methods are based on the spectral response characteristics of salinized soils. Researchers have combined the spectral information obtained from different remote sensing data with non-remote sensing parameters to establish inverse models for regional soil salinity monitoring. However, due to the high correlation between water and salt in soil salinization, salt moves with water, and soil salinity is easily shifted with changes in moisture. In the monitoring of salinity, not only the salt but also the water will be considered to further develop water–salt synergistic monitoring so as to achieve the purpose of high monitoring accuracy and effective monitoring.

For the scale studies of salinized soils, researchers have mostly conducted a single scale or chosen the administrative district as the scale indicator for analysis. Some results have been obtained for soil salinity monitoring at different scales, such as field and regional scales. However, the correlation between different scales and their transformation have been less studied. Various reasons, such as surface heterogeneity, complexity of the salinization process, and significant variability in the dominant factors affecting the spatial variability of soil salinity at different scales [176], can lead to large differences in the autocorrelation of the same variable at different scales. There are limitations in studying salinization at a single scale.

The variability of soil salinity shows obvious scale effects with spatial scales. If the analysis is performed only at large scales, it may cause the spatial structural features at small scales to be obscured, making it difficult to analyze the structural features of soil spatial variability in depth. Soil salinity research based on multi-scale methods as well as scale-transformation methods can solve this problem well, which will provide new research ideas for soil salinity monitoring.

7. Conclusions

The proper management of saline soil resources in China is related to national food security and ecological stability. The leaders of the Party and the State attach great importance to the management and utilization of saline soils. As an important land resource in China, saline soils of different types, vast areas, and great potential provide unique research conditions for our researchers. Efficient and accurate monitoring of salinization information along with the management and development of unused saline soils provides more room for development to expand the country's arable land resources and expand the path of agricultural development.

With the continuous progress of remote sensing technology, research on soil salinization information extraction methods has developed over a long period. In order to obtain more accurate inversion data to characterize the interrelationship between soil salinity status and its influencing factors, the construction of remote sensing monitoring models has gradually become a research hotspot in the field of soil salinity monitoring. In future research, the exploration of soil salinization information extraction will become more extensive, will generate more data, and will be more accurate in judgment [176]. In general, according to current scientific needs and national demands, soil salinization research in China will play an important role for national food security [177], arable land security [178], saline land improvement [179], land use protection [180], ecology, and sustainable agricultural development. In the face of today's increasingly serious soil salinization situation, it is important to seek more efficient, reliable, accurate, and economical soil salinization monitoring technologies.

Author Contributions: Conceptualization, methodology, formal analysis, investigation, data curation, and writing—original draft preparation, Y.M.; writing—review and editing, supervision, project administration, and funding acquisition, N.T. All authors have read and agreed to the published version of the manuscript.

Funding: This research was carried out with the financial support provided by the National Natural Science Foundation of China "Study on the optimal scale for soil salinization water–salt remote sensing monitoring", grant number 41761077.

Institutional Review Board Statement: Not applicable.

Informed Consent Statement: Not applicable.

Data Availability Statement: The data presented in this study are openly available in the China National Knowledge Infrastructure at https://www.cnki.net (accessed on 12 January 2023), and the Web of Science at https://www.webofscience.com/wos/woscc/basic-search (accessed on 17 March 2023).

Acknowledgments: We extend our heartfelt gratitude to the anonymous reviewers of this manuscript for their constructive comments and helpful suggestions, which strengthened the manuscript.

Conflicts of Interest: The authors declare no conflict of interest.

References

1. Guo, S.; Ruan, B.; Chen, H.; Guan, X.; Wang, S.; Xu, N.; Li, Y. Characterizing the spatiotemporal evolution of soil salinization in Hetao Irrigation District (China) using a remote sensing approach. *Int. J. Remote Sens.* **2018**, *39*, 6805–6825. [CrossRef]
2. Ding, J.; Wu, M.; Tiyip, T. Study on soil salinization information in arid region using remote sensing technique. *Agric. Sci. China* **2011**, *10*, 404–411. [CrossRef]
3. Wang, Z.; Zhang, F.; Zhang, X.; Chan, N.W.; Kung, H.-t.; Ariken, M.; Zhou, X.; Wang, Y. Regional suitability prediction of soil salinization based on remote-sensing derivatives and optimal spectral index. *Sci. Total Environ.* **2021**, *775*, 145807. [CrossRef] [PubMed]
4. Wu, Y.; Liu, G.; Su, L.; Yang, J. Accurate evaluation of regional soil salinization using multi-source data. *Spectrosc. Spectr. Anal.* **2018**, *38*, 3528–3533.
5. Li, J.; Pu, L.; Han, M.; Zhu, M.; Zhang, R.; Xiang, Y. Soil salinization research in China: Advances and prospects. *J. Geogr. Sci.* **2014**, *24*, 943–960. [CrossRef]
6. Metternicht, G.I.; Zinck, J.A. Remote sensing of soil salinity: Potentials and constraints. *Remote Sens. Environ.* **2003**, *85*, 1–20. [CrossRef]
7. Yang, J.; Yao, R.; Wang, X.; Xie, W.; Zhang, X.; Zhu, W.; Zhang, L.; Sun, R. Research on salt-affected soils in China: History, status quo and prospect. *Acta Pedol. Sin.* **2022**, *59*, 10–27. [CrossRef]
8. Nawar, S.; Buddenbaum, H.; Hill, J.; Kozak, J. Modeling and mapping of soil salinity with reflectance spectroscopy and Landsat data using two quantitative methods (PLSR and MARS). *Remote Sens.* **2014**, *6*, 10813–10834. [CrossRef]
9. Zhang, Y.; Hu, K.; Li, B.; Zhou, L.; Zhu, J. Spatial distribution pattern of soil salinity and saline soil in yinchuan plain of china. *Trans. Chin. Soc. Agric. Eng.* **2009**, *25*, 19–24. [CrossRef]
10. Wen, W.; Timmermans, J.; Chen, Q.; van Bodegom, P.M. A review of remote sensing challenges for food security with respect to salinity and drought threats. *Remote Sens.* **2021**, *13*, 6. [CrossRef]
11. Zhang, H. Analysis of the distribution and evolution characteristics of saline soils in China. *Agric. Technol.* **2022**, *73*, 104–107. [CrossRef]
12. Zhang, J.L. Understanding of saline soil formation processes. *Chin. J. Soil Sci.* **1957**, 49–51. [CrossRef]
13. Kan, M.A.W.; Yadav, J.S.P.; Qiao, H.Q. Characteristics of saline land and afforestation problems. *For. Sci. Technol.* **1962**, *22*, 2–4. [CrossRef]
14. Zen, X.X. Overview of soil salinization in Henan Province and experience in its improvement and utilization. *Chin. J. Soil Sci.* **1961**, *1*, 28–35. [CrossRef]
15. Yang, J.S.; Zhu, S.Q. International Symposium on Saline Soil Dynamics held in Nanjing. *Soils* **1990**, *1*, 57.
16. Xu, Y.Z.; Shi, T.K. Study on the saline soil foundation treatments. *Ind. Constr.* **1991**, *3*, 7–12+61.
17. Guo, S.L.; Zhu, X.Y.; Li, F.R. A brief discussion on saline soils and their improvement in the inland saline zone of Hexi. *Soil Fertil. Sci. China* **1990**, *1*, 4–7.
18. Ding, G.W.; Zhao, C.X. The prevention and amelioration of salt-affected land on the loess plateau. *J. Arid Land Resour. Environ.* **1991**, *4*, 49–60. [CrossRef]
19. Zhang, Y.H.; Fen, H.C.; Yi, Z.R.; Li, Y.Y.; Gui, L.G.; Wang, T.N.; Wang, P.; Huang, J.C. Research and demonstration of highly efficient utilization technology model for saline land agriculture in Ningxia Irrigation Area. *Inst. Agric. Resour. Environ. Ningxia Acad. Agric. For. Sci.* **2013**, *11*, 28.
20. Liang, X.; Gy, A.; Lf, B.; Db, B.; Jing, W. Soil properties and the growth of wheat (*Triticum aestivum* L.) and maize (*Zea mays* L.) in response to reed (phragmites communis) biochar use in a salt-affected soil in the Yellow River Delta. *Agric. Ecosyst. Environ.* **2020**, *303*, 107124. [CrossRef]
21. National Development and Reform Commission. Development and reform of agricultural economics. No. [2014]594. Guidance on strengthening saline land management. *Xinjiang Water Resour.* **2014**, *3*, 28–29.
22. Xiao, K.B. Experimental Study on Mechanism of Halophyteremediation in Alkali-Saline Soil in the North Region of Yin Chuan City Ningxia Province. Ph.D. Thesis, Northwest Agriculture and Forestry University, Xianyang, China, 2013.
23. Li, H.Y. Study on the Effect of Biological Measures on the Improvement of Saline Degraded Grassland in Songnen Plain. Ph.D. thesis, Northeast Agricultural University, Harbin, China, 2014.
24. Yang, J.S.; Yao, R.J. Management and efficient agricultural utilization of salt-affected soil in China. *Bull. Chin. Sci.* **2015**, *30*, 162–170.
25. Wen, K.J.; Liu, J.; Wu, L.P. Evaluation of sustainability for ecological restoration of saline soil using planting hole control body. *Environ. Sci. Technol.* **2013**, *36*, 200–205.

26. Ma, X.L.; Meng, J.S.; Fu, D.P. Study on environmental geological problems and prevention countermeasures of typical ecologically vulnerable areas, Take the Bashang area in Hebei as an example. *Ground Water* **2020**, *42*, 122–126. [CrossRef]

27. Du, X.J.; Hu, S.W. Research progress of saline-alkali land at home and abroad over the past 30 years based on biliometric analysis. *J. Anhui Agric. Sci.* **2021**, *49*, 236–239.

28. Yang, G.S.; Wang, X.P.; Yao, R.J. Halt soil salinization, Boost soil productivity. *Science* **2021**, *73*, 30–34+2+4.

29. Zhang, Y.; Xiao, H.; Nie, X.; Zhongwu, L.I.; Deng, C.; Zhou, M. Evolution of research on soil erosion at home and abroad in the past 30 years-based on bibliometric analysis. *Acta Pedol. Sin.* **2020**, *57*, 797–810. [CrossRef]

30. Park, S.; Lee, B.; Lee, J.; Kim, T. S nutrition alleviates salt stress by maintaining the assemblage of photosynthetic organelles in Kentucky bluegrass (*Poa pratensis* L.). *Plant Growth Regul.* **2016**, *79*, 367–375. [CrossRef]

31. Corwin, D.L.; Lesch, S.M. Apparent soil electrical conductivity measurements in agriculture. *Comput. Electron. Agric.* **2005**, *46*, 11–43. [CrossRef]

32. Gong, L.; Han, L.; Ren, M.; Gui, D. Spatial variability of soil water-salt in a typical oasis on the upper reaches of the Tarim River. *J. Soil Water Conserv.* **2012**, *26*, 251–255. [CrossRef]

33. Wang, D.L.; Shu, Y.G.; University, G.J. Research progress in determination methods for soil water content. *J. Mt. Agric. Biol.* **2017**, *36*, 061–065. [CrossRef]

34. Al-Ali, Z.M.; Bannari, A.; Rhinane, H.; El-Battay, A.; Shahid, S.A.; Hameid, N. Validation and comparison of physical models for soil salinity mapping over an arid landscape using spectral reflectance measurements and Landsat-OLI data. *Remote Sens.* **2021**, *13*, 494. [CrossRef]

35. Liu, H.; Chu, G.; Zhao, F.; Huang, Q.; Wang, F. Study on the variation and trend analysis of soil secondary salinization of cotton field under long-term drip irrigation condition in northern Xinjiang. *Soil Fertil. Sci. China* **2010**, *4*, 12–17. [CrossRef]

36. Deng, K.; Ding, J.; Yang, A.; Wang, S. Modeling of the spatial distribution of soil profile salinity based on the electromagnetic induction technique. *Acta Ecol. Sin.* **2016**, *36*, 6387–6396. [CrossRef]

37. Xie, W.; Yang, J.; Yao, R.; Wang, X. Spatial and temporal variability of soil salinity in the Yangtze River estuary using electromagnetic induction. *Remote Sens.* **2021**, *13*, 1875. [CrossRef]

38. Yasenjiang, K.; Yang, S.; Nigara, T.; Zhang, F. Hyperspectral estimation of soil electrical conductivity based on fractional order differentially optimised spectral indices. *Acta Ecol. Sin.* **2019**, *39*, 7237–7248. [CrossRef]

39. Li, Y.; Zhao, G.; Chang, C.; Wang, Z.; Wang, L.; Zheng, J. Soil salinity retrieval model based on OLI and HSI image fusion. *Trans. Chin. Soc. Agric. Eng.* **2017**, *33*, 173–180. [CrossRef]

40. Ivushkin, K.; Bartholomeus, H.; Bregt, A.K.; Pulatov, A. Satellite thermography for soil salinity assessment of cropped areas in Uzbekistan. *Land Degrad. Dev.* **2017**, *28*, 870–877. [CrossRef]

41. Alqasemi, A.S.; Ibrahim, M.; Al-Quraishi, A.M.F.; Saibi, H.; Al-Fugara, A.k.; Kaplan, G. Detection and modeling of soil salinity variations in arid lands using remote sensing data. *Open Geosci.* **2021**, *13*, 443–453. [CrossRef]

42. Perri, S.; Suweis, S.; Holmes, A.; Marpu, P.R.; Entekhabi, D.; Molini, A. River basin salinization as a form of aridity. *Proc. Natl. Acad. Sci. USA* **2020**, *117*, 17635–17642. [CrossRef]

43. Aldabaa, A.A.A.; Weindorf, D.C.; Chakraborty, S.; Sharma, A.; Li, B. Combination of proximal and remote sensing methods for rapid soil salinity quantification. *Geoderma* **2015**, *239*, 34–46. [CrossRef]

44. Zhang, F. The Study on the Saline Soil Spectrum, Spatial Characteristic and Composition in the Arid Regions: A Case Study in the Delta Oasis of Weigan and Kuqa Rivers. Master's Thesis, Xinjiang University, Ürümqi, China, 2007.

45. Zhang, S. Application and Improvement of electrical conductivity measurements in soil salinity. *Chin. J. Soil Sci.* **2014**, *45*, 754–759. [CrossRef]

46. Zarai, B.; Walter, C.; Michot, D.; Pmontoroi, J.; Hachicha, M. Integrating multiple electromagnetic data to map spatiotemporal variability of soil salinity in Kairouan region, Central Tunisia. *J. Arid Land* **2022**, *14*, 186–202. [CrossRef]

47. Sun, G.; Zhu, Y.; Gao, Z. Spatiotemporal Patterns and key driving factors of soil salinity in dry and wet years in an arid agricultural area with shallow groundwater table. *Agriculture* **2022**, *12*, 1243. [CrossRef]

48. Du, X.J.; Yan, B.W.; Xu, K.; Wang, S.Y. Research progress on water-salt transport theories and models in saline-alkali soil. *Chin. J. Soil Sci.* **2021**, *52*, 713–721. [CrossRef]

49. Xiu, X.F.; Pu, L.J. Evolution and prospects in modeling of water and salt transport in soils. *Sci. Geogr. Sin.* **2016**, *36*, 1565–1572. [CrossRef]

50. Guo, J.Y. Pollution assessment and hyperspectral monitoring of heavy metal farmland soils in Ebinur Lake Basin. *Xinjiang Univ.* **2022**, *3*, 1–84. [CrossRef]

51. Tang, R.; Zhang, X.H.; Wang, W.Y.; Shi, X.Y.; Li, P. Determination and analysis of heavy metals Cd-Zn-Cu in salinized soils of Heilonggang watershed. *J. Xingtai Univ.* **2020**, *35*, 182–185.

52. Zhang, X.; Huang, B.; Liu, F. Information extraction and dynamic evaluation of soil salinization with a remote sensing method in a typical county on the Huang-Huai-Hai Plain of China. *Pedosphere* **2020**, *30*, 496–507. [CrossRef]

53. Qi, G.; Chang, C.; Yang, W.; Gao, P.; Zhao, G. Soil salinity inversion in coastal corn planting areas by the satellite-UAV-ground integration approach. *Remote Sens.* **2021**, *13*, 3100. [CrossRef]

54. Zhou, X.; Zhang, F.; Liu, C.; Kung, H.-t.; Johnson, V.C. Soil salinity inversion based on novel spectral index. *Environ. Earth Sci.* **2021**, *80*, 501. [CrossRef]

55. Li, Y.; Ding, J.; Sun, Y.; Wang, G.; Wang, L. Remote sensing monitoring models of soil salinization based on the three dimensional feature space of MSAVI-WI-SI. *Res. Soil Water Conserv.* **2015**, *22*, 113–117+121. [CrossRef]

56. Wang, M.; Mo, H.; Chen, H. Study on model method of inversion of soil salt based on multispectral image. *Chin. J. Soil Sci.* **2016**, *47*, 1036–1041. [CrossRef]

57. Umut, H.; Mamat, S.; Yiliyas, N.; Wang, J. Soil salinity inversion model based on WorldView-2 images. *Trans. Chin. Soc. Agric. Eng.* **2017**, *33*, 200–206. [CrossRef]

58. Samiee, M.; Ghazavi, R.; Pakparvar, M.; Vali, A.A. Mapping spatial variability of soil salinity in a coastal area located in an arid environment using geostatistical and correlation methods based on the satellite data. *Desert* **2018**, *23*, 233–242. [CrossRef]

59. Chen, J.; Wang, X.; Zhang, Z.; Han, J.; Yao, Z.; Wei, G.J. Soil salinization monitoring method based on UAV-satellite remote sensing scale-up. *Trans. Chin. Soc. Agric.* **2019**, *53*, 226–238. [CrossRef]

60. Zaytungul, Y.; Mamat, S.; Abdusalam, A.; Zhang, D. Soil salinity inversion in Yutian Oasis based on PALSAR radar data. *Resour. Sci.* **2018**, *40*, 2110–2117. [CrossRef]

61. Li, Y.; Guo, Q.; Wan, B.; Qing, H.; Wang, D.; Xu, K.; Song, S.; Sun, Q.; Zhao, X.; Yang, M.; et al. Current status and prospect of three-dimensional dynamic monitoring of natural resources based on LiDAR. *Natl. Remote Sens. Bull.* **2021**, *25*, 381–402. [CrossRef]

62. An, D.; Xing, Q.; Zhao, G. Hyperspectral remote sensing of soil salinity for coastal saline soil in the Yellow River Delta based on HICO bands. *Acta Oceanol. Sin.* **2018**, *40*, 51–59. [CrossRef]

63. Bian, J.; Zhang, Z.; Chen, J.; Chen, H.; Cui, C.; Li, X.; Chen, S.; Fu, Q. Simplified evaluation of cotton water stress using high resolution unmanned aerial vehicle thermal imagery. *Remote Sens.* **2019**, *11*, 267. [CrossRef]

64. Zhang, Z.; Tan, C.; Xu, C.; Chen, S.; Han, W.; Li, Y. Retrieving soil moisture content in field maize root zone based on UAV multispectral remote sensing. *Trans. Chin. Soc. Agric. Mach.* **2019**, *50*, 246–257. [CrossRef]

65. Chen, S.; Chen, Y.; Chen, J.; Zhang, Z.; Fu, Q.; Bian, J.; Cui, T.; Ma, Y. Retrieval of cotton plant water content by UAV-based vegetation supply water index (VSWI). *Int. J. Remote Sens.* **2020**, *41*, 4389–4407. [CrossRef]

66. Zhang, Z.; Wei, G.; Yao, Z.; Tan, C.; Wang, X.; Han, J. Soil salt inversion model based on UAV multispectral remote sensing. *Trans. Chin. Soc. Agric. Mach.* **2019**, *50*, 151–160. [CrossRef]

67. Ding, J.; Wang, F. Environmental modeling of large-scale soil salinity information in an arid region: A case study of the low and middle altitude alluvial plain north and south of the Tianshan Mountains, Xinjiang. *Acta Geogr. Sin.* **2017**, *72*, 64–78. [CrossRef]

68. Li, X.; Zhang, F.; Wang, Z. Present situation and development trend of remote sensing monitoring model for soil salinization. *Remote Sens. Nat. Resour.* **2022**, *34*, 11–21. [CrossRef]

69. Khan, N.M.; Rastoskuev, V.V.; Sato, Y.; Shiozawa, S. Assessment of hydrosaline land degradation by using a simple approach of remote sensing indicators. *Agric. Water Manag.* **2005**, *77*, 96–109. [CrossRef]

70. Allbed, A.; Kumar, L.; Aldakheel, Y.Y. Assessing soil salinity using soil salinity and vegetation indices derived from IKONOS high-spatial resolution imageries: Applications in a date palm dominated region. *Geoderma* **2014**, *230*, 1–8. [CrossRef]

71. Douaoui, A.E.K.; Nicolas, H.; Walter, C. Detecting salinity hazards within a semiarid context by means of combining soil and remote-sensing data. *Geoderma* **2006**, *134*, 217–230. [CrossRef]

72. Wang, F.; Ding, J.; Wei, Y.; Zhou, Q.; Yang, X.; Wang, Q. Sensitivity analysis of soil salinity and vegetation indices to detect soil salinity variation by using Landsat series images: Applications in different oases in Xinjiang, China. *Acta Ecol. Sin.* **2017**, *37*, 5007–5022.

73. Ar, H.; Batchily, K.; Vanleeuwen, W.; Liu, H.J. A comparison of vegetation indices global set of TM images for Eos-MODIS. *Remote Sens. Environ. Interdiscip. J.* **1997**, *59*, 440–451. [CrossRef]

74. Abbas, A.; Khan, S.; Hussain, N.; Hanjra, M.A.; Akbar, S. Characterizing soil salinity in irrigated agriculture using a remote sensing approach. *Phys. Chem. Earth* **2013**, *55–57*, 43–52. [CrossRef]

75. Taylor, G.R.; Mah, A.H.; Kruse, F.A.; Kierein-Young, K.S.; Hewson, R.D.; Bennett, B.A. Characterization of saline soils using airborne radar imagery. *Remote Sens. Environ.* **1996**, *57*, 127–142. [CrossRef]

76. Khan, N.M.; Rastoskuev, V.V.; Shalina, E.V.; Sato, Y. Mapping salt-affected soils using remote sensing indicators—A simple approach with the use of GIS IDRISI. In Proceedings of the 22nd Asian Conference on Remote Sensing, Ulaanbaatar, Mongolia, 3–5 October 2022.

77. Fourati, H.T.; Bouaziz, M.; Benzina, M.; Bouaziz, S.J. Modeling of soil salinity within a semi-arid region using spectral analysis. *Arab. J. Geosci.* **2015**, *8*, 11175–11182. [CrossRef]

78. Birth, G.S.; Mcvey, G.R.J. Measuring the color of growing turf with a reflectance spectrophotometer. *Agron. J.* **1968**, *60*, 640–643. [CrossRef]

79. Mcfeeters, S.K. The use of the Normalized Difference Water Index (NDWI) in the delineation of open water features. *Int. J. Remote Sens.* **2007**, *17*, 1425–1432. [CrossRef]

80. Scudiero, E.; Skaggs, T.H.; Corwin, D.L. Regional scale soil salinity evaluation using Landsat 7, western San Joaquin Valley, California, USA. *Geoderma Reg.* **2014**, *2–3*, 82–90. [CrossRef]

81. Taghizadeh-Mehrjardi, R.; Minasny, B.; Sarmadian, F.; Malone, B.P. Digital mapping of soil salinity in Ardakan region, central Iran. *Geoderma* **2014**, *213*, 15–28. [CrossRef]

82. Alhammadi, M.S.; Glenn, E.P. Detecting date palm trees health and vegetation greenness change on the eastern coast of the United Arab Emirates using SAVI. *Int. J. Remote Sens.* **2008**, *29*, 1745–1765. [CrossRef]

83. Fang, X.; Gao, J.; Xie, C.; Zhu, F.; Huang, L.; He, Y. Review of crop canopy spectral information detection technology and methods. *Spectrosc. Spectr. Anal.* **2015**, *35*, 1949–1955. [CrossRef]

84. Huete, A.; Didan, K.; Miura, T.; Rodriguez, E.P.; Gao, X.; Ferreira, L.G. Overview of the radiometric and biophysical performance of the MODIS vegetation indices. *Remote Sens. Environ.* **2002**, *83*, 195–213. [CrossRef]

85. Jordan, C.F. Derivation of leaf area index from light quality of the forest floor. *Ecology* **1969**, *50*, 663–666. [CrossRef]

86. Tucker, C.J. Red and photographic infrared linear combinations for monitoring vegetation. *Remote Sens. Environ.* **1979**, *8*, 127–150. [CrossRef]

87. Roujean, J.L.; Breon, F.M. Estimating PAR absorbed by vegetation from bidirectional reflectance measurements. *Remote Sens. Environ.* **1995**, *51*, 375–384. [CrossRef]

88. Wu, W.; Ahmad, S.M.; Boubaker, D.; Eddy, D.P. Mapping soil salinity changes using remote sensing in Central Iraq. *Geoderma Reg.* **2014**, *2–3*, 21–31. [CrossRef]

89. Bunkei, M.; Wei, Y.; Chen, J.; Yuyichi, O.; Qiu, G. Sensitivity of the enhanced vegetation index (evi) and normalized difference vegetation index (ndvi) to topographic effects: A case study in high-density cypress forest. *Sensors* **2007**, *11*, 2636. [CrossRef]

90. Gitelson, A.A.; Kaufman, Y.J.; Merzlyak, M.N. Use of a green channel in remote sensing of global vegetation from EOS-MODIS. *Remote Sens. Environ.* **1996**, *58*, 289–298. [CrossRef]

91. Gitelson, A.A.; Merzlyak, M.N. Remote sensing of chlorophyll concentration in higher plant leaves. *Adv. Space Res.* **1998**, *22*, 689–692. [CrossRef]

92. Liu, H.Q.; Huete, A. A feedback based modification of the NDVI to minimize canopy background and atmospheric noise. *IEEE Trans. Geosci. Remote Sens.* **1995**, *33*, 457–465. [CrossRef]

93. Jiang, Z.; Huete, A.R.; Didan, K.; Miura, T. Development of a two-band enhanced vegetation index without a blue band. *Remote Sens. Environ.* **2008**, *112*, 3833–3845. [CrossRef]

94. Sishodia, R.P.; Ray, R.L.; Singh, S.K. Applications of remote sensing in precision agriculture: A review. *Remote Sens.* **2020**, *12*, 3136. [CrossRef]

95. Brunner, P.; Li, H.T.; Kinzelbach, W.; Li, W.P. Generating soil electrical conductivity maps at regional level by integrating measurements on the ground and remote sensing data. *Int. J. Remote Sens.* **2007**, *28*, 3341–3361. [CrossRef]

96. Peng, G.; Pu, R.; Biging, G.S.; Larrieu, M.R. Estimation of forest leaf area index using vegetation indices derived from Hyperion hyperspectral data. *IEEE Trans. Geosci. Remote Sens.* **2003**, *41*, 1355–1362. [CrossRef]

97. Sims, D.A.; Gamon, J.A. Relationships between leaf pigment content and spectral reflectance across a wide range of species, leaf structures and developmental stages. *Remote Sens. Environ.* **2002**, *81*, 337–354. [CrossRef]

98. Haboudane, D.; Miller, J.R.; Pattey, E.; Zarco-Tejada, P.J.; Strachan, I.B. Hyperspectral vegetation indices and novel algorithms for predicting green LAI of crop canopies: Modeling and validation in the context of precision agriculture. *Remote Sens. Environ.* **2004**, *90*, 337–352. [CrossRef]

99. Gitelson, A.A.; Gritz, Y.; Merzlyak, M.N. Relationships between leaf chlorophyll content and spectral reflectance and algorithms for non-destructive chlorophyll assessment in higher plant leaves. *J. Plant Physiol.* **2003**, *160*, 271–282. [CrossRef]

100. Huete, A.R. A soil-adjusted vegetation index (SAVI). *Remote Sens. Environ.* **1988**, *25*, 295–309. [CrossRef]

101. Sripada, R.P.; Heiniger, R.W.; White, J.G.; Meijer, A.D. Aerial color infrared photography for determining early in-season nitrogen requirements in corn. *Agron. J.* **2006**, *98*, 968–977. [CrossRef]

102. Baret, S.F. Optimization of soil-adjusted vegetation indices. *Remote Sens. Environ.* **1996**, *55*, 95–107. [CrossRef]

103. Okin, G.S.; Clarke, K.D.; Lewis, M.M. Comparison of methods for estimation of absolute vegetation and soil fractional cover using MODIS normalized BRDF-adjusted reflectance data. *Remote Sens. Environ.* **2013**, *130*, 266–279. [CrossRef]

104. Fernandez, B.N.; Siebe, C.; Cram, S.; Palacio, J.L. Mapping soil salinity using a combined spectral response index for bare soil and vegetation: A case study in the former lake Texcoco, Mexico. *J. Arid Environ.* **2006**, *65*, 644–667. [CrossRef]

105. Gitelson, A.A.; Stark, R.; Grits, U.; Rundquist, D.; Kaufman, Y.; Derry, D. Vegetation and soil lines in visible spectral space: A concept and technique for remote estimation of vegetation fraction. *Int. J. Remote Sens.* **2002**, *23*, 2537–2562. [CrossRef]

106. Wang, J.; Peng, J.; Li, H.; Yin, C.; Liu, W.; Wang, T.; Zhang, H. Soil salinity mapping using machine learning algorithms with the Sentinel-2 MSI in arid areas, China. *Remote Sens.* **2021**, *13*, 305. [CrossRef]

107. Song, C.; Ren, H.; Huang, C.; Botany, I.O.; Sciences, C. Estimating soil salinity in the yellow river delta, eastern chin—An integrated approach using spectral and terrain indices with the generalized additive model. *Pedosphere* **2016**, *26*, 626–635. [CrossRef]

108. Chen, H.; Xu, X.; Zou, C. Study of a gis-supported remote sensing method and a model for monitoring soil moisture at depth. In *Ecosystems' Dynamics, Agricultural Remote Sensing and Modeling, and Site-Specific Agriculture*; SPIE: Bellingham, WA, USA, 2003; Volume 5153, pp. 147–152. [CrossRef]

109. Jiang, H.; Shu, H. Optical remote-sensing data based research on detecting soil salinity at different depth in an arid-area oasis, Xinjiang, China. *Earth Sci. Inform.* **2019**, *12*, 43–56. [CrossRef]

110. Ma, G.; Ding, J.; Han, L.; Zhang, Z. Digital mapping of soil salinization in arid area wetland based on variable optimized selection and machine learning. *Trans. Chin. Soc. Agric. Eng.* **2020**, *36*, 124–131. [CrossRef]

111. Jiang, Y.; Wang, R.; Li, Y.; Li, C.; Peng, Q.; Wu, X. Hyper-spectral retrieval of soil nutrient content of various land-cover types in Ebinur Lake Basin. *Chin. J. Eco-Agric.* **2016**, *24*, 1555–1564. [CrossRef]

112. Rossel, R.A.V.; Behrens, T. Using data mining to model and interpret soil diffuse reflectance spectra. *Geoderma* **2010**, *158*, 46–54. [CrossRef]
113. Cui, S.; Zhou, K.; Ding, R. Extraction of plant abnormal information in mining area based on hyperspectral. *Spectrosc. Spectr. Anal.* **2019**, *39*, 241–249.
114. Yumiti, M.; Wang, X. Hyperspectral estimation of soil organic matter content based on continuous wavelet transformation. *Spectrosc. Spectr. Anal.* **2022**, *42*, 1278–1284. [CrossRef]
115. Guan, X.; Wang, S.; Gao, Z.; Lv, Y. Dynamic prediction of soil salinization in an irrigation district based on the support vector machine. *Math. Comput. Model.* **2013**, *58*, 719–724. [CrossRef]
116. Tan, K.; Ye, Y.Y.; Du, P.; Zhang, Q. Estimation of heavy metal concentrations in reclaimed mining soils using reflectance spectroscopy. *Spectrosc. Spectr. Anal.* **2014**, *34*, 3317–3322. [CrossRef]
117. Tan, K.J. Hyperspectral remote sensing image classification based on support vector machine. *J. Infrared Millim. Waves* **2008**, *27*, 123–128. [CrossRef]
118. Fei, Z.; Tashpolat, T.; Jianli, D.; Yuan, T.; Ilyas, N.; Xueping, H.A. Extracting of soil salinization by SVM and accuracy evaluation based on texture characteristic. *Arid Land Geogr.* **2009**, *32*, 57–66. [CrossRef]
119. Yiliyas, J. Remote Sensing Monitoring of Soil Salinization based on Fusion and Classification of Radar and TM Image. Ph.D. Thesis, Xinjiang University, Ürümqi, China, 2008.
120. Wang, X.; Li, Y.H.; Wang, R.Y.; Shi, R.F.; Xu, S.T. Remote sensing inversion of surface soil organic matter at jointing stage of winter wheat based on unmanned aerial vehicle multispectral. *Chin. J. Appl. Ecol.* **2020**, *31*, 2399–2406. [CrossRef]
121. Chen, S.; Gao, C.; Xu, B.; Jin, Y.; Li, J.; Ma, H.; Zhao, F.; Guo, J.; Yang, X. Quantitative inversion of soil salinity and analysis of its spatial pattern in agricultural area in Shihezi of Xinjiang. *Geogr. Res.* **2014**, *33*, 2135–2144.
122. Feng, W.; Wang, X.; Han, J. Research on soil salinization monitoring based on scale conversion of satellite and UAV remote sensing data. *J. Shandong Univ. Sci. Technol. (Nat. Sci.)* **2020**, *11*, 87–93.
123. Feng, X.; Liu, Q. Regional soil salinity monitoring based on multi-source collaborative remote sensing data. *Trans. Chin. Soc. Agric. Mach.* **2018**, *49*, 127–133. [CrossRef]
124. Wang, F.; Yang, S.; Ding, J.; Wei, Y.; Ge, X.; Liang, J. Environmental sensitive variable optimization and machine learning algorithm using in soil salt prediction at oasis. *Trans. Chin. Soc. Agric. Eng.* **2018**, *34*, 102–110. [CrossRef]
125. Mutanga, O.; Adam, E.; Cho, M.A. High density biomass estimation for wetland vegetation using WorldView-2 imagery and random forest regression algorithm. *Int. J. Appl. Earth Obs. Geoinf.* **2012**, *18*, 399–406. [CrossRef]
126. Kennedy, W.; Dieu, T.; Bui, Y. A comparative assessment of support vector regression, artificial neural networks, and random forests for predicting and mapping soil organic carbon stocks across an Afromontane landscape. *Ecol. Indic.* **2015**, *52*, 394–403. [CrossRef]
127. Hu, J. Estimation of Soil Salinity in Arid Area based on Multi-Source Remote Sensing. Ph.D. Thesis, Zhejiang University, Hangzhou, China, 2019.
128. Meng, L.; Ding, J.; Wang, J.; Ge, X. Spatial distribution of soil salinity in Ugan-Kuqa River delta oasis based on environmental variables. *Trans. Chin. Soc. Agric. Eng.* **2020**, *36*, 175–181. [CrossRef]
129. Ding, J.; Yao, Y.; Wang, F. Quantitative remote sensing of soil salinization in arid regions based on three dimensional spectrum eigen spaces. *Acta Pedol. Sin.* **2013**, *50*, 853–861. [CrossRef]
130. Guo, B.; Yang, F.; Han, B.; Fan, Y.; Chen, S.; Yang, W.; Jiang, L. A model for the rapid monitoring of soil salinization in the Yellow River Delta using Landsat 8 OLI imagery based on VI-SI feature space. *Remote Sens. Lett.* **2019**, *10*, 796–805. [CrossRef]
131. Ding, J.; Qu, J.; Sun, Y.; Zhang, Y. The retrieval model of soil salinization information in arid region based on MSAVI-WI feature space:A case study of the delta oasis in Weigan-Kuqa watershed. *Geogr. Res.* **2013**, *32*, 223–232.
132. Wang, F.; Ding, J.; Wu, M. Remote sensing monitoring models of soil salinization based on NDVI-SI feature space. *Trans. Chin. Soc. Agric. Eng.* **2010**, *26*, 168–173. [CrossRef]
133. Bian, L.; Wang, J.; Guo, B.; Cheng, K.; Wei, H. Remote sensing extraction of soil salinity in Yellow River Delta Kenli County based on Feature Space. *Remote Sens. Technol. Appl.* **2020**, *35*, 211–218.
134. Feng, J.; Ding, J.L.; Wei, W.Y. A Study of Soil Salinization in Weigan and Kuqa Rivers Oasis Based on Albedo-MSAVI Feature Space. *China Rural Water Hydropower* **2018**, *2*, 147–152.
135. Ha, X.; Ding, J.; Tashpolat, T.; Luo, J.; Zhang, F. SI-Albedo space-based extraction of salinization information in arid area. *Acta Pedol. Sin.* **2009**, *46*, 381–390. [CrossRef]
136. Bi, D.; He, Q.; Li, J.L.; Lu, Z.; Zhou, C.; Meng, L.; Jiang, H. Study on a dust pollution process in Taklamakan Desert based on vertical observation by UAV. *Acta Sci. Circumstantiae* **2022**, *42*, 298–310.
137. Shiga, Y.; Greene, R.; Scott, K.M. Recognising terrestrially-derived salt (NaCl) in SE australian dust. *Aeolian Res.* **2011**, *2*, 215–220. [CrossRef]
138. Olaya-Abril, A.; Parras-Alcántara, L.; Lozano-García, B.; Obregón-Romero, R. Soil organic carbon distribution in Mediterranean areas under a climate change scenario via multiple linear regression analysis. *Sci. Total Environ.* **2017**, *592*, 134–143. [CrossRef] [PubMed]
139. Wang, J.; Zhenhai, L.I.; Zhu, H.; Xingang, X.U.; Gan, P.; Wang, H.; Yang, G.J. Estimation of salt content in saline alkali soil based on field-derived spectra of salinized soils. *J. Shandong Univ. Sci.* **2017**, *36*, 17–24. [CrossRef]

140. Kahaer, Y.; Tashpolat, N.; Shi, Q.; Liu, S. Possibility of Zhuhai-1 hyperspectral imagery for monitoring salinized soil moisture content using fractional order differentially optimized spectral indices. *Water* **2020**, *12*, 3360. [CrossRef]

141. Wang, H.F.; Zhang, Z.T.; Qiu-Ping, F.U.; Chen, S.B.; Bian, J.; Cui, T. Inversion of soil moisture content based on multispectral remote sensing data of low altitude UAV. *Water Sav. Irrig.* **2018**, *1*, 90–94.

142. Scudiero, E.; Skaggs, T.H.; Corwin, D.L. Comparative regional-scale soil salinity assessment with near-ground apparent electrical conductivity and remote sensing canopy reflectance. *Ecol. Indic.* **2016**, *70*, 276–284. [CrossRef]

143. Peng, S.; Li, C.; Huang, J.; Wang, F.J. The prediction of soil moisture model based on partial least-squares regression. *J. Agric. Mech.* **2010**, *32*, 45–49. [CrossRef]

144. Lombardo, L.; Saia, S.; Schillaci, C.; Mai, P.M.; Huser, R. Modeling soil organic carbon with quantile regression: Dissecting predictors' effects on carbon stocks. *Geoderma Int. J. Soil Sci.* **2018**, *318*, 148–159. [CrossRef]

145. Jia, P.; Shang, T.; Zhang, J.; Yuan, S. Inversion of soil pH during the dry and wet seasons in the Yinbei region of Ningxia, China, based on multi-source remote sensing data. *Geoderma Reg.* **2021**, *25*, e00399. [CrossRef]

146. Anderson, R.H.; Basta, N.T. Application of ridge regression to quantify marginal effects of collinear soil properties on phytoaccumulation of arsenic, cadmium, lead, and zinc. *Environ. Toxicol. Chem.* **2009**, *28*, 619–628. [CrossRef]

147. Fox, G.A.; Metla, R. Soil property analysis using principal components analysis, soil line, and regression models. *Soil Sci. Soc. Am. J.* **2005**, *69*, 1782–1788. [CrossRef]

148. Shen, Y.; Zhang, X.P.; Liang, A.Z.; Shi, X.H.; Fan, R.Q.; Yang, X. Multiplicative scatter correction and stepwise regression to build NIRS model for analysis of soil organic carbon content in black soil. *Syst. Sci. Compr. Stud. Agric.* **2010**, *26*, 174–180. [CrossRef]

149. Chen, H.; Wang, S.; Guan, X.; Gao, L. Hyperspectra based models for soil electrical conductivity estimation-A case study from sandy loam soil in Shahaoqu District of Hetao Irrigation Area. *J. Arid Land Resour. Environ.* **2014**, *28*, 172–177. [CrossRef]

150. Sidike, A.; Zhao, S.; Wen, Y. Estimating soil salinity in Pingluo County of China using QuickBird data and soil reflectance spectra. *Int. J. Appl. Earth Obs. Geoinf.* **2014**, *26*, 156–175. [CrossRef]

151. Erkin, N.; Zhu, L.; Gu, H.; Tusiyiti, A. Method for predicting soil salinity concentrations in croplands based on machine learning and remote sensing techniques. *J. Appl. Remote Sens.* **2019**, *13*, 034520. [CrossRef]

152. Tripathi, A.; Tiwari, R.K. A simplified subsurface soil salinity estimation using synergy of SENTINEL-1 SAR and SENTINEL-2 multispectral satellite data, for early stages of wheat crop growth in Rupnagar, Punjab, India. *Land Degrad. Dev.* **2021**, *32*, 3905–3919. [CrossRef]

153. Wang, Y.; Qi, Y.; Chen, Y.; Xie, F. Prediction of soil organic matter based on multi-resolution remote sensing data and random forest algorithm. *Acta Pedol. Sin.* **2016**, *53*, 342–354. [CrossRef]

154. Wang, J.; Ding, J.; Zhang, Z.; Chen, W. SWAT model parameters correction based on multi-source remote sensing data in saline soil in Ebinur Lake Watershed. *Trans. Chin. Soc. Agric.* **2017**, *33*, 139–144. [CrossRef]

155. Wang, X.; Zhang, F.; Kung, H.-t.; Johnson, V.C.; Latif, A. Extracting soil salinization information with a fractional-order filtering algorithm and grid-search support vector machine (GS-SVM) model. *Int. J. Remote Sens.* **2020**, *41*, 953–973. [CrossRef]

156. Barman, U.; Choudhury, R.D. Soil texture classification using multi class support vector machine. *Inf. Process. Agric.* **2020**, *7*, 318–332. [CrossRef]

157. Wang, Z.; Zhang, X.; Zhang, F.; Ngai Weng, C.; Kung, H.-t.; Liu, S.; Deng, L. Estimation of soil salt content using machine learning techniques based on remote-sensing fractional derivatives, a case study in the Ebinur Lake Wetland National Nature Reserve, Northwest China. *Ecol. Indic.* **2020**, *119*, 106869. [CrossRef]

158. Zhong, Y.Q.; Chen, Y.X.; Shi, H.B.; Wei, Z.M. An artificial neural network forecast model for regional soil water-salt regime. *Irrig. Drain.* **2002**, *04*, 40–44. [CrossRef]

159. Corbane, C.; Syrris, V.; Sabo, F.; Politis, P.; Melchiorri, M.; Pesaresi, M.; Soille, P.; Kemper, T. Convolutional neural networks for global human settlements mapping from Sentinel-2 satellite imagery. *Neural Comput. Appl.* **2021**, *33*, 6697–6720. [CrossRef]

160. Sedaghat, A.; Bayat, H.; Sinegani, A.A.S. Estimation of soil saturated hydraulic conductivity by artificial neural networks ensemble in smectitic soils. *Eurasian Soil Sci.* **2016**, *49*, 347–357. [CrossRef]

161. Dawei, H.U.; Xinming, B.; Shuyu, W.; Weiguo, F.U. Study on spatial distribution of farmland soil heavy metals in Nantong City based on BP- ANN modeling. *J. Saf. Environ.* **2007**, *7*, 91–95. [CrossRef]

162. Wang, X.; Zhang, F.; Ding, J.; Kung, H.T.; Latif, A.; Johnson, V.C. Estimation of soil salt content (SSC) in the Ebinur Lake Wetland National Nature Reserve (ELWNNR), Northwest China, based on a Bootstrap-BP neural network model and optimal spectral indices. *Sci. Total Environ.* **2018**, *615*, 918. [CrossRef]

163. Mahajan, G.R.; Das, B.; Gaikwad, B.; Murgaonkar, D.; Desai, A.; Morajkar, S.; Patel, K.P.; Kulkarni, R.M. Monitoring properties of the salt-affected soils by multivariate analysis of the visible and near-infrared hyperspectral data. *Catena* **2021**, *198*, 105041. [CrossRef]

164. Zhang, J.; Li, X.; Yang, R.; Liu, Q.; Zhao, L.; Dou, B. An extended kriging method to interpolate near-surface soil moisture data measured by wireless sensor networks. *Sensors* **2017**, *17*, 1390. [CrossRef]

165. Eldeiry, A.A.; Garcia, L.A. Detecting soil salinity in alfalfa fields using spatial modeling and remote sensing. *Soil Sci. Soc. Am. J.* **2008**, *72*, 201–211. [CrossRef]

166. Wang, K.; Jiang, Z. Accuracy analysis of kriging with local regression residuals on soil cation exchange capacity. *Acta Agric. Univ. Jiangxiensis* **2013**, *35*, 195–203. [CrossRef]

167. Ting, D.U.; Jiao, J.Z.; Xie, Y.W. Method exploration for quantitative evaluation of salinization using landsat satellite image: A case study of Guazhou-Dunhuang area. *Hubei Agric. Sci.* **2018**, *57*, 51–55. [CrossRef]
168. Guo, B.; Zang, W.; Zhang, R. Soil salizanation information in the Yellow River Delta based on feature surface models using Landsat 8 OLI data. *IEEE Access* **2020**, *8*, 94394–94403. [CrossRef]
169. Wu, Y.; Wang, W.; Zhao, S.; Liu, S. Dielectric properties of saline soils and an improved dielectric model in C-Band. *IEEE Trans. Geosci. Remote Sens.* **2015**, *53*, 440–452. [CrossRef]
170. Liu, Y.; Pan, X.; Wang, C.; Li, Y.; Shi, R. Can subsurface soil salinity be predicted from surface spectral information?—From the perspective of structural equation modelling. *Biosyst. Eng.* **2016**, *152*, 138–147. [CrossRef]
171. Purandara, B.K.; Sujitha, V.; Shivapur, A.V.; Tyagi, J.V. Modelling of soil moisture movement and solute transport in parts of malaprabha command. *J. Geol. Soc. India* **2021**, *97*, 293–296. [CrossRef]
172. Suchithra, M.S.; Pai, M.L. Improving the prediction accuracy of soil nutrient classification by optimizing extreme learning machine parameters. *Inf. Process. Agric.* **2020**, *7*, 72–82. [CrossRef]
173. Liu, Y.; Pan, X.; Wang, C.; Li, Y.; Shi, R.; Li, Z. Prediction of saline soil moisture content based on differential spectral index: A case study of coastal saline soil. *Soils* **2016**, *48*, 381–388. [CrossRef]
174. Scudiero, E.; Skaggs, T.H.; Corwin, D.L. Regional-scale soil salinity assessment using Landsat ETM plus canopy reflectance. *Remote Sens. Environ.* **2015**, *169*, 335–343. [CrossRef]
175. Li, Z.; Li, Y.; Xing, A.; Zhuo, Z.; Zhang, S.; Zhang, Y.; Huang, Y. Spatial prediction of soil salinity in a semiarid oasis: Environmental sensitive variable selection and model comparison. *Chin. Geogr. Sci.* **2019**, *29*, 784–797. [CrossRef]
176. Yan, A.; Jiang, P.; Sheng, J.; Wang, X.; Wang, Z. Spatial variability of surface soil salinity in Manas River basin. *Acta Pedol. Sin.* **2014**, *51*, 410–414. [CrossRef]
177. Yuan, G.; Chen, D.; Yangyang, X.U.; Meng, D.; Zhang, Y.; Wang, X. Summary of methods for extracting soil salinization information. *J. North China Univ. Water Resour. Electr. Power (Nat. Sci. Ed.)* **2022**, *43*, 95–101. [CrossRef]
178. Shen, R.; Wang, C.; Sun, B. Soil related scientific and technological problems in implementing strategy of storing grain in land and technology. *Bull. Chin. Acad. Sci.* **2018**, *33*, 135–144. [CrossRef]
179. Zhu, W.; Yang, J.; Yao, R.; Wang, X.; Xie, W.; Shi, Z. Buried layers change soil water flow and solute transport from the Yellow River Delta, China. *J. Soils Sediments* **2021**, *21*, 1598–1608. [CrossRef]
180. Zheng, H.; Wang, X.; Chen, L.; Wang, Z.; Xia, Y.; Zhang, Y.; Wang, H.; Luo, X.; Xing, B. Enhanced growth of halophyte plants in biochar-amended coastal soil: Roles of nutrient availability and rhizosphere microbial modulation. *Plant Cell Environ.* **2018**, *41*, 517–532. [CrossRef] [PubMed]

MDPI
St. Alban-Anlage 66
4052 Basel
Switzerland
www.mdpi.com

Sustainability Editorial Office
E-mail: sustainability@mdpi.com
www.mdpi.com/journal/sustainability